# Clostridia in Gastrointestinal Disease

Editor

**S. Peter Borriello, Ph.D.**
Research Fellow
Division of Communicable Diseases
Clinical Research Center
Harrow, Middlesex
England

CRC Press
Taylor & Francis Group
Boca Raton London New York

CRC Press is an imprint of the
Taylor & Francis Group, an **informa** business

First published 1985 by CRC Press
Taylor & Francis Group
6000 Broken Sound Parkway NW, Suite 300
Boca Raton, FL 33487-2742

Reissued 2018 by CRC Press



**Library of Congress Cataloging in Publication Data**
Main entry under title:

Clostridia in gastrointestinal disease.

Bibliography: p.
Includes index.
1. Intestines—Diseases—Etiology. 2. Clostridium diseases. 3. Intestines—Microbiology. 4. Intestines—Diseases—Animal models. I. Borriello, S. Peter.
[DNLM: 1. Clostridium—pathogenicity. 2. Clostridium Infections—complications. 3. Gastrointestinal

Diseases—etiology. WC 368 C645]
RC862.C47C56 1985 616.3'3 84-11425
ISBN 0-8493-5656-3

A Library of Congress record exists under LC control number: 84011425

Publisher's Note
The publisher has gone to great lengths to ensure the quality of this reprint but points out that some imperfections in the original copies may be apparent.

Disclaimer
The publisher has made every effort to trace copyright holders and welcomes correspondence from those they have been unable to contact.

ISBN 13: 978-1-315-89161-3 (hbk)
ISBN 13: 978-1-351-07071-3 (ebk)

Visit the Taylor & Francis Web site at http://www.taylorandfrancis.com and the
CRC Press Web site at http://www.crcpress.com

# FOREWORD

Although numerically the clostridia form only a very small component of the total gut flora the genus produces one of the most potent collection of toxins known and represents one of the most lethal collection of potential opportunist pathogens harboured in the gut. They are ideally situated to take advantage of any compromise in the hosts normal defense mechanisms to cause gastrointestinal disease. The recent discovery of *Clostridium difficile* as a cause of antibiotic-associated diarrhea and colitis has served to refocus peoples attention of the pathogenic potential of the clostridia. It also demonstrates our ignorance of the importance of clostridia in gut disease as stated clearly in Chapter 1, "Although antimicrobials have been producing diarrhea for more than 40 years, the role of *C. difficile* . . . has been revealed only within the past 4 years. This single fact unequivocally demonstrates the ignorance of medical science as regards the importance of intestinal clostridia . . . "

The aim of this book is to bring together the information available on established clostridial diseases of the gastrointestinal tract, including the more recent observations with respect to mechanisms of action and to critically review the data available which implicate clostridia in gastrointestinal diseases of unknown etiology such as infantile necrotizing enterocolitis and large bowel cancer. Information on the wide range of gut diseases in animals, both natural and laboratory induced, in which clostridia have been shown to be involved or are implicated, has been included, as in many instances these observations serve to help delineate the etiologies of human disease.

**S. P. Borriello**

# THE EDITOR

**S. Peter Borriello, Ph.D., MiBiol.,** is a lecturer in the Division of Communicable Diseases at the Clinical Research Centre at Harrow in Middlesex, England. He graduated with an honors degree in Microbiology at University College, University of London in 1975 and received his Ph.D. degree from the University of London registered through St. Thomas' Hospital Medical School in 1981. Formerly a member of the scientific staff of the Public Health Laboratory Service, Bacterial Metabolism Research Laboratory, London, he took up his present post in 1979. His major area of interest is the bacterial flora of the gastrointestinal tract in health and disease, including in particular antibiotic-associated diarrhea and colitis and colon cancer. His recent research has included the discovery of the cause of a colitis in rabbits and a new cause of antibiotic-associated diarrhea in man. Current interests also include microbial metabolism of dietary components, microbial interactions, and the etiology of sexually transmitted diseases of unknown cause.

Dr. Borriello is a member of the Institute of Biology, American Society of Microbiology, the Society for Intestinal Microbial Ecology and Disease, the Anaerobe Discussion Group and the Pathological Society of Great Britain and Ireland. He has been a member of the Anaerobe Discussion Group steering committee since 1980 and is Chairman of the Publications Committee for the Society for Intestinal Microbial Ecology and Disease. Dr. Borriello also serves as an outside reviewer for a number of journals and has edited or co-edited two other books as well as having published over 50 scientific articles.

## CONTRIBUTORS

**Stephen S. Arnon, M.D.**
Senior Investigator
Infant Botulism Research Project
California Department of Health Services
Berkeley, California

**Edward Balish, Ph.D.**
Professor
Departments of Surgery and Medical Microbiology
University of Wisconsin Medical School
Madison, Wisconsin

**S. Peter Borriello, Ph.D.**
Lecturer
Division of Communicable Diseases
Clinical Research Centre
Harrow, Middlesex, England

**R. J. Carman, Ph.D.**
Research Microbiologist
Department of Animal Unit
Charing Cross Hospital Medical School
London, England

**Sydney M. Finegold, M.D.**
Chief, Infectious Disease Section
VA Wadsworth Medical Center
Professor of Medicine
Professor of Microbiology and Immunology
UCLA School of Medicine
Los Angeles, California

**W. Lance George, M.D.**
Assistant Chief
Infectious Disease Section
Director, Clinical Microbiology Laboratory
VA Wadsworth Medical Center
Associate Professor of Medicine
UCLA School of Medicine
Los Angeles, California

**Michael J. Hill, Ph.D.**
Public Health Laboratory Service
PHLS Centre for Applied Microbiology and Research
Bacterial Metabolism Research Laboratory
Salisbury, Wiltshire, England

**Robert M. Kliegman, M.D.**
Assistant Professor
Department of Pediatrics
Case Western Reserve University
Rainbow Babies Children's Hospital
Cleveland, Ohio

**H. Elliott Larson, M.D., F.R.C.P.**
Member, Scientific Staff
Division of Communicable Diseases
Clinical Research Centre
Harrow, Middlesex, England

**Ashley B. Price, M.A., B.M., B.Ch., F.R.C. Pathol.**
Consultant Pathologist
Department of Pathology
Northwick Park Hospital
Harrow, Middlesex, England

**M. F. Stringer, Ph.D.**
Senior Microbiologist
Food Hygiene Laboratory
Public Health Laboratory Service
London, England

**P. D. Walker, Ph.D.**
Head, Bacteriology R&D Department
Wellcome Research Laboratories
Beckenham, Kent, England

**Charles E. Yale, M.D.**
Professor and Vice Chairman
Department of Surgery
University of Wisconsin Medical School
Madison, Wisconsin

# TABLE OF CONTENTS

Chapter 1

# CLOSTRIDIA IN THE HUMAN GASTROINTESTINAL FLORA

**W. Lance George and Sydney M. Finegold**

## TABLE OF CONTENTS

## I. INTRODUCTION

Clostridia are frequently present in the normal flora of the human gastrointestinal (GI) tract. Certain species of *Clostridium* (present either as normal flora or as exogenous "invaders") in the GI tract may be of great consequence to the host — both as potential sources of extra-GI infection and gut-related illnesses; the latter group of diseases will be reviewed in subsequent chapters. The purpose of this chapter is to review clostridia as components of the "normal" human GI flora and to review factors which may affect the presence of clostridia in the gut.

## II. VARIATIONS IN DIFFERENT STUDIES OF BOWEL FLORA

Moore and Holdeman[1] have reported that the "total number of different kinds of bacteria in the intestinal tract at any one time probably exceeds 400 to 500 species, but most are represented by less than $10^8$ cells per gram of feces (less than 1/1000 of the bacterial population)". An investigator attempting to study the fecal flora is confronted with the need to isolate species of bacteria that may be present in appreciable numbers ($10^4$ to $10^{10}$ per gram), but that may be outnumbered as much as 10 million to 1 by other components of the flora. Selective media for clostridia (and other bacteria) are quite important in this regard.[2] It is not known, however, whether most of the species of *Clostridium* that may be present in the gut will grow on the various selective media available. Heat shock is commonly used as a selective means to recover clostridia from feces because spores are generally resistant to such treatment; however, the resistance of spores to heat is somewhat variable, and vegetative cells of clostridia usually do not survive heating. The effect of heating may therefore reduce the viable counts of a given species of *Clostridium* or even eradicate it altogether.

Limitations in methodology in some of the studies that we reviewed include the following:

1. Failure to use selective media.
2. Inoculation of media only with high dilutions of the specimen (if the lowest dilution used is $10^7$, then the lowest count of any *Clostridium* that could be detected would be $10^7$ or $10^8$).
3. Use of less than optimal anaerobic transport of specimens.
4. Delays in processing of the specimen.
5. Freezing of the specimen prior to study (freezing is likely to kill vegetative cells, although spores would most likely survive).
6. Failure to homogenize the specimen (the distribution of organisms in feces is irregular).
7. Failure to base counts on the "dry weight" of the sample (to correct for variable moisture content).[2]

These variables are quite significant and undoubtedly have a substantial impact on the data reported.

Variations between the different populations studied may also have a significant impact on the results of bowel flora studies; these variations may either be within a given study or between studies to be compared. Variations include age of the subjects, type of diet, general state of health, presence of GI disease (e.g., ulcerative colitis and sprue), and country of residence. Such information, whenever available, was included in our review.

Improved methods for identification of clostridia (and other anaerobes) have been developed over the past 10 to 15 years; it is difficult to determine the reliability of identification techniques that were used prior to 1970. For example, *C. tetani* was isolated rather frequently from feces in several studies performed in the early part of this century, yet was not recovered from apparently similar populations studied in the past 10 years (*vide infra*). This rather striking difference may represent either a change in the composition of fecal flora, or may be a reflection of incorrect identification in earlier studies. We have mentioned a number of limitations in various studies; the purpose of such comments is not to denigrate the fine work performed by many investigators, but rather to point out factors which we feel may be important to the reader.

Despite the above-mentioned limitations, a great deal of information exists regarding the presence of clostridia in the human GI tract. Many investigators use the term "normal bowel flora", which is a rather elusive concept. An organism that is persistently present in the gut of a given individual could be said to be normal flora. On the other hand, an organism that is present only occasionally could be considered to be transient flora. Clostridial spores are rather resistant to the harsh conditions in the upper GI tract (i.e., gastric acid and enzymes in gut secretions); therefore, it is possible that when clostridia are recovered from bowel contents or feces, they may represent transient flora more frequently than is the case with nonspore-forming bacteria.

## III. TOTAL COUNTS OF CLOSTRIDIA AND SPECIES OF *CLOSTRIDIUM* AT VARIOUS SITES IN THE GASTROINTESTINAL TRACT

### A. Healthy Infants and Children

Very few studies of the gut flora of infants and children, other than those of feces, have been done. Challacombe et al.[3] were unable to recover clostridia from either the gastric or duodenal juices of 13 infants studied. Brook et al.[4] recovered *C. perfringens* on three occasions, and *C. bifermentans* and *C. limosum* on one occasion each from the gastric contents of 35 newborn infants studied within 10 min of delivery.

#### *1. Feces — Total Counts of Clostridium*

Although the human GI tract is sterile at birth, it rapidly becomes colonized by bacteria. The incidence and total counts of clostridia in feces of infants at various times during the first 2 months of life are shown in Table 1. From these data, it is evident that the gut frequently becomes colonized within the first 1 to 2 days of life. Such colonization often persists for extended periods.

Most studies of older infants indicate that clostridia can be recovered from the feces of 80 to 100% of infants between 2 and 12 months of age; counts are usually $10^5$ or higher per gram of feces (Table 2).

#### *2. Feces — Species of Clostridium*

The species of *Clostridium* isolated from the feces of healthy infants are shown in Table 3. A total of 21 different species have been reported. A limitation of these data is the relatively small number of patients studied. The most detailed studies of infant flora appear to be those of Snyder[25] who in 1940 recovered clostridia (6 different species) from all of 22 infants studied, and of Stark and Lee,[14] who recovered clostridia (12 different species) from all of 10 infants.

*Clostridium* species recovered from either feces or infection derived from the gut flora of infants who were ill are shown in Table 4. Most of these infants had necrotizing enterocolitis (NEC) and probably received antimicrobial therapy prior to having cultures performed. The role of clostridia in neonatal NEC (to be discussed in a subsequent chapter) is unclear; the data from infants with NEC are included in Table 4 because they may simply reflect the normal presence of clostridia in the gut. Additional studies of normal infants and infants with NEC are clearly needed.

**Table 1**
**PRESENCE OF CLOSTRIDIA IN FECES OF NORMAL INFANTS UP TO 2 MONTHS OF AGE**

| Age | No. of subjects | % Yielding clostridia | Range of total counts of clostridia[a] | Lowest count of clostridia detectable by methods employed[a] | Range of total counts of bacteria[a] | Country of residence | Type of feeding | Ref. |
|---|---|---|---|---|---|---|---|---|
| 0—11 hr | 30 | ~6 | ~4 | 4 | ~4—6 (40% positive) | Guatemala | Breast | 5 |
| 12—23 hr | 30 | ~50 | ~6—10 | 4 | ~8—11 (85% positive) | Guatemala | Breast | 5 |
| 24 hr | 23 | 0 | NA[b] | NA | NA | England | Breast plus cow's milk | 6 |
| 24—35 hr | 30 | ~55 | ~8—11 | 4 | ~9—11 (100% positive) | Guatemala | Breast | 5 |
| 18—51 hr | 14 | 29 | 2—9 | 2 | ~9.2—10.8 | France | NA | 7 |
| 1.5—2 days | 30 | ~83 | ~9—11 | 4 | ~9—11 (90% positive) | Guatemala | Breast | 5 |
| 2 days | 23 | 26 | NA | NA | NA | England | Breast plus cow's milk | 6 |
| 2 days | 30 | ~70 | ~7—11 | 4 | ~8—11 (90% positive) | Guatemala | Breast | 5 |
| 3 days | 30 | ~33 | ~8—10 | 4 | ~9—11 (85% positive) | Guatemala | Breast | 5 |
| 3 days | 23 | 70 | NA | NA | NA | England | Breast plus cow's milk | 6 |
| 4 days | 30 | ~27 | ~6—10 | 4 | ~10—11 (95% positive) | Guatemala | Breast | 5 |
| 5 days | 28 | 61 | NA | NA | NA | England | Breast plus "artificial milk" | 8 |
| 5 days | 75 | 73 | NA | NA | NA | England | Bottle | 8 |
| 6 days | 23 | 78 | NA | NA | NA | England | Breast plus cow's milk | 6 |
| 1—6 days | 17—19 | 22—39 | NA | NA | NA | U.S. | Breast | 9 |
| 1—7 days | 19 | 42 | NA | NA | NA | Canada | NA | 10 |
| 5—7 days | 30 | ~25 | ~9—11 | 4 | ~10—11 (100% positive) | Guatemala | Breast | 5 |
| 7 days | 7 | 86 | 3.3—6.8 | 3 | NA | Australia | Breast | 11 |
| 7 days | 7 | 57 | 5.2—11.1 | 3 | NA | Australia | Bottle (some received limited amounts of breast milk during 1st 1—2 weeks) | 11 |

| | | | | | | | | |
|---|---|---|---|---|---|---|---|---|
| 7 days | 10 | NA | ~7.2[c] | 3 | ~8.5[c] | England | Bottle | 12 |
| 7 days | 10 | NA | <3[c] | 3 | ~10.2[c] | England | Breast | 13 |
| 7 days | 8 | NA | ~4.7[c] | 3 | ~9.6[c] | England | Cow's milk | 13 |
| 7 days | 8 | NA | ~5.5[c] | 3 | ~8.6[c] | England | Formula | 13 |
| 7 days | 7 | NA | ~4.8[c] | 3 | ~9.9[c] | England | Breast plus dried cow's milk during 1st week of life | 13 |
| 0—7 days | 10 | 100 | 4—6 | 2 | NA | Australia | Breast or bottle | 14 |
| 0—10 days | 10 | 70 | NA | NA | NA | U.S. | NA | 15, 16 |
| "Newborn" | 13 | 31 | NA | NA | NA | U.S. | NA | 17 |
| "Neonates" | 17 | 56 | NA | NA | NA | England | NA | 18 |
| 2 weeks | 12 | ~8 | ~8 | 8 | ~10—11 (100% positive) | Guatemala | Breast | 5 |
| 2 weeks | 5 | NA | ~6.3[c] | 3 | ~9.1[c] | England | Breast plus dried cow's milk during 1st week of life | 12 |
| 2 weeks | 4 | NA | ~7.2[c] | 3 | ~8.5[c] | England | Bottle | 12 |
| 2 weeks | 10 | NA | ~3.2[c] | 3 | ~9.8[c] | England | Breast | 13 |
| 2 weeks | 9 | NA | ~7.9[c] | 3 | ~9.7[c] | England | Cow's milk | 13 |
| 2 weeks | 7 | NA | ~6.0[c] | 3 | ~8.7[c] | England | Formula | 13 |
| 2 weeks | 11 | NA | ~6.1[c] | 3 | ~9.5[c] | England | Breast milk plus dried cow's milk during 1st week of life | 13 |
| 3 weeks | 12 | 0 | | 8 | ~10—11 | Guatemala | Breast | 5 |
| 3 weeks | 7 | NA | ~3.7[c] | 3 | ~9.2[c] | England | Breast plus dried cow's milk during 1st week of life | 12 |
| 3 weeks | 4 | NA | ~7.1[c] | 3 | ~9.0[c] | England | Bottle | 12 |
| 3 weeks | 8 | NA | <3[c] | 3 | ~9.9[c] | England | Breast | 13 |
| 3 weeks | 9 | NA | ~7.9[c] | 3 | ~10.2[c] | England | Cow's milk | 13 |
| 3 weeks | 5 | NA | ~6.3[c] | 3 | ~8.7[c] | England | Formula | 13 |
| 3 weeks | 13 | NA | ~5.7[c] | 3 | ~10.2[c] | England | Breast plus dried cow's milk during 1st week of life | 13 |
| 4 weeks | 12 | ~17 | ~8—10 | 8 | ~9—11 | Guatemala | Breast | 5 |
| 4 weeks | 7 | 86 | 3.3—7.1(6.1)[c] | 3 | NA | Australia | Breast | 11 |
| 4 weeks | 7 | 86 | 9.1—10.2(9.4)[c] | 3 | NA | Australia | Bottle (some received limited amounts of breast milk during 1st 1—2 weeks) | |
| 4 weeks | 5 | NA | ~2.7[c] | 3 | ~9[c] | England | Breast plus dried cow's milk during 1st week of life | 12 |
| 4 weeks | 4 | NA | ~7.3[c] | 3 | ~8.6[c] | England | Bottle | 12 |

**Table 1 (continued)**
**PRESENCE OF CLOSTRIDIA IN FECES OF NORMAL INFANTS UP TO 2 MONTHS OF AGE**

| Age | No. of subjects | % Yielding clostridia | Range of total counts of clostridia[a] | Lowest count of clostridia detectable by methods employed[a] | Range of total counts of bacteria[a] | Country of residence | Type of feeding | Ref. |
|---|---|---|---|---|---|---|---|---|
| 4 weeks | 9 | NA | ~3.2[c] | 3 | ~10.2[c] | England | Breast | 13 |
| 4 weeks | 6 | NA | ~7.1[c] | 3 | ~9.8[c] | England | Cow's milk | 13 |
| 4 weeks | 5 | NA | ~6.5[c] | 3 | ~9.2[c] | England | Formula | 13 |
| 4 weeks | 10 | NA | ~6.1[c] | 3 | ~9.3[c] | England | Breast plus dried cow's milk during 1st week of life | 13 |
| 2—4 1/2 weeks | 4 | 100 | ~7.9[c] | NA | ~9.6—10.3 | England | NA | 19 |
| 5 weeks | 4 | NA | ~5.0[c] | 3 | ~9.2[c] | England | Breast plus dried cow's milk during 1st week of life | 12 |
| 5 weeks | 3 | NA | ~7.2[c] | 3 | ~9.3[c] | England | Bottle | 12 |
| 5 weeks | 8 | NA | ~3.0[c] | 3 | ~10.2[c] | England | Breast | 13 |
| 5 weeks | 9 | NA | ~7.3[c] | 3 | ~9.7[c] | England | Cow's milk | 13 |
| 5 weeks | 6 | NA | ~7.0[c] | 3 | ~8.8[c] | England | Formula | 13 |
| 5 weeks | 9 | NA | ~7.2[c] | 3 | ~9.5[c] | England | Breast plus dried cow's milk during 1st week of life | 13 |
| 6 weeks | 5 | NA | <3[c] | 3 | ~10.0[c] | England | Breast | 13 |
| 6 weeks | 7 | NA | ~7.2[c] | 3 | ~9.8[c] | England | Cow's milk | 13 |
| 6 weeks | 6 | NA | ~6.9[c] | 3 | ~8.6[c] | England | Formula | 13 |
| 6 weeks | 7 | NA | ~5.9[c] | 3 | ~10.2[c] | England | Breast plus dried cow's milk during 1st week of life | 13 |
| 1—2 months | 14 | 64 | ~3.1[c] | 3 | ~9.5[c] | France | Formula and cereal | 20 |

[a] Counts expressed as number of bacteria $\log_{10}$ per gram of feces.
[b] NA indicates information not available.
[c] Indicates mean value rather than range.

## Table 2
## PRESENCE OF CLOSTRIDIA IN FECES OF NORMAL INFANTS AND CHILDREN, AGE 2 MONTHS TO 16 YEARS

| Age | No. of subjects | % Yielding clostridia | Range of total counts of clostridia[a] | Lowest count of clostridia detectable by methods employed[a] | Range of total counts of bacteria[a] | Country of residence | Type of feeding | Ref. |
|---|---|---|---|---|---|---|---|---|
| <4 months | 16 | 100 | ~6.9—10(8.7)[b] | 2 | ~10.2[b] | England | Formula-fed | 21 |
| 2 days—14 weeks | 8 | 100 | 3.6—8.4 | ~2 | NA | England | | 22 |
| 1 week—4 months | 5 | 80 | NA[c] | NA | NA | U.S. | | 23 |
| 2—4 months | 19 | 11 | (10)[b] | 8 | 11.5[b] | Guatemala | Breast-fed | 24 |
| ~9 months | 5 | 100 | 4.2—5.5(5.1)[b] | 3 | ~10[b] | Australia | Breast-fed infants, at time of "late weaning" | 11 |
| 9 months | 7 | 86 | 4.5—6.4(5.6)[b] | 3 | ~10[b] | Australia | Formula-fed | 11 |
| 1—12 months | 22 | 100 | NA | NA | NA | U.S. | | 25 |
| 4—12 months | 17 | 88 | ~8—10.5(~9.3)[b] | 2 | ~10.3[b] | England | Formula-fed | 21 |
| 1 year | 7 | 100 | 4.3—6.1(5.2)[b] | 3 | ~10.2[b] | Australia | Breast fed, after weaning | 11 |
| 1 year | 7 | 100 | 3.5—6.1(4.9)[b] | 3 | ~10.5[b] | Australia | Formula-fed | 11 |
| 1—20 months | 29 | 28 | (5.2)[b] | 2 | ~10.3[b] | India | | 26 |
| 1—4 years | 12 | 83 | 6.7—8.7(7.6)[b] | 2 | ~10.2[b] | England | | 21 |
| 2—3 years | 12 | 17 | 10 | 8 | 11[b] | Guatemala | | 24 |
| 4—8 years | 12 | 67 | ~4—8.7(~5.5)[b] | 2 | ~10[b] | England | | 21 |
| 6—16 years | 265[d] | 48 | ~2.7—4.8(3.8)[b] | NA | ~10[b] | East Germany | | 27 |

[a] Counts expressed as number of bacteria $\log_{10}$ per gram of feces.
[b] Numbers in ( ) = mean count.
[c] NA indicates information not available.
[d] 432 samples from 265 children.

**Table 3**
**SPECIES OF CLOSTRIDIA ISOLATED FROM FECES OF HEALTHY INFANTS**

| Species of *Clostridium* | No. of subjects studied | % Yielding species of *Clostridium* indicated | Counts of *Clostridium*[a] | Prior antimicrobial therapy | Comments | Ref. |
|---|---|---|---|---|---|---|
| *C. acetobutylicum* | 8 | 13 | 2.6 | NA[b] | Controls for NEC[c] study | 22 |
| *C. barati* | 10 | 10 | NA | No | Infants sampled 8 times during 1st year of life | 14 |
| *C. beijerinckii* | 8 | 13 | NA | NA | Controls for NEC[c] study | 22 |
| *C. bifermentans* | 22 | 41 | NA | No | Infants, age 1—12 months | 25 |
| *C. butyricum* | 13 | 8 | NA | NA | Asymptomatic controls for NEC[c] study | 17 |
| | 4 | 100 | 4—6 | No | Low birth weight infants | 19 |
| | 17 | 35 | NA | NA | Controls for NEC[c] study | 18 |
| | 8 | 88 | 3.6—8.4 | NA | Controls for NEC[c] study | 22 |
| | 10 | 70 | NA | No | Infants sampled 8 times during 1st year of life | 14 |
| *C. cadaveris* | 10 | 10 | NA | No | Infants sampled 8 times during 1st year of life | 14 |
| | 22 | 18 | NA | No | Infants, age 1—12 months | 25 |
| *C. cochlearium* | 23 | 9 | NA | No | Neonates, 6-days old | 6 |
| *C. difficile* | 44 | 11 | NA | No | Formula- and breast-fed infants, age 10 days to 8 months | 28 |
| | 13 | 8 | NA | No | Breast-fed infants, age 1—5 months | 29 |
| | 17 | 71 | NA | No | Formula-fed infants, age 1—5 months | 29 |
| | 105 | 10[d] | NA | No | Healthy newborn infants | 30[d] |
| | 13 | 0[d] | NA | No | Healthy neonates | 17[d] |
| | 4 | 25 | 7.0 | No | Low birth weight infants | 19 |
| | 10 | 40 | NA | No | Neonates | 15 |
| | 10 | 20 | NA | NA | Controls for NEC[c] study | 18 |
| | 45 | 24 | NA | No | Healthy neonates | 31 |
| | 23 | 17 | NA | No | Neonates, 6-days old | 6 |
| | 8 | 88 | 4.2—5.0 | NA | Controls for NEC[c] study | 22 |
| | 22 | 45 | NA | No | Infants, age 1 month—1 year | 25 |
| | 10 | 90 | 3.1—7.1 | No | Infants sampled at 7 intervals during 1st year of life | 32 |
| | 45 | 29 | 4.6 | No | Healthy neonates | 33 |

| | | | | | | |
|---|---|---|---|---|---|---|
| *C. hastiforme* | 4 | 25 | 5 | No | Low birth weight infants | 19 |
| *C. innocuum* | 10 | 20 | NA | No | Infants sampled 8 times during 1st year of life | 14 |
| *C. lentoputrescens* | 10 | 10 | NA | No | Infants sampled 8 times during 1st year of life | 14 |
| *C. paraputrificum* | 13 | 8 | NA | NA | Asymptomatic controls for NEC[c] study | 17 |
| | 10 | 60 | NA | No | Neonates in the 1st 10 days of life | 15, 16 |
| | 23 | 13 | NA | No | Neonates, 6 days old | 6 |
| | 22 | 18 | NA | No | Healthy neonates, 1—12 months of age | 25 |
| | 10 | 90 | NA | No | Infants sampled 8 times during 1st year of life | 14 |
| *C. perfringens* | 13 | 8 | NA | NA | Asymptomatic controls for NEC[c] study | 17 |
| | 4 | 0 | NA | NA | 6 other species of *Clostridium* recovered | 19 |
| | 5 | 40 | NA | No | | 23 |
| | 17 | 35 | NA | NA | Controls for NEC[c] study | 18 |
| | 23 | 70 | NA | No | Neonates, 6 days old | 6 |
| | 8 | 13 | 3.1 | NA | Controls for NEC[c] study | 22 |
| | 22 | 100 | NA | No | Neonates studied from 1—12 months of age | 25 |
| | 10 | 60 | NA | No | 10 infants studied during 1st year of life at 8 different times; 11 other species of clostridia isolated | 14 |
| | 17—19 | 22—39 | NA | No | Breast-fed infants studied daily for the 1st 6 days of life | 9 |
| *C. putrificum* | 4 | 50 | 4—6 | No | Low birth weight infants | 19 |
| *C. ramosum* | 10 | 70 | NA | No | Infants sampled 8 times during 1st year of life | 14 |
| *C. rectum* | 10 | 10 | NA | No | Infants sampled 8 times during 1st year of life | 14 |
| | 4 | 25 | 8 | No | Low birth weight infant | 19 |
| *C. sartagoformum* | 10 | 60 | NA | No | Infants sampled 8 times during 1st year of life | 14 |
| *C. scatologenes* | 4 | 50 | 7—8 | No | Low birth weight infants | 19 |
| *C. septicum* | 4 | 25 | 5 | No | Low birth weight infants | 19 |
| *C. sporogenes* | 4 | 50 | 3—6 | No | Low birth weight infants | 19 |
| | 22 | 9 | NA | No | Infants, age 1—12 months | 25 |
| *C. tertium* | 5 | 40 | NA | No | | 23 |
| | 23 | 9 | NA | No | Neonates, 6 days old | 6 |
| | 22 | 95 | NA | No | Infants, age 1—12 months | 25 |
| | 10 | 30 | NA | No | Infants sampled 8 times during 1st year of life | 14 |

[a] Counts expressed as number of bacteria $\log_{10}$ per gram of feces.
[b] NA indicates information not available.
[c] NEC = necrotizing enterocolitis.
[d] Indicates that presence of *C. difficile* was determined by cytotoxin assay of feces rather than by culture.

**Table 4**
**SPECIES OF CLOSTRIDIA ISOLATED FROM ILL INFANTS**

| Species of *Clostridium* | No. of subjects studied | % Yielding species of *Clostridium* indicated | Counts of *Clostridium*[a] | Site of isolation | Underlying illness | Prior antimicrobial therapy | Comments | Ref. |
|---|---|---|---|---|---|---|---|---|
| *C. beijerinckii* | 6 | 17 | 5.8—6.3 | Feces | NEC[b] | Probable | | 22 |
| *C. bifermentans* | 10 | 20 | NA[c] | Feces | Congenital large bowel obstruction | No | Initial cultures negative; 20% positive by 2nd postoperative day | 34 |
| *C. butyricum* | 11 | 64 | NA | Feces or blood | NEC | Probable | *C. butyricum* recovered from blood of 7 infants; 2 of the 7 also had *C. butyricum* recovered from feces | 35 |
| | 3 | 100 | NA | Feces | NEC | Probable | | 18 |
| | 10 | 60 | NA | Feces | NEC | Probable | | 10 |
| | 19 | 37 | NA | Feces | Not specified | Probable | Controls for NEC study | 10 |
| | 6 | 83 | 1.7—8.5 | Feces | NEC | Probable | | 22 |
| *C. carnis* | 1 | 100 | NA | Blood | Gastroenteritis | NA | Infant, 8-weeks old | 36 |
| *C. cochlearium* | 9 | 11 | NA | Feces | Esophageal atresia | Yes | Preoperative cultures negative; 11% positive by 5th postoperative day | 34 |
| | 5 | 20 | NA | Feces | Small intestinal atresia | Yes | Preoperative cultures negative; 20% positive by 10th postoperative day | 34 |
| *C. difficile*[d] | 51 | 55[e] | NA | Feces | Not specified | Most had received antibiotics | | 30 |
| | 20 | 25[e] | NA | Feces | NEC,[b] suspected or confirmed | Probable | | 17 |
| | 121 | 14[e] | NA | Feces | Not specified | NA | | 37 |
| | 9 | 33 | NA | Feces | Esophageal atresia | Yes | Initial cultures negative; 33% positive by 7th postoperative day | 34 |

| | | | | | | | | |
|---|---|---|---|---|---|---|---|---|
| | 5 | 20 | NA | Feces | Small intestinal atresia | Yes | Initial cultures negative; 20% positive by 10th postoperative day | 34 |
| | 5 | 100 | NA | Feces | NEC | Yes | 40% Positive preoperatively; 100% positive by 2nd postoperative day | 34 |
| | 10 | 20 | NA | Feces | Congenital large bowel obstruction | No | Initial cultures negative; 20% positive by 2nd postoperative day | 34 |
| | 6 | 67 | 3.9—6.3 | Feces | NEC | Probable | | 22 |
| | 32 | 53 | NA | Feces | Various illnesses other than NEC | Most had received antibiotics | | 38 |
| *C. paraputrificum* | 20 | 20 | NA | Feces | NEC, suspected or confirmed | Probable | | 17 |
| | 10 | 10 | NA | Feces | NEC | Probable | | 10 |
| | 19 | 11 | NA | Feces | Not specified | Probable | Controls for NEC study | 10 |
| *C. perfringens* | 20 | 15 | NA | Feces | NEC, suspected or confirmed | Probable | | 17 |
| | 8 | 13 | NA | Feces | NEC | Probable | | 35 |
| | 3 | 33 | NA | Feces | NEC | Probable | | 18 |
| | 51 | 14 | NA | Feces or peritoneal fluid | NEC | Probable | | 39 |
| | 9 | 78 | NA | Rectum | Esophageal atresia | Yes | Initial cultures negative; 78% positive by 7th postoperative day | 34 |
| | 5 | 2 | NA | Rectum | Small intestinal atresia | Yes | 1st positive culture obtained on 5th postoperative day | 34 |
| | 5 | 80 | NA | Rectum | NEC | Yes | 20% Positive preoperatively; 80% by 2nd postoperative day | 34 |
| | 10 | 70 | NA | Rectum | Congenital large bowel obstruction | No | 70% Positive by 5th postoperative day | 34 |
| | 6 | 17 | 7.1 | Feces | NEC | Probable | 5 Other species of *Clostridium* recovered | 22 |
| | 19 | 5 | NA | Feces | Not specified | Probable | Controls for NEC study | 10 |

**Table 4 (continued)**
**SPECIES OF CLOSTRIDIA ISOLATED FROM ILL INFANTS**

| Species of *Clostridium* | No. of subjects studied | % Yielding species of *Clostridium* indicated | Counts of *Clostridium*[a] | Site of isolation | Underlying illness | Prior antimicrobial therapy | Comments | Ref. |
|---|---|---|---|---|---|---|---|---|
| *C. rectum* | 6 | 17 | NA | Feces | NEC | Probable | | 22 |
| *C. sporogenes* | 9 | 11 | NA | Feces | Esophageal atresia | Yes | Initial cultures negative; 11% positive by 3rd post-operative day | 34 |
| | 5 | 40 | NA | Feces | Small intestinal atresia | Yes | 20% Positive at surgery; 40% positive by 3rd post-operative day | 34 |
| *C. tertium* | 20 | 15 | NA | Feces | NEC, suspected or confirmed | Probable | | 17 |
| | 9 | 22 | NA | Feces | Esophageal atresia | Yes | Initial cultures negative; 22% positive by 7th post-operative day | 34 |
| | 5 | 20 | NA | Feces | NEC | Yes | 1st positive culture on the 1st postoperative day | 34 |
| *C. tyrobutyricum* | 6 | 17 | 6.2 | Feces | NEC | Probable | | 22 |

[a] Counts expressed as number of bacteria $\log_{10}$ per gram of feces.
[b] NEC = necrotizing enterocolitis.
[c] NA indicates information not available.
[d] Does not include cases of antimicrobial agent-associated diarrhea.
[e] Indicates that presence of *C. difficile* was determined by cytotoxin assay of feces rather than by culture.

**Table 5**
**RECOVERY OF CLOSTRIDIA FROM THE STOMACH OF ADULTS**

| No. of subjects | % Yielding clostridia | Total counts of bacteria[a] | Comments | Ref. |
|---|---|---|---|---|
| 15 | 80 | NA[b] | All subjects studied within 4 days of gastrectomy | 40 |
| 49 | 0 | NA | Patients with duodenal or gastric ulcer, or gastric carcinoma | 41 |
| 7 | 14 | NA | Gastric mucosa from patients with gastric carcinoma | 42 |
| 12 | 0 | 6.6—8.2[c](7.2)[d,e] | Normal subjects | 43 |
| 33 | 0 | 1.9—8.9[c](5.5)[d,f] | Adults with tropical sprue | 43 |
| 5 | 0 | NA | Achlorhydric patients | 44 |
| 149 | 0 | 0 | Fasting individuals (3 had previous partial gastrectomy) | 45 |
| 34 | 0 | NA | Normal subjects or patients with achlorhydria, or pernicious anemia | 46 |
| 31 | 0 | NA | Fasting patients with Crohn's disease or normals | 47 |
| 9 | 0 | NA | Normal subjects and patients with ileostomy | 48 |
| 33 | 0 | NA | Patients with various GI conditions | 49 |

[a] Counts expressed as number of bacteria $\log_{10}$ per mℓ of secretions.
[b] NA indicates information not available.
[c] Range of counts.
[d] Numbers in ( ) = mean count.
[e] 6 of 12 specimens yielded bacteria or yeast on culture.
[f] 24 of 33 specimens yielded bacteria or yeast on culture.

### B. Adults

#### *1. Stomach — Total Counts of Clostridium*

The results of 10 studies (377 subjects) of gastric flora in which anaerobic cultures were performed are shown in Table 5. Clostridia were recovered in only 2 studies;[40,42] surprisingly, 80% of the 15 subjects in one study[40] yielded clostridia (*C. perfringens*). It is possible that achlorhydria due to recent gastrectomy or the presence of a recent gastric incision were important factors; however, none of the achlorhydric subjects studied by Drasar et al.[44] and Gianella et al.[46] yielded clostridia from the stomach.

#### *2. Duodenum — Total Counts of Clostridium*

Clostridia were not recovered from the duodenal contents of any of the 61 subjects from the 5 studies shown in Table 6. These data are consistent with the usual absence of clostridia in gastric contents. Clostridia are sometimes present in foodstuffs (e.g., spices) which are not necessarily cooked prior to ingestion. Because clostridial spores are resistant to killing by harsh conditions (e.g., low pH), it is somewhat surprising that they are not detectable in gastroduodenal contents (other than in post-gastrectomy patients).

#### *3. Jejunum — Total Counts of Clostridium*

Table 7 shows 14 studies of the microbial flora at various levels in the jejunum. Of 405 subjects studied, only 4.7% yielded clostridia; the species of *Clostridium* recovered were not reported.

**Table 6**
**RECOVERY OF CLOSTRIDIA FROM THE DUODENUM OF ADULTS**

| No. of subjects | % Yielding clostridia | Total counts of bacteria[a] | Comments | Ref. |
|---|---|---|---|---|
| 36 | 0 | 1.3—7.0[b] | Specimens collected by transduodenal needle aspiration at surgery | 50 |
| 4 | 0 | NA[c] | Duodenal mucosa from 2 normals and 2 patients with gastric carcinoma | 42 |
| 5 | 0 | 3.0—3.7[d] | Duodenal contents of 4 normal subjects and 1 patient with ulcerative colitis | 51 |
| 5 | 0 | NA | Normal subjects | 48 |
| 11 | 0 | NA | Specimens collected by transduodenal needle aspiration at surgery | 52 |

[a] Counts expressed as number of bacteria $\log_{10}$ per mℓ of secretions.
[b] 17% of samples yielded bacteria on culture.
[c] NA indicates information not available.
[d] 80% of samples yielded bacteria on culture.

*4. Ileum — Total Counts of Clostridium*

Clostridia, usually in low counts, were recovered from the ileum of 23% of 230 subjects studied (Table 8). The increased frequency of clostridia in the lower vs. upper small intestine is in keeping with the increased complexity of the flora and counts of bacteria in the distal small bowel noted by Drasar and Hill.[63]

*5. Proximal and Mid-colon — Total Counts of Clostridium*

Very few data are available as regards to total counts of clostridia in the proximal and midcolon. In general, only about 50% of patients harbor clostridia in these segments of the colon and the counts are usually $10^3$ to $10^5$ per gram (Table 9). Clostridia can also be found associated with the colonic mucosa[65] (Table 10).

*6. Feces — Total Counts of Clostridium*

Considerable data are available as regards to clostridia in feces (Table 11). When the most recent studies are considered, it is evident that clostridia can be recovered from the feces of 80 to 100% of subjects. In studies in which good selective techniques for clostridia are employed, clostridia can be recovered from greater than 90% of subjects[21,66-71] with mean counts of approximately 8.5 to 9.5 $\log_{10}$ organisms per gram of feces. The lower incidences and lower counts of clostridia in some of the older studies most likely represent suboptimal isolation techniques. In general, clostridia represent 1/100 to 1/1000 of the total fecal flora of adults.

*7. Vermiform Appendix — Species of Clostridium*

A number of investigators studied the clostridial flora of normal and inflamed appendixes between 1900 and 1935 (Table 12). The species most frequently recovered were *C. perfringens, C. pseudotetanicum, C. septicum,* and *C. tertium.* The frequency of isolation of clostridia is appreciably greater in the region of the cecum than in the distal small bowel (see Tables 8 and 9).

## Table 7
## RECOVERY OF CLOSTRIDIA FROM JEJUNUM

| No. of subjects | % Yielding clostridia | Range of total counts of clostridia[a] | Specimens yielding any bacteria (%) | Range of total counts of bacteria[a] | Site in jejunum (as reported) | Comments | Ref. |
|---|---|---|---|---|---|---|---|
| 12 | 0 | 0 | 67 | ~2.3—7.1(~5.0)[b] | Upper jejunum | Subjects from southern India | 43 |
| 33 | 3 | ~2.1 | 88 | ~2.2—8.2(~5.4) | Upper jejunum | Patients with tropical sprue from southern India | 43 |
| 12 | 8 | ~6.1 | NA[c] | NA | Upper jejunum | Patients with Crohn's disease | 47 |
| 10 | 30 | 0—7.0 | NA | NA | Upper Jejunum | Normal subjects | 53 |
| 19 | 0 | 0 | NA | NA | Upper jejunum | Normal subjects | 47 |
| 11 | 9 | 6.2 | NA | (~7.3) | Jejunum | Normal subjects | 54 |
| 10 | 0 | 0 | NA | (~6.3) | Jejunum | Patients with tropical sprue | 54 |
| 7 | 14 | ~5 | NA | NA | Mucosal-associated jejunal flora | | 42 |
| 25 | 16 | <2—6.7 | 76 | <2—7.1 | Jejunum | Sample obtained by transjejunal needle aspiration at the time of surgery | 55 |
| 25 | 0 | 0 | NA | NA | Jejunum | Normal subjects | 44 |
| 34 | 3 | 1.0 | 29 | 1—10 | Mid-jejunum | Sample obtained by transjejunal needle aspiration at the time of surgery | 50 |
| 5 | 0 | 0 | 100 | 3.5—3.8 | Mid-jejunum | Normal subjects | 51 |
| 8 | 0 | NA | 37 | 2—7.8(4.8) | Jejunum | Samples obtained at time of ileojejunal bypass surgery | 56 |
| 5 | 0 | NA | 100 | NA | Mid-jejunum | Normal subjects | 48 |
| 10 | 0 | NA | NA | NA | Jejunum | Patients with achlorhydria | 44 |
| 11 | 9 | 1.6 | NA | NA | Mid-intestine | Patients with Crohn's disease | 47 |
| 11 | 0 | 0 | NA | NA | Mid-intestine | Normal subjects | 47 |
| 11 | 9 | 5.3 | NA | NA | Jejunum | Patients with Crohn's disease | 57 |
| 14 | 0 | 0 | NA | NA | Jejunum | Following gastric surgery | 58 |
| 43 | 0 | 0 | NA | NA | Jejunum | Following Polya partial gastrectomy | 58 |
| 15 | 7 | 3.0 | NA | NA | Jejunum | Patients with jejunal diverticula | 58 |

**Table 7 (continued)**
**RECOVERY OF CLOSTRIDIA FROM JEJUNUM**

| No. of subjects | % Yielding clostridia | Range of total counts of clostridia[a] | Specimens yielding any bacteria (%) | Range of total counts of bacteria[a] | Site in jejunum (as reported) | Comments | Ref. |
|---|---|---|---|---|---|---|---|
| 6 | 0 | 0 | NA | NA | Jejunum | Patients with regional enteritis | 58 |
| 33 | 0 | 0 | NA | NA | Jejunum | Patients with miscellaneous GI conditions | 49 |
| 12 | 8 | ~5 | 100 | ~2.1—8.1(~4.9) | Lower jejunum | Normal subjects in southern India | 43 |
| 33 | 9 | 2—4.5(3) | 94 | ~2.7—7.6(~5.3) | Lower jejunum | Patients from southern India with tropical sprue | 43 |

[a] Counts expressed as number of bacteria $\log_{10}$ per milliliter of jejunal contents.
[b] Numbers in ( ) = mean count.
[c] NA indicates information not available.

## Table 8
## RECOVERY OF CLOSTRIDIA FROM ILEUM

| No. of subjects | % Yielding clostridia | Total counts of clostridia[a] | Specimens yielding any bacteria (%) | Total counts of bacteria[a] | Site in ileum (as reported) | Comments | Ref. |
|---|---|---|---|---|---|---|---|
| 12 | 17 | 2.6, 3.9 | 100 | ~3.1—8.7(5.9)[b] | Upper ileum | Subjects from southern India | 43 |
| 33 | 21 | 2.5—5.0(3.5) | 97 | ~2.8—9.6(5.8) | Upper ileum | Patients with tropical sprue from southern India | 43 |
| 39 | 8 | ~3.6, 5.0, 7.4 | 44 | NA[c] | Mid-ileum | Samples collected at time of operation | 50 |
| 4 | 0 | 0 | NA[c] | NA | Ileum | | 48 |
| 6 | 67 | 5.7—7.0 | NA | NA | Ileum (various sites) | Patients with tropical sprue | 59 |
| 8 | 0 | 0 | NA | NA | Ileum | | 60 |
| 10 | 30 | 2.7, 2.7, 3.7 | 100 | ~2.1—8.5(~5.8) | Lower ileum | Subjects from southern India | 43 |
| 33 | 42 | ~2.2—3.7 | 100 | ~2.2—9.8(~6.6) | Lower ileum | Patients with tropical sprue from southern India | 43 |
| 12 | 0 | 0 | A | NA | Terminal ileum | | 61 |
| 25 | 24 | <2—5.7 | 72 | <2—10 | Terminal ileum | | 55 |
| 8 | 25 | 3.7, 6.2 | 63 | ~2.1—7.8 | Terminal ileum | Samples obtained at the time of ileo-jejunal bypass surgery | 56 |
| 10 | 20 | ~1.1, 8.1 | NA | NA | Terminal ileum | Patients with Crohn's disease | 47 |
| 12 | 17 | 4.2, 4.8 | NA | NA | Terminal ileum | | 47 |
| 8 | 75 | 0—7(3.9) | NA | NA | Terminal ileum | | 59 |
| 4 | 0 | 0 | NA | NA | Ileostomy effluent | | 48 |
| 6 | 17 | 3.3 | | (8.3) | Ileostomy effluent | | 62 |

[a] Counts expressed as number of bacteria $\log_{10}$ per milliliter of jejunal contents.
[b] Number in ( ) = mean count.
[c] NA indicates information not available.

## Table 9
## RECOVERY OF CLOSTRIDIA FROM PROXIMAL AND MID-LARGE BOWEL

| No. of subjects | % Yielding clostridia | Counts of clostridia[a] | Total counts of bacteria | Site | Ref. |
|---|---|---|---|---|---|
| 7 | 29 | <4—9 (7.5)[b] | NA[c] | Appendiceal mucosa | 42 |
| 6 | 50 | ~3.5—5.2 | NA | Cecal contents | 61 |
| 6 | 50 | ~2.1—5.1 | NA | Transverse colon contents | 61 |
| 2 | 100 | 4.7, 5.0 | 8.4, 8.5 | Transverse colostomy effluent | 62 |
| 25 | 8 | ~4.0 | NA | Colonic mucosa | 64 |

[a] Counts expressed as number of bacteria $\log_{10}$ per milliliter of contents.
[b] Numbers in ( ) = mean count.
[c] NA indicates information not available.

## Table 10
## RECOVERY OF CLOSTRIDIA FROM COLONIC MUCOSAL SAMPLES[a]

| Subject no. | Colon biopsy (site) | Clostridia recovered (counts) |
|---|---|---|
| 1[b] | Rectosigmoid | *C. paraputrificum* (3.2)[c], *C. ramosum* (3.6) *Clostridium* species (3.7) *Clostridium* species (4.0) |
| 2 | Rectosigmoid | *C. glycolicum* (2.9) |
| 3 | Rectosigmoid | *C. chauvoei* (3.6) |
| 4 | Rectosigmoid | *C. bifermentans* (3.5), *C. perfringens* serotype 52 (3.2), *C. perfringens* serotype 11 (2.7) |
| 5 | Rectosigmoid | *C. paraputrificum* (4.5) |
| 6[d] | Rectosigmoid | *C. bifermentans* (3.9) |
| 7[d] | Rectosigmoid | *C. cochlearium* (4.9) |
| 8 | Proximal | *C. tertium* (1.6) |
| | Transverse | *C. paraputrificum* (3.9), *C. perfringens* serotype 17, 50 (2.9) |
| | Distal | *C. paraputrificum* (4.0), *C. perfringens* serotype 17, 50 (3.2), *C. tertium* (3.4) |
| 9 | Rectosigmoid | *C. perfringens* serotype 34(4.8), *Clostridium* species (4.8) |
| 10 | Rectosigmoid | *C. perfringens* (nontypable) (3.1) |
| 11 | Rectosigmoid | *Clostridium* species (2.7) |
| 12 | Proximal | *Clostridium* species (2.3) |
| | Distal | *C. perfringens* (nontypable) (1.7), *C. perfringens* serotype 30 (1.4) |
| 13 | Proximal | *C. butyricum* (2.1), *C. cochlearium* (2.1) |
| | Distal | *C. carnis* (2.1) |

[a] Of the 13 subjects shown above, Clostridia were recovered from mucosal samples in a 24-subject study.
[b] Subject 1 to 4 were normal volunteers; subjects 5 to 13 had ulcerative colitis.
[c] Numbers in ( ) = count of organism ($\log_{10}$ /g of mucosa).
[d] Indicates subjects receiving salazopyrine.

Modified from Borriello, S. P., Clostridia Flora of the Gastrointestinal Tract in Health and Disease, Ph.D. thesis, London, 1981.

### 8. Feces — Species of Clostridium

*Clostridium* species that have been recovered from feces are shown in Tables 13 and 14. The data shown in Table 13 are from total fecal flora studies of 141 subjects studied in our

# Table 11
## RECOVERY OF CLOSTRIDIA FROM FECES OF ADULTS

| No. of subjects | % Yielding clostridia | Total counts of clostridia[a] | Lowest count of clostridia detectable by methods employed[a] | Total counts of bacteria[a] | Country of residence | Comments | Ref. |
|---|---|---|---|---|---|---|---|
| 20 | 100 | ~6—11 (9.6)[b] | 4 or 5 | (11.0) | U.S. | Subjects with malignancy other than that of the GI tract | 66 |
| 15 (20 specimens) | 100 | ~7—11.4 (9.7) | 3 | 10—12 (11.4) | U.S. | Japanese subjects who ate traditional Japanese diet | 67 |
| 18 (20 specimens) | 100 | ~3.4—11.1(9.5) | 3 | 11—11.8 (11.5) | U.S. | Japanese subjects eating an Occidental diet | 67 |
| 6 | 100 | ~6—10.5 | 3 | (11.5) | U.S. | | 68 |
| 175 | 99.4 | <2—7.7 | NA[c] | NA | 7 Different countries | | 69 |
| 16 | 94 | ~5.8—9.2 (~7.4) | ?2 | (9.7) | England | | 21 |
| 14 | 93 | ~5—11 (8.4) | 3 | (11.6) | U.S. | 7th Day Adventist subjects who consumed limited quantities of meat | 70 |
| 13 | 92 | ~4—12 (8.8) | 3 | (11.7) | U.S. | 7th Day Adventists who were strict vegetarians | 70 |
| 10 | 90 | (~9.1) | NA | (11.4) | Greece | Patients with GI malignancy | 71 |
| 26 | 89 | ~5—11 (9.4) | 4 or 5 | (11.4) | U.S. | | 66 |
| 17 | 88 | ~7—12 (9.8) | 4 or 5 | (11.4) | U.S. | Patients with GI malignancy | 66 |
| 12 | 83 | ~1.7—7.1 (~3.8) | ~2 | ~8.9—10.5(~9.8) | India | | 43 |
| 6 | 83 | ~3.2—5.8 (3.6) | 2 | (~9.6) | U.S. | | 61 |
| 33 | 79 | ~2—7.2 (~4.3) | 2 | NA | France | | 20 |
| 31 | 77 | ~1—>10 | NA | NA | U.S. | | 72 |
| 10 | 70 | (~8.0) | NA | (~11.2) | Greece | | 71 |
| 9 | 67 | (8.9) | 5 | (9.8) | Sweden | | 73 |
| 47 (specimens) | 64 | (4.7) | NA | | U.S. | Lecithinase-positive clostridia | 74 |
| 12 | 58 | (9.3) | 8 | (~10.5) | Guatemala | | 5 |
| 200 | ~48 | 0—10 | NA | NA | U.S. | | 75 |
| 25 | NA | ~4—11 (10.4) | 3 | (11.6) | U.S. | | 76 |
| 25 | NA | ~4—11 (10.0) | 3 | (11.6) | U.S. | Patients with colonic polyps | 76 |
| 40 | NA | (4.5) | NA | (~10.1) | England | | 77 |
| 18 | NA | (2.7) | NA | (~10) | England | Subjects were strict vegetarians | 77 |

**Table 11 (continued)**
**RECOVERY OF CLOSTRIDIA FROM FECES OF ADULTS**

| No. of subjects | % Yielding clostridia | Total counts of clostridia[a] | Lowest count of clostridia detectable by methods employed[a] | Total counts of bacteria[a] | Country of residence | Comments | Ref. |
|---|---|---|---|---|---|---|---|
| 40 | NA | (4.4) | NA | NA | England | | 78 |
| 48 | NA | (4.0) | NA | NA | Uganda | | 78 |
| 10 | NA | (2.7) | ?3 | (10.4) | U.S. | | 79 |
| 25 | NA | 0—5 (2.6) | 2 | (~10.7) | England | | 44 |
| 10 | NA | (8.4) | NA | NA | U.S. | | 80 |
| 5 | NA | (5.5) | 3 | NA | U.S. | | 81 |

[a] Counts expressed as number of bacteria $\log_{10}$/g of feces.
[b] Numbers in ( ) = mean counts.
[c] NA indicates information not available.

**Table 12**
**RECOVERY OF CLOSTRIDIA FROM THE NORMAL OR DISEASED VERMIFORM APPENDIX**

| Species of *Clostridium* | No. of subjects | % Yielding *Clostridium* sp. indicated | Comments | Ref. |
|---|---|---|---|---|
| *C. bifermentans* | NA[a] | 24 | Normal appendices | 82 |
| | 160 | 2 | Periappendiceal pus | 83 |
| *C. butyricum* | NA | 35 | Normal appendices | 82 |
| *C. fallax* | 160 | 4 | Periappendiceal pus | 83 |
| *C. histolyticum* | 160 | 1 | Periappendiceal pus | 83 |
| *C. perfringens* | NA | 90 | Luminal contents of appendixes removed at operation | 84 |
| | NA | 79 | Normal appendices | 82 |
| | 14 | 43 | Appendicitis | 85 |
| | 160 | 34 | Periappendiceal pus | 83 |
| | 64 | 25 | Peritonitis secondary to appendicitis | 86 |
| | 172 | 11 | Diseased appendices (either wall or lumen) | 87 |
| | 102 | 9 | Appendicitis | 88 |
| | 33 | 9 | Mucosa of inflamed appendixes | 89 |
| | 28 | 7 | Contents of normal appendixes | 87 |
| *C. pseudotetanicum* | 8 | 63 | Normal appendices | 90 |
| | 138 | 43 | Appendicitis | 85 |
| | 102 | 17 | Appendicitis | 88 |
| *C. ramosum* | 102 | 24 | Appendicitis | 88 |
| | 160 | 9 | Periappendiceal pus | 83 |
| *C. septicum* | 8 | 63 | Normal appendices | 90 |
| | 138 | 36 | Appendicitis | 85 |
| | 102 | 10 | Appendicitis | 88 |
| | 160 | 1 | Periappendiceal pus | 83 |
| *C. sphenoides* | NA | 22 | Normal appendices | 82 |
| *C. sporogenes* | 64 | 3.1 | Peritonitis secondary to appendicitis | 86 |
| | 160 | 0.6 | Periappendiceal pus | 83 |
| *C. tertium* | NA | 37 | Normal appendices | 82 |
| | 64 | 4.6 | Peritonitis secondary to appendicitis | 86 |
| *Clostridium* other than *C. perfringens* | 28 | 11 | Contents of normal appendixes | 87 |
| | 172 | 8 | Diseased appendices (either wall or lumen) | 87 |
| Species not given | 11 | 18 | Contents of normal appendices | 91 |

[a] NA indicates information not available.

laboratory.[67,70,76,125] Since the original publications, some of these strains of *Clostridium* have been reclassified based on newer techniques and recent taxonomic changes. Of the 141 subjects in our study, 46 recognized species of clostridia were recovered. In addition, we recovered 71 other isolates of *Clostridium* that could not be speciated by current methods; these 71 strains could be separated into 42 different groups of *Clostridium* species by biochemical testing and gas-liquid chromatographic analysis of metabolic end products. The total number of all isolates of *Clostridium* was 489 (average of 3.5 isolates per subject).

# Table 13
## SUMMARY OF CLOSTRIDIA RECOVERED FROM FECES OF SUBJECTS IN A COLON CANCER STUDY[a]

| | Patient Category | | | | | | | | | | | | |
|---|---|---|---|---|---|---|---|---|---|---|---|---|---|
| | Strict vegetarian (13)[b] | | Vegetarian, some meat (14) | | Japanese diet (15) | | Colonic polyp (25) | | Colon cancer (12) | | Western diet (62) | | Total (141) |
| Species | % Pos. | Range(mean)[c] | % Pos. | Range(mean) | % Pos. | Range(mean) | % Pos. | Range(mean) | % Pos. | Range(mean) | % Pos. | Range(mean) | % Pos. |
| *C. acetobutylicum* | 0 | | 0 | | 0 | | 12 | 7.9—10.7(9.0) | 0 | | 2 | 7.6 | 2.8 |
| *C. aminovalericum* | 8 | 11.9 | 14 | 6.3—7.9(7.1) | 13 | 5.8—6.2(6.0) | 28 | 6.0—10.1(8.0) | 0 | | 13 | 5.7—10.6(7.9) | 14.2 |
| *C. aurantibutyricum* | 0 | | 0 | | 0 | | 0 | | 0 | | 3 | 8.3 | 1.4 |
| *C. barkeri* | 0 | | 0 | | 0 | | 0 | | 0 | | 8 | 9.1—10.5(9.8) | 3.5 |
| *C. beijerinckii* | 0 | | 14 | 4.1—10.7(7.4) | 0 | | 4 | 9.4 | 8 | 10.0 | 8 | 5.6—12.8(9.3) | 6.4 |
| *C. bifermentans* | 0 | | 0 | | 20 | 4.3—8.2(6.3) | 60 | 3.5—11.5(9.3) | 25 | 5.0—9.7(7.5) | 32 | 3.3—12.6(8.8) | 29.1 |
| *C. butyricum* | 0 | | 21 | 3.8—10.2(6.4) | 7 | 5.4 | 8 | 6.4—6.5(6.4) | 0 | | 2 | 10.4 | 5.0 |
| *C.cadaveris* | 0 | | 0 | | 0 | | 0 | | 8 | 8.7 | 2 | 5.6 | 1.4 |
| *C. carnis* | 0 | | 0 | | 0 | | 20 | 8.3—10.7(9.9) | 0 | | 5 | 6.3—7.8(7.3) | 5.7 |
| *C. celatum* | 0 | | 0 | | 0 | | 0 | | 17 | 5.7—8.8(7.2) | 0 | | 1.4 |
| *C. chauvoei* | 0 | | 0 | | 7 | 4.7 | 0 | | 0 | | 5 | 5.9—11.1(8.8) | 2.8 |
| *C. cellobioparum* | 0 | | 0 | | 0 | | 12 | 7.8—9.7(8.7) | 0 | | 4 | 5.7—6.5(6.1) | 3.5 |
| *C. clostridiiforme* | 8 | 11.7 | 0 | | 7 | 6.0 | 16 | 7.2—10.7(8.6) | 8 | 4.9 | 8 | 9.6—11.1(10.3) | 8.5 |
| *C. cochlearium* | 0 | | 7 | 4.8 | 0 | | 0 | | 0 | | 5 | 5.7—10.8(7.7) | 2.8 |
| *C. difficile* | 0 | | 0 | | 0 | | 0 | | 0 | | 2 | 4.8 | 0.7 |
| *C. fallax* | 0 | | 0 | | 0 | | 0 | | 0 | | 3 | 7.9—9.6(8.7) | 1.4 |
| *C. felsineum* | 0 | | 0 | | 7 | 8.7 | 0 | | 0 | | 0 | | 0.7 |
| *C. ghoni* | 0 | | 0 | | 0 | | 0 | | 0 | | 2 | 10.4 | 0.7 |
| *C. glycolicum* | 0 | | 7 | 5.1 | 13 | 5.6—8.8(7.2) | 4 | 7.9 | 0 | | 6 | 5.0—6.3(5.5) | 5.7 |
| *C. haemolyticum* | 0 | | 0 | | 0 | | 0 | | 0 | | 3 | 7.3—10.8(9.1) | 1.4 |
| *C. indolis* | 0 | | 7 | 10.1 | 0 | | 0 | | 0 | | 2 | 11.0 | 1.4 |
| *C. innocuum* | 8 | 7.9 | 43 | 6.9—11.4(9.9) | 53 | 6.7—11.4(9.2) | 24 | 5.8—9.9(8.4) | 25 | 4.5—9.9(7.1) | 21 | 6.5—9.2(8.2) | 26.2 |
| *C. irregularis* | 0 | | 0 | | 0 | | 0 | | 0 | | 2 | 7.0 | 0.7 |
| *C. lentoputrescens* | 0 | | 0 | | 0 | | 0 | | 0 | | 2 | 5.7 | 0.7 |
| *C. limosum* | 0 | | 7 | 5.3 | 0 | | 0 | | 0 | | 2 | 4.4 | 1.4 |
| *C. malenominatum* | 0 | | 0 | | 0 | | 4 | 4.8 | 8 | 4.6 | 2 | 9.8 | 1.4 |
| *C. mangenoti* | 0 | | 0 | | 0 | | 8 | 7.9—10.7(9.3) | 0 | | 8 | 3.6—10.1(6.6) | 5.0 |
| *C. nexile* | 0 | | 7 | 3.9 | 0 | | 0 | | 0 | | 2 | 9.7 | 1.4 |
| *C. oceanicum* | 0 | | 0 | | 0 | | 0 | | 0 | | 2 | 7.6 | 0.7 |
| *C. oroticum* | 23 | 10.4—12.0(11.0) | 0 | | 7 | 10.5 | 4 | 4.4 | 0 | | 6 | 6.3—10.4(7.5) | 6.4 |
| *C. paraperfringens* | 15 | 4.1 | 7 | 3.5 | 13 | 6.2—6.8(6.5) | 8 | 3.6—8.8(6.2) | 0 | | 3 | 7.7—8.7(8.2) | 6.4 |
| *C. paraputrificum* | 8 | 4.5 | 7 | 5.4 | 13 | 7.4—8.0(7.7) | 12 | 4.4—8.3(5.8) | 0 | | 11 | 4.1—11.8(7.6) | 9.9 |
| *C. pasteurianum* | 0 | | 0 | | 0 | | 0 | | 0 | | 2 | 10.0 | 0.7 |
| *C. perenne* | 0 | | 0 | | 0 | | 0 | | 0 | | 3 | 6.4—7.5(7.0) | 1.4 |
| *C. perfringens* | 0 | | 7 | 4.0 | 73 | 4.0—10.0(7.6) | 32 | 3.9—10.8(7.3) | 33 | 6.7—11.5(8.2) | 55 | 3.8—12.5(6.9) | 41.1 |

| | | | | | | | | | | | | | |
|---|---|---|---|---|---|---|---|---|---|---|---|---|---|
| *C. plagarum* | 0 | | 0 | | 0 | | 0 | | 0 | | 2 | 5.3 | 0.7 |
| *C. pseudotetanicum* | 0 | | 0 | | 7 | 5.1 | 4 | 4.2 | 0 | | 2 | 4.0 | 2.1 |
| *C. putrefaciens* "B" | 15 | 4.4—6.8(5.6) | 0 | | 0 | | 12 | 9.9—10.9(10.5) | 17 | 4.7—8.4(6.5) | 6 | 4.7—10.1(7.1) | 7.8 |
| *C. ramosum* | 69 | 7.1—11.1(9.4) | 36 | 6.4—11.4(9.3) | 67 | 7.6—10.5(9.1) | 44 | 6.0—10.8(9.4) | 25 | 8.5—10.5(9.6) | 60 | 4.5—12.5(8.9) | 53.2 |
| *C. sartagoformum* | 0 | | 7 | 9.9 | 0 | | 8 | 7.7—10.0(8.8) | 8 | 8.3 | 5 | 8.3—10.9(9.6) | 5.0 |
| *C. septicum* | 0 | | 0 | | 0 | | 0 | | 8 | 9.8 | 3 | 6.9—7.3(7.1) | 2.1 |
| *C. sordellii* | 0 | | 0 | | 7 | 8.5 | 8 | 9.1—11.1(10.1) | 8 | 12.3 | 2 | 9.9 | 3.5 |
| *C. sphenoides* | 8 | 9.5 | 7 | 5.7 | 13 | 4.7—6.7(5.7) | 4 | 9.9 | 8 | 3.8 | 5 | 4.1—10.3(7.0) | 6.4 |
| *C. sporosphaeroides* | 0 | | 0 | | 13 | 4.1—10.2(7.1) | 4 | 6.9 | 0 | | 6 | 5.5—10.9(8.4) | 5.0 |
| *C. subterminale* | 0 | | 0 | | 0 | | 0 | | 0 | | 2 | 7.5 | 0.7 |
| *C. tertium* | 0 | | 0 | | 13 | 4.9—5.9(5.4) | 0 | | 0 | | 8 | 3.9—10.7(7.1) | 5.0 |
| *Clostridium* species[d] | 15 | 4.0—10.0(7.0) | 43 | 5.7—11.0(8.5) | 47 | 4.7—11.2(8.3) | 60 | 6.1—10.9(9.0) | 33 | 5.7—8.1(6.8) | 34 | 4.4—12.5(8.2) | 55 |
| Total counts of *Clostridium* | 100 | 2.1—12.2(8.8) | 100 | 2.1—11.5(8.4) | 100 | 7.7—11.5(9.7) | 100 | 7.8—11.5(10.4) | 100 | 3.8—12.4(8.3) | 100 | 6.5—13.1:(10.2) | 100 |

[a] Total flora studies were done on 141 subjects (Seventh Day Adventists who were strict vegetarians; Seventh Day Adventists who consumed only small quantities of meat; Japanese who consumed an oriental diet that included fish but no beef; patients with colonic polyps; patients with nonobstructing colon cancer; and healthy subjects who consumed a western diet that contained relatively large quantities of beef). Some of these data have been published previously (see text).

[b] Number in ( ) is number of subjects studied.

[c] Range and mean counts of bacteria expressed as number of organisms $\log_{10}$/g feces (dry weight).

[d] In addition to the named species shown above, a total of 42 distinct groups of clostridia (71 isolates) were recovered from the 141 subjects. These 42 groups of *Clostridium* species do not conform to any currently recognized species of *Clostridium* as determined by biochemical testing and gas-liquid chromatographic analysis of metabolic end products.

## Table 14
## SPECIES OF *CLOSTRIDIUM* ISOLATED FROM FECES — LITERATURE REVIEW

| Species of *Clostridium* | No. of subjects | % Yielding species of *Clostridium* indicated | Comments | Ref. |
|---|---|---|---|---|
| *C. acetobutylicum* | 175 | 0.6 | | 69 |
| *C. aminovalericum* | 175 | 1.1 | | 69 |
| | 10 | 10 | Count = 6.3[a] | 71 |
| *C. aurantibutyricum* | 175 | 0.6 | | 69 |
| | 10 | 10 | Patient with colon cancer; count = 9.2 | 71 |
| *C. beijerinckii* | 175 | 1.1 | | 69 |
| *C. bifermentans* | 175 | 24 | | 69 |
| | 72 | 13.8 | Various conditions | 92 |
| | 10 | 10 | Count = 7.6 | 71 |
| *C. butyricum* | 175 | 1.1 | | 69 |
| | 10 | 10 | Patient with colon cancer; count = 9.5 | 71 |
| *C. cadaveris* | 175 | 5.1 | | 69 |
| *C. chauvoei* | 175 | 4.0 | | 69 |
| *C. difficile* | 431 | 10.9 | | 93 |
| | 175 | 0.6 | | 69 |
| | 60 | 0 | | 33 |
| *C. fallax* | 72 | 4.3 | Various conditions | 92 |
| *C. felsineum* | 175 | 6.9 | | 69 |
| *C. glycolicum* | 175 | 0.6 | | 69 |
| *C. indolis* | 175 | 5.7 | | 69 |
| | 10 | 10 | Patient with colon cancer; count = 7.3 | 71 |
| *C. innocuum* | 175 | 5.1 | 69 | |
| | 10 | 10 | Count = 8.8 | 71 |
| *C. irregularis* | 175 | 0.6 | | 69 |
| *C. limosum* | 175 | 1.7 | | 69 |
| *C. malenominatum* | 175 | 1.7 | | 69 |
| *C. novyi* | 72 | 2.8 | Various conditions | 92 |
| *C. oroticum* | 175 | 2.3 | | 69 |
| *C. paraputrificum* | 175 | 45 | | 69 |
| | 10 | 10 | Count = 6.2 | 71 |
| | 10 | 50 | Patients with colon cancer; mean count = 8.2 | 71 |
| *C. pasteurianum* | 175 | 1.7 | | 69 |
| *C. perfringens* | 500 | 26 | Heat-resistant *C. perfringens* only | 94 |
| | 364 | 55 | From post-mortem specimens in Hong Kong; heat resistant *C. perfringens* only | 95 |
| | 175 | 51 | | 69 |
| | 100 | 70 | From Australian aborigines | 96 |
| | 83 | 63 | From patients in Hong Kong; heat-resistant *C. perfringens* only | 95 |
| | 72 | 90 | Miscellaneous conditions | 92 |
| | 58 | 100 | Japanese subjects; counts = 3.5—8.9 | 97 |
| | 57 | 9 | Subjects from England; heat-resistant *C. perfringens* only | 95 |

**Table 14 (continued)**

| Species of *Clostridium* | No. of subjects | % Yielding species of *Clostridium* indicated | Comments | Ref. |
|---|---|---|---|---|
| | 50 | 68 | Mean count = 5.2 | 96 |
| | 43 | 84 | | 98 |
| | 10 | 80 | | 99 |
| | 10 | 10 | Count = 7.5 | 71 |
| | 10 | 20 | Patients with colon cancer; mean counts = 8.1 | 71 |
| *C. plagarum* | 175 | 0.6 | | 69 |
| *C. propionicum* | 10 | 10 | Count = 7.5 | 71 |
| *C. pseudotetanicum* | 175 | 1.7 | | 69 |
| *C. putrefaciens* | 175 | 10 | | 69 |
| *C. putrificum* | 175 | 1.7 | | 69 |
| | 72 | 27.8 | Miscellaneous conditions | 92 |
| | 10 | 10 | Count = 5.3 | 71 |
| | 10 | 20 | Patients with colon cancer; mean count = 7.2 | 71 |
| *C. ramosum* | 175 | 14 | | 69 |
| *C. rectum* | 175 | 11.4 | | 69 |
| *C. sartagoformum* | 175 | 5.1 | | 69 |
| *C. septicum* | 72 | 2.8 | Miscellaneous conditions | 92 |
| *C. sphenoides* | 175 | 4.0 | | 69 |
| | 10[b] | 10 | Patients with colon cancer; count = 5.5 | 71 |
| *C. sporogenes* | 175 | 8.6 | | 71 |
| | 72 | 37.5 | Miscellaneous conditions | 92 |
| *C. sporosphaeroides* | 175 | 10.3 | | 69 |
| | 10 | 10 | Count = 7.5 | 71 |
| *C. sticklandii* | 175 | 0.6 | | 69 |
| *C. tertium* | 72 | 20.8 | Miscellaneous conditions | 92 |
| *C. tetani* | 487 | 24.6 | | 100 |
| | 204 | 0 | | 101 |
| | 78 | 34.7 | | 102 |
| *C. tyrobutyricum* | 10 | 10 | Count = 6.3 | 71 |

[a] Counts expressed as number of bacteria $\log_{10}$/g feces.
[b] Organism was reported as "*C. spheroides*"; presumably *C. spheroides* was meant.

Approximately 15% of isolates in our study did not conform to known species of*Clostridium*. The most frequent isolates were *C. ramosum* (53% of subjects), *C. perfringens* (41%), *C. bifermentans* (29%), *C. innocuum* (26%), *C. aminovalericum* (14%), and *C. paraputrificum* (10%). The remaining species shown in Table 13 were present in fewer than 10% of subjects. The most striking differences in frequency between the 6 subject groups were those of *C. perfringens, C. bifermentans,* and *C. innocuum*. Only 4% of the 27 Seventh Day Adventists (strict vegetarians or "vegetarians" eating small quantities of meat) harbored *C. perfringens* whereas 32 to 73% of subjects in the other 4 groups harbored *C. perfringens*. None of the Seventh Day Adventists harbored *C. bifermentans* as opposed to 20 to 60% of subjects in the other groups. The incidence of *C. innocuum* was appreciably lower in the strict vegetarians than in the other subject groups.

Our literature review (Table 14) revealed that 39 different *Clostridium* species have been recovered from feces by other investigators. *C. perfringens* was the most frequent isolate although the incidence varied from 10 to 100% in the studies reviewed. *C. bifermentans, C. paraputrificum, C. tertium,* and *C. tetani* were also present relatively frequently; the last two organisms have been isolated infrequently or not at all in more recent studies, however.

The above data indicate that a great variety of clostridia, often in significant counts, are

**Table 15**
**FACTORS THAT MAY AFFECT THE COMPOSITION OF HUMAN INTESTINAL FLORA**

| | |
|---|---|
| Age | Intrinsic bowel disease (colon cancer, Crohn's disease, etc.) |
| Diet | |
| Gut motility | Cathartics |
| Anatomic abnormalities of the gut (obstruction, diverticulae, fistulae, etc.) | Diarrhea |
| | Antimicrobial therapy |
| Bowel surgery | "Accidental" ingestion of bacteria in foods, etc. |

present in feces. The potential significance of the clostridial flora of the large bowel for both health and disease processes is immense.

## IV. FACTORS THAT AFFECT THE PRESENCE OF CLOSTRIDIA IN THE HUMAN BOWEL FLORA

The microbial flora of the gut is frequently subjected to exogenous influences that may alter its composition. The data presented in Tables 5 to 8 reveal that clostridia are infrequently isolated from the stomach and small intestine; the available data are generally not sufficient to permit conclusions as regards alterations in gastric and small intestinal flora. Hence, most of our knowledge of changes in the gut flora is from studies of feces.

Limited data are available regarding the stability of fecal flora over time. Many investigators consider a change in bacterial counts of less than 2 $\log_{10}$ (100-fold difference) to be insignificant; we would consider a change of less than $10^4$ organisms per gram of no significance, particularly in the case of organisms for which there is not an efficient selective medium available. Zubrzycki and Spaulding[103] sequentially studied the fecal flora of 4 normal subjects for periods of 17 to 21 days; total counts of clostridia in 1 subject varied less than 2 $\log_{10}$ organisms per gram of feces, whereas the differences between highest and lowest clostridial counts in each of the other 3 were approximately 2.3, 2.5, and 2.6 $\log_{10}$ organisms per gram of feces. Mitsuoka et al.[104] studied the fecal flora of 6 subjects on 6 occasions during a 2.5-month period; 1 subject had less than 2 $\log_{10}$ organisms per gram of feces change in counts of clostridia, whereas changes in clostridial counts in the other 5 ranged from approximately 2.4 to 5.4 $\log_{10}$ organisms per gram of feces. Thus, the total counts of clostridia in feces may vary over relatively short periods of time.

Borriello[65] performed weekly fecal flora studies on 7 subjects for a 6-week period. The maximum change in total counts of clostridia ranged from 0.3 to 2.6 $\log_{10}$ organisms per gram of feces; only 1 of the subjects had a change of 2 $\log_{10}$ organisms or greater, however. The counts of *C. perfringens* in this study varied appreciably over the 6-week interval; 3 subjects had changes of from 4.1 to 4.5 $\log_{10}$ organisms per gram of feces, and 3 others had changes in counts of 2.0 to 3.8 $\log_{10}$ organisms per gram of feces. These data indicate that the changes in total counts of certain species of clostridia may be appreciably more striking than are the changes in total counts of *Clostridium*.

Assessment of factors that affect clostridia in the feces may be approached in two ways: one can compare the flora of one distinct group of individuals to that of another (e.g., vegetarians vs. nonvegetarians), or one can determine the flora in a given individual and then assess the effect of a specified manipulation (e.g., a change of diet or administration of antimicrobials) upon the "baseline" flora. A reasonable amount of data exist for both approaches.

Factors that may produce changes in the bowel flora have been reviewed by Drasar and Hill[63] and are summarized in Table 15.

### A. Age

Ellis-Pegler et al.[21] studied infants, children, and adults, and found that mean fecal counts of clostridia were approximately 9 $\log_{10}$ organisms per gram of feces in infants less than 4 months of age; the mean counts fell progressively in older age groups to a level of approximately 5.6 $\log_{10}$ organisms per gram of feces in children 4 to 8 years of age. The mean count in adults was approximately 7.4 $\log_{10}$ organisms per gram of feces.

Viscidi et al.[33] recovered *C. difficile* from 29% of 45 neonates, 9% of healthy children aged 3 to 24 months, and none from 60 adults. Rietra et al.[37] reported the incidence of cytopathic effect (presumably due to *C. difficile* cytotoxin) in stools of infants and children; 14% of 138 infants younger than 5 months had a fecal cytopathic effect compared to only 0.7% of children age 2 to 15 years. Nakamura et al.[93] studied 431 Japanese subjects who had not received antibiotics and did not have diarrhea; 15.4% of fecal culture of 149 medical students (age 22 to 28 years), 15.9% of 69 inpatients with cerebrovascular disease (age 50 to 80 years), and 7% of 213 healthy residents of a home for the aged (age 60 to 90 years) yielded *C. difficile*. Viscidi et al.[33] did not recover *C. difficile* from the feces of any of the 60 healthy adult Americans, and George et al.[105] recovered *C. difficile* from 3% of 137 healthy adults. The reason for the discrepancies between the Japanese and American adults are unclear.

### B. Diet

#### *1. Infants*

Diet has a significant impact on the flora of infants. Stark and Lee[11] studied 7 breast-fed and 7 formula-fed infants from birth to 1 year of age (Table 1). Clostridia were recovered from the feces of a majority of both groups during the 1st week of life. By the 4th week of life, none of the breast-fed infants yielded clostridia; the same results were obtained until the time of weaning (approximately 9 months of age). In contrast, 6 of the 7 formula-fed infants became permanently "colonized" prior to introduction of solid foods (9 months of age). Bullen and colleagues[12,13] noted that clostridia were absent or present in low counts in breast-fed neonates, whereas neonates who received formula, cow's milk, or breast milk along with cow's milk usually had clostridia in their feces (range of mean counts was approximately 5 to 8 $\log_{10}$ organisms per gram of feces). Bullen et al.[12] also noted that clostridia first appeared in the feces of breast-fed neonates when a supplement (dried cow's milk) was added to the diet. These data indicate that breast milk tends to prevent the establishment of clostridia in the gut of infants, and that the addition of any supplement to the diet results in colonization by clostridia. The explanation for this interesting finding is not known.

Cooperstock[29] found *C. difficile* in the feces of 8% of 13 breast-fed and 71% of 17 bottle-fed infants. Brettle and Wallace[28] and Stark et al.,[32] on the other hand, found that the incidence of colonization of breast- and bottle-fed infants were roughly equivalent.

#### *2. Adults*

Although the total counts of clostridia in the feces of adults may vary somewhat between different groups of subjects (Table 11), the most striking differences are those in the frequency with which certain species occur in various parts of the world.

Drasar et al.[69] cultured feces of 175 adults from 7 different countries; for the 11 species of *Clostridium* that were most frequently isolated, there were major differences in incidence in the various countries (Table 16). The data regarding subjects from Scotland is particularly striking; *C. bifermentans, C. cadaveris, C. felsineum, C. indolis, C. paraputrificum,* and *C. sporogenes* were present in these subjects in an appreciably higher frequency than in subjects from most or all of the other countries. On the other hand, only one eighth of subjects in Hong Kong yielded *C. perfringens*, whereas one half to three fourths of subjects

**Table 16**
**DISTRIBUTION OF CLOSTRIDIA IN FECES OF INDIVIDUALS FROM VARIOUS COUNTRIES**

| | % Of subjects positive from | | | | | | |
|---|---|---|---|---|---|---|---|
| Species | England (36)[a] | Scotland (11) | Wales (65) | U.S.A. (8) | Uganda (19) | Japan (11) | Hong Kong (25) |
| *C. bifermentans* | 19 | 73 | 17 | 25 | 21 | 18 | 8 |
| *C. cadaveris* | 3 | 55 | 0 | 0 | 11 | 0 | 0 |
| *C. felsineum* | 11 | 46 | 0 | 0 | 11 | 0 | 4 |
| *C. indolis* | 6 | 36 | 0 | 0 | 16 | 9 | 0 |
| *C. paraputrificum* | 61 | 91 | 43 | 75 | 26 | 36 | 12 |
| *C. perfringens* | 53 | 55 | 60 | 75 | 53 | 64 | 12 |
| *C. putrefaciens* | 6 | 0 | 22 | 13 | 0 | 0 | 0 |
| *C. ramosum* | 17 | 0 | 12 | 38 | 0 | 27 | 16 |
| *C. rectum* | 11 | 0 | 14 | 0 | 11 | 0 | 20 |
| *C. sporogenes* | 17 | 46 | 3 | 13 | 0 | 9 | 0 |
| *C. sporosphaeroides* | 11 | 0 | 22 | 0 | 0 | 0 | 0 |

[a] Number in ( ) = number of subjects.

Modified from Drasar, B. S., Goddard, P., Heaton, S., Peach, S., and West, B., *J. Med. Microbiol.*, 9, 63, 1976.

from the other six countries harbored *C. perfringens* in their feces. A possible correlation between the incidence of colon cancer and the presence of *C. paraputrificum* in feces has been studied by a number of investigators;[67,69-71,76] this topic is reviewed in a subsequent chapter. Finegold et al.[70] noted significant differences in the incidence of *C. septicum* and *C. tertium* (both are members of the *C. paraputrificum* group) in Seventh Day Adventists (all consumed small quantities of meat or were vegetarians) vs. individuals consuming a western diet (this group consumes a relatively large amount of meat); differences in incidence of other clostridia were not present.

A variety of dietary manipulations has been studied and has generally been found to have little impact on the fecal flora. Hentges et al.[80] studied 10 volunteers who ingested prescribed diets sequentially; the counts of clostridia were 4.9 (initial control diet), 7.5 (meatless diet), 6.1 (high beef diet), and 8.9 $\log_{10}$ organisms per gram (final control diet following high beef diet). The counts of clostridia were statistically different for only the initial control vs. final control diets. Drasar et al.[106] found that addition of fiber (bran) to the diet of volunteers had no appreciable effect on fecal counts of clostridia. Several studies[79,107,108] of the effect of an elemental or chemically defined diet on fecal flora failed to detect an effect of the diet on total fecal counts of clostridia. Attebery et al.[109] studied the effect of a chemically defined diet on the fecal flora of three subjects; two subjects acquired *C. innocuum* in high counts whereas the third lost *C. innocuum* while receiving the defined diet. *C. perfringens* and *C. ramosum* disappeared from one subject each during the diet, and *C. ramosum*, *C. tetanomorphum*, and an unidentified *Clostridium* species were acquired by the third subject. From the above data, it is evident that short-term dietary changes have little influence on total counts of clostridia in feces; the data from Attebery et al.,[109] however, suggest that diet may result in changes in the individual species of *Clostridium* that are present. The results of the study by Drasar et al.[69] (Table 16) support this concept. Factors (not studied) other than diet that might influence the species of *Clostridium* found in feces include general state of health, GI immunity to various *Clostridium* species, hygiene, and types and quantities of *Clostridium* ingested (accidentally) with a particular diet. In areas where hygiene is poor, the ingestion of certain *Clostridium* species during infancy and childhood might be a factor

Table 17
PRESENCE OF CLOSTRIDIA IN THE FECES OF NEONATES WITH CONGENITAL OBSTRUCTION OF THE GASTROINTESTINAL TRACT[a]

| | Type of illness | | |
|---|---|---|---|
| **Organism** | **Esophageal atresia(9)[b]** | **Small bowel obstruction(5)[b]** | **Large bowel obstruction(10)** |
| *C. bifermentans* | 0[c] | 0 | 40 |
| *C. cochlearium* | 11 | 20 | 0 |
| *C. difficile* | 33 | 20 | 40 |
| *C. perfringens* | 78 | 20 | 70 |
| *C. sporogenes* | 9 | 40 | 0 |
| *C. tertium* | 0 | 0 | 0 |

[a] Culture of rectal swab specimens taken postoperatively; only two patients had clostridia recovered from cultures of bowel contents at time of surgery.
[b] Indicates that patients received prophylactic antibiotics.
[c] Numbers indicate percent of subjects yielding the species indicated.

in the ability (or lack of ability) of certain species subsequently to establish residence in the gut. It is reasonable to speculate that certain culinary practices might influence the fecal flora. Uncooked fish (sashimi) is a significant part of the diet in Japan, yet raw fish is rarely, if ever, ingested in Occidental populations. To our knowledge, colonization of sashimi by clostridia has not been studied.

## C. Abnormalities of Gut Anatomy and Motility

Motility of the gut is important in determining the distribution of bacteria in the bowel;[63] factors that impair motility may lead to proliferation of bacteria, including clostridia, within the bowel lumen. The exception to this is the case of newborn infants who have congenital obstruction of the GI tract. Rotimi and Duerden[34] studied 24 newborns with esophageal atresia, small intestinal atresia, or congenital large bowel obstruction within 2 days of birth. None of the babies had clostridia recovered from rectal swabs cultured preoperatively, and only two had clostridia recovered from bowel contents cultured at the time of surgery. Within 10 days of surgery, several species of *Clostridium* could be recovered from rectal swabs of a number of the infants (Table 17).

Bishop and Anderson[110] studied the small intestinal flora of 25 infants and children with bowel obstruction. A variety of bacteria were recovered, including nine isolates of *C.perfringens* and one isolate each of *C. fallax*, *C. butyricum*, and *C. paraputrificum*. Although the small intestinal flora of normal infants and children has not been well studied, the data for adults (Tables 7 and 8) suggests that the presence of these clostridia in the small intestine may be a consequence of bowl obstruction.

Farrar et al.[111] studied eight subjects with "stagnation-inducing lesions" of the small bowel (blind loop syndrome or small bowel diverticulosis); one of the eight (who had jejunal diverticulosis) had a *Clostridium* species and two coliform bacteria recovered from jejunal fluid. Of eight patients with small intestinal diverticulosis studied by Goldstein et al.[112] anaerobic cultures of small bowel contents were performed for two, and clostridia, along with other anaerobes and facultative organisms, were recovered from one. McKenna et al.[113] recovered a heavy growth of a *Clostridium* (along with several other organisms) from small bowel contents of a patient with jejunal diverticulosis. Polter et al.[114] studied a patient with blind loop syndrome (Billroth II anastomosis and jejunal diverticula) on several occasions.

Of 9 specimens of upper small bowel contents, 2 yielded clostridia in counts of 8 $\log_{10}$ organisms per gram of material (total bacterial count was 9.1 $\log_{10}$ organisms per gram) and 8.3 $\log_{10}$ organisms per gram of material (total bacterial count was 9.0 $\log_{10}$ organisms per gram).

### D. Effect of Cathartics and Diarrhea Upon Bowel Flora

Levison and Kaye[81] studied the effect of milk of magnesia upon fecal flora in five volunteers and found that the counts of clostridia in one rose significantly following administration of the cathartic. Nichols et al.[61] studied the bowel flora in six patients who received preoperative mechanical cleansing (only) of the bowel and six control subjects. The mean total counts of clostridia in cecum, transverse colon, and feces were not significantly different between treated patients and controls.

Mata et al.[115] did not recover clostridia from any of the 8 patients with severe diarrhea due to *Shigella dysenteriae*, whereas 5 of the 12 healthy volunteers had mean counts of clostridia that ranged from 9 to 10 $\log_{10}$ organisms per gram of feces. Albert et al.,[26] on the other hand, found that there was not a significant difference between the counts of clostridia (based on wet weight) in 49 patients with diarrhea vs. 29 controls.

### E. Effect of Antimicrobials on Clostridia in Feces

Administration of certain antimicrobial agents may have a profound impact on the human fecal flora. Despite the clinical significance of this, there are relatively few detailed studies of the effects of various antimicrobials upon fecal flora.

#### *1. Penicillins*

Cohn[75] found that phenoxymethyl penicillin (2.4 million units daily for 3 days) promptly reduced fecal counts of *Clostridium* from a median count of approximately $10^{10}$ organisms per gram of feces to 0 in 5 patients. Heimdahl and Nord,[116] on the other hand, did not find appreciable changes in fecal counts of *Clostridium* in 10 patients who received 1600 mg of phenoxymethyl penicillin daily for 7 days. Benzyl-penicillin is said to produce a slight decrease in fecal clostridia,[117] but detailed studies are lacking.

Finegold et al.[118] found that 1 to 2 g of ampicillin reduced counts of clostridia from 8 to 9 $\log_{10}$ organisms per gram of feces to 0 by the 7th day of therapy in 3 patients; hetacillin (a related compound) had no effect in 1 patient. Sutter and Finegold[119] found that cyclacillin administration (2 g/day) in 3 subjects had variable effects upon clostridia in feces; *C. perfringens* was eradicated in 2 patients, but was acquired by the 3rd during therapy. *C. ramosum* was eradicated in one subject and was unaffected in the other two. Other *Clostridium* species were unaffected. Heimdahl et al.[120] found no affect of bacampicillin on fecal clostridia in 6 subjects receiving 1200 mg of drug daily for 7 days. From thes limited data it is difficult to predict the effects of the various penicillins on fecal clostridia; additional studies are obviously needed.

#### *2. Cephalosporins*

Sutter and Finegold[119] found that cephalexin (2 g/day) had variable effects on clostridia in 3 subjects; *C. perfringens* was eradicated in 1 patient and *C. ramosum* was acquired in another. Other *Clostridium* species were reduced in one subject and unaffected in the other two.

Mulligan et al.[126] found that *Clostridium* species, other than *C. difficile*, were reduced by a factor of 2 to 6.6 $\log_{10}$ organisms per gram of feces in 3 patients and unaffected in 3 others treated with cefoxitin. Of note, *C. difficile* was acquired during or after therapy in counts of 4.6 to 8.9 $\log_{10}$ organisms per gram of feces in 5 of the 6 cefoxitin-treated subjects.

Mulligan et al.[121] also studied the effect of cefoperazone on the fecal flora of four patients. Species of *Clostridium* (other than *C. difficile*) were present in counts of 9 to 10.6 $\log_{10}$

organisms per gram of feces before therapy, and were eradicated in 3 and reduced by a factor of 4.5 $\log_{10}$ organisms per gram of feces in the other patient. *C. difficile* was acquired in 1 subject following therapy (count of 8.8 $\log_{10}$ organisms per gram of feces) and was initially present in 1 other, but was eradicated during therapy.

Allen et al.[122] found that low-dose moxalactam (1.5 g/day for 9 days) did not produce a decrease in total counts of clostridia of 4 $\log_{10}$ organisms per gram of feces or greater in any of the 5 subjects. Of 5 other subjects, 2 who received high-dose moxalactam (6 g/day for 9 days), however, did have a decrease in counts of clostridia greater than 4 $\log_{10}$ organisms per gram of feces; 1 subject transiently acquired *C. difficile* (count of $10^{3.6}$/gram of feces) during low-dose moxalactam therapy. This isolate was inhibited by 16 μg of moxalactam per milliliter in vitro. Allen et al.[122] also implied that counts of *C. ramosum* and *C. innocuum* increased during therapy to as high as 11 $\log_{10}$ organisms per gram of feces. Moxalactam was "not particularly active" against *C. innocuum, C. ramosum,* and certain other *Clostridium* species recovered before, during, and after therapy; 14 strains of *C. innocuum* isolated before, during, or after therapy required greater than 64 μg/mℓ of moxalactam for inhibition of growth.

*3. Lincosamides*

Finegold et al.[123] studied the effect of lincomycin on fecal flora and found that 2.4 g/day (intramuscularly) did not affect clostridia in 1 patient. The counts of clostridia in 1 of 3 patients who received 1.5 to 2.0 g/day of lincomycin by mouth decreased by a factor of $10^3$ per gram of feces. Clostridia, which initially were present in counts of $10^8$ organisms per gram of feces were eradicated in the other 2 subjects.

Heimdahl and Nord[116] studied clindamycin (600 mg/day by mouth for 7 days) in 10 subjects. A clindamycin-resistant *C. perfringens* was isolated from both the saliva and feces during therapy of 1 subject; the minimal inhibitory concentration (MIC) of clindamycin was 16 μg/mℓ for this organism. Two other subjects had "increased" numbers of resistant *C. innocuum* (MIC was 64 μg of clindamycin per milliliter) and 1 other was transiently colonized with *C. difficile* (MIC was 8 μg of clindamycin per milliliter). Sutter and Finegold[119] noted substantial changes in fecal flora of 5 subjects who received 1.2 g of clindamycin daily by mouth; anaerobes other than *Eubacterium* and *Clostridium* were eliminated. *C. perfringens* was eradicated in three of the subjects and *C. ramosum* in one other. The *Clostridium* species found in high counts during therapy were *C. difficile, C. innocuum, C. oroticum,* and *C. ramosum*.

*4. Tetracyclines*

Finegold et al.[72] noted that 2 of 5 patients given various tetracycline salts had increases in clostridia of 5 and 6 $\log_{10}$ organisms per gram of feces. Cohn[75] studied several different tetracyclines and found that the effect on clostridia in feces was quite variable. Doxycycline had no effect on total counts of clostridia in one subject studied by Finegold et al.[118]

*5. Erythromycin*

Cohn[75] reported that fecal clostridia were eliminated in all of 3 patients receiving 2 g of erythromycin per day.

*6. Metronidazole*

Krook[73] gave metronidazole (800 mg/day by mouth for 14 days) to 9 subjects. Clostridial counts were increased by greater than 1 $\log_{10}$ organisms per gram of feces in 4 subjects, decreased by greater than 1 $\log_{10}$ organisms per gram of feces in 4, and unchanged in the remaining subject. Lewis et al.[68] treated 6 subjects with several different doses of metronidazole (10, 20, and 30 mg/kg/day); total counts of clostridia were reduced by the highest dose and *C. perfringens* was eradicated in all 4 subjects in whom it was present initially. Metronidazole has been used with good effect in a number of cases of *C. difficile*-induced colitis, but has also induced the disease on at least one occasion.

*7. Other Agents*

Data are extremely limited as regards the effect of other antimicrobials on bowel flora. The polymyxins, aminoglycosides, and vancomycin are poorly excreted by the liver and undoubtedly have a minimal impact on clostridia in the gut. When given by the oral route these agents may, however, have a major impact on clostridia. Finegold et al.[72] found that oral administration of neomycin eradicated clostridia in three subjects studied. Hewitt and Finegold[124] found that clostridia were eradicated from the feces of 2 subjects who received oral kanamycin (6 g/day). Fecal counts in a 3rd subject receiving oral kanamycin for 2 weeks increased from 3 to 9 $\log_{10}$ organisms per gram of feces on the 3rd day of drug administration; at 7 days clostridia were not detected, but were again present in a count of 3 $\log_{10}$ organisms per gram of feces on the 14th day of drug administration.

Orally administered vancomycin is currently the agent of choice for treatment of *C. difficile*-induced diarrhea. Oral bacitracin has also been effective in treating *C. difficile*-associated disease (see Chapter 7).

Our knowledge of the impact of various antimicrobials upon clostridia in the gut is extremely limited. The clinical implications of this impact are significant both as regards selection of antimicrobial-resistant strains during therapy that may become involved in subsequent infections, and as regards the development of antimicrobial-induced diarrhea. Although *C. difficile* has clearly been shown (but only in the past 3 to 4 years) to be a cause of antimicrobial agent-induced diarrhea or colitis, this organism cannot be implicated in approximately 2/3 of the cases of diarrhea related to antimicrobial therapy. Moreover, Viscidi et al.[33] and Mulligan and colleagues[121] (*vide supra*) have documented the presence of *C. difficile* in the feces of antibiotic recipients who do not have diarrhea.

## V. SUMMARY

Clostridia are an important component of the human GI flora. The GI tract of the infant is rapidly colonized by clostridia soon after birth; this colonization persists except in infants whose diet consists solely of breast milk.

In adults (and presumably children), clostridia are infrequently present in the stomach, duodenum, jejunum, and proximal ileum. Both the incidence and counts of clostridia are higher in the terminal ileum than in the more proximal small bowel. In feces, a large number of species of *Clostridium* may be found, sometimes in counts that represent a significant proportion of the total fecal flora.

Marked differences exist in the particular species of *Clostridium* found in feces between various patient populations. The reasons for these differences are not clear; although short-term changes in diet have not been shown to alter the total counts of *Clostridium* in feces, the available data suggests that diet may have an influence on the species present in the colon.

A variety of other factors may affect both the types and numbers of clostridia in feces; of these, antimicrobial therapy is probably the most significant. Antimicrobials that are incompletely absorbed form the gut lumen or that are excreted in an active form by the liver may have a profound effect on fecal clostridia. Although antimicrobials have been producing diarrhea for more than 40 years, the role of *C. difficile* as one cause of this diarrhea has been revealed only within the past 4 years. This single fact unequivocally demonstrates the ignorance of medical science as regards the importance of intestinal clostridia in both health and disease.

## REFERENCES


1. **Moore, W. E. C. and Holdeman, L. V.,** Human fecal flora: the normal flora of 20 Japanese-Hawaiians, *Appl. Microbiol.*, 27, 961, 1974.
2. **Sutter, V. L., Citron, D. M., and Finegold, S. M.,** *Wadsworth Anaerobic Bacteriology Manual*, 3rd ed., C. V. Mosby, St. Louis, 1980, 131.
3. **Challacombe, D. N., Richardson, J. M., and Anderson, C. M.,** Bacterial microflora of the upper gastrointestinal tract in infants without diarrhea, *Arch. Dis. Child.*, 49, 264, 1974.
4. **Brook, I., Barrett, C. T., Brinkman, C. R., III, Martin, W. J., and Finegold, S. M.,** Aerobic and anaerobic bacterial flora of the maternal cervix and newborn gastric fluid and conjuntiva: a prospective study, *Pediatrics*, 63, 451, 1979.
5. **Mata, L. J. and Urrutia, J. J.,** Intestinal colonization of breast-fed children in a rural area of low socio-economic level, *Ann. N.Y. Acad. Sci.*, 176, 93, 1971.
6. **Rotimi, V. O. and Duerden, B. I.,** The development of the bacterial flora in normal neonates, *J. Med. Microbiol.*, 14, 51, 1981.
7. **Patte, C., Tancrede, C., Raibaud, P., and Ducluzeau, R.,** Premières étapes de la colonisation bactérienne du tube digestif du nouveau-né, *Ann. Microbiol. (Inst. Pasteur)*, 130A, 69, 1979.
8. **Kindley, A. D., Roberts, P. J., and Tulloch, W. H.,** Neonatal necrotizing enterocolitis, *Lancet*, 1, 649, 1977.
9. **Upton, M. F.,** The anaerobic intestinal flora of normal breast-fed and artificially fed infants, *Am. J. Dis. Child.*, 37, 1221, 1929.
10. **Laverdière, M., Robert, A., Chicoine, R., Salet, D., and Rosenfeld, R.,** Clostridia in necrotizing enterocolitis, *Lancet*, 2, 377, 1978.
11. **Stark, P. L. and Lee, A.,** The microbial ecology of the large bowel of breast-fed and formula-fed infants during the first year of life, *J. Med. Microbiol.*, 15, 189, 1982.
12. **Bullen, C. L., Tearle, P. V., and Willis, A. T.,** Bifidobacteria in the intestinal tract of infants: an in-vivo study, *J. Med. Microbiol.*, 9, 325, 1976.
13. **Bullen, C. L., Tearle, P. V., and Stewart, M. G.,** The effect of "humanized" milks and supplemented breast feeding on the faecal flora of infants, *J. Med. Microbiol.*, 10, 403, 1977.
14. **Stark, P. L. and Lee, A.,** Clostridia isolated from the feces of infants during the first year of life, *J. Pediatr.*, 100, 362, 1982.
15. **Hall, I. C. and O'Toole, E.,** Intestinal flora in new-born infants, *Am. J. Dis. Child.*, 49, 390, 1935.
16. **Hall, I. C. and Snyder, M. L.,** Isolation of an obligately anaerobic bacillus from the feces of newborn infants and from other human sources, and its probable identity with the "Köpfchenbakterien" of Escherich, Rodella's "Bacillus III", and Bacillus paraputrificus (Bienstock), *J. Bacteriol.*, 28, 181, 1934.
17. **Cashore, W. J., Peter, G., Stonestreet, B. S., and Oh, W.,** Clostridia colonization and clostridial toxin in neonatal necrotizing enterocolitis, *J. Pediatr.*, 98, 308, 1981.
18. **Kelsey, M. C. and Vince, A. J.,** Clostridia in neonatal feces, *Lancet*, 2, 100, 1979.
19. **Gothefors, L. and Blenkharn, I.,** *Clostridium butyricum* and necrotizing enterocolitis, *Lancet*, 1, 52, 1978.
20. **Leclerc, H. and Moriamez, J. C.,** Etude quantitative de la flore fécale de l'adulte et du nourrisson alimenté artificiellement, *Pathol. Biol.*, 28, 217, 1980.
21. **Ellis-Pegler, R. B., Crabtree, C., and Lambert, H. P.,** The faecal flora of children in the United Kingdom, *J. Hyg.*, 75, 135, 1975.
22. **Smith, M. F., Borriello S. P., Clayden, G. S., and Casewell, M. W.,** Clinical and bacteriological findings in necrotizing enterocolitis: a controlled study, *J. Infect.*, 2, 23, 1980.
23. **Hall, I. C. and Matsumura, K.,** Recovery of *Bacillus tertius* from stools of infants, *J. Infect. Dis.*, 35, 502, 1924.
24. **Mata, L. J., Carrillo, C., and Villatoro, E.,** Fecal microflora in healthy persons in a preindustrial region, *Appl. Microbiol.*, 17, 596, 1969.
25. **Snyder, M. L.,** The normal fecal flora of infants between two weeks and one year of age, *J. Infect. Dis.*, 66, 1, 1940.
26. **Albert, M. J., Bhat, P., Rajan, D., Maiya, P. P., Pereira, S. M., and Baker, S. J.,** Faecal flora of south Indian infants and young children in health and with acute gastroenteritis, *J. Med. Microbiol.*, 11, 137, 1978.
27. **Haenel, H.,** Human normal and abnormal gastrointestinal flora, *Am. J. Clin. Nutr.*, 23, 1433, 1970.
28. **Brettle, R. P. and Wallace, E.,** *Clostridium difficile* from stools of normal children, *Lancet*, 1, 1193, 1982.
29. **Cooperstock, M. S.,** *Clostridium difficile*, enterocolitis, and Hirschsprung's disease, *Lancet*, 1, 800, 1982.
30. **Donta, S. T. and Myers, M. G.,** *Clostridium difficile* toxin in asymptomatic neonates, *J. Pediatr.*, 100, 431, 1982.

31. **Mardh, P.-A., Helin, I., Colleen, I., Öberg, M., and Holst, E.,** *Clostridium difficile* toxin in faecal specimens of healthy children and children with diarrhea, *Acta Paediatr. Scand.*, 71, 275, 1982.
32. **Stark, P. L., Lee, A., and Parsonage, B. D.,** Colonization of the large bowel by *Clostridium difficile* in healthy infants; quantitative study, *Infect. Immun.*, 35, 895, 1982.
33. **Viscidi, R., Willey, S., and Bartlett, J. G.,** Isolation and toxigenic potential of *Clostridium difficile* isolates from various patient populations, *Gastroenterology*, 81, 5, 1981.
34. **Rotimi, V. O. and Duerden, B. I.,** The bacterial flora of neonates with congenital abnormalities of the gastrointestinal tract, *J. Hyg.*, 88, 69, 1982.
35. **Howard, F. M., Flynn, D. M., Bradley, J. M., Noone, P., and Szawatkowski, M.,** Outbreak of necrotizing enterocolitis caused by *Clostridium butyricum*, *Lancet*, 2, 1099, 1977.
36. **Wort, A. J. and Ozere, R. L.,** Characteristics of a strain of *Clostridium carnis* causing septicaemia in a young infant, *J. Clin. Pathol.*, 29, 1011, 1976.
37. **Rietra, P. J. G. M., Slaterus, K. W., Zanen, H. C., and Meuwissen, S. G. M.,** Clostridial toxin in faeces of healthy infants, *Lancet*, 2, 319, 1978.
38. **Sherertz, R. J. and Sarubbi, F. A.,** The prevalence of *Clostridium difficile* and toxin in a nursery population: a comparison between patients with necrotizing enterocolitis and an asymptomatic group, *J. Pediatr.*, 100, 435, 1982.
39. **Kliegman, R. M., Fanaroff, A. A., Izant, R., and Speck, W. T.,** Clostridia as pathogens in neonatal necrotizing enterocolitis, *J. Pediatr.*, 95, 287, 1979.
40. **Howie, J. W., Duncan, I. B. R., and Mackie, L. M.,** Growth of *Clostridium welchii* in the stomach after partial gastrectomy, *Lancet*, 2, 1018, 1953.
41. **Nichols, R. L. and Smith, J. W.,** Intragastric microbial colonization in common disease states of the stomach and duodenum, *Ann. Surg.*, 182, 557, 1975.
42. **Nelson, D. P. and Mata, L. J.,** Bacterial flora associated with the human gastrointestinal mucosa, *Gastroenterology*, 58, 56, 1970.
43. **Bhat, P., Shantakumari, S., Rajan, D., Mathan, V. I., Kapadia, C. R., Swarnabai, C., and Baker, S. J.,** Bacterial flora of the gastrointestinal tract in southern Indian control subjects and patients with tropical sprue, *Gastroenterology*, 62, 11, 1972.
44. **Drasar, B. S., Shiner, M., and McLeod, G. M.,** Studies on the intestinal flora. I. The bacterial flora of the gastrointestinal tract in healthy and achlorhydric persons, *Gastroenterology*, 56, 71, 1969.
45. **Franklin, M. A. and Skoryna, S. C.,** Studies on natural gastric flora. I. Bacterial flora of fasting human subjects, *Can. Med. Assoc. J.*, 95, 1349, 1966.
46. **Gianella, R. A., Broitman, S. A., and Zamcheck, N.,** Gastric acid barrier to ingested microorganisms in man: studies *in vivo* and *in vitro*, *Gut*, 13, 251, 1972.
47. **Vince, A., Dyer, N. H., O'Grady, F. W., and Dawson, A. M.,** Bacteriological studies in Crohn's disease, *J. Med. Microbiol.*, 5, 219, 1972.
48. **Gorbach, S. L., Spanknebel, G., Weinstein, L., Plaut, A. G., Nahas, L., and Levitan, R.,** Studies of intestinal microflora. VIII. Effect of lincomycin on the microbial population of the human intestine, *J. Infect. Dis.*, 120, 298, 1969.
49. **Gray, J. D. A. and Shiner, M.,** Influence of gastric pH on gastric and jejunal flora, *Gut*, 8, 574, 1967.
50. **Thadepalli, H., Lou, M. A., Bach, V. T., Matsui, T. K., and Mandal, A. K.,** Microflora of the human small intestine, *Am. J. Surg.*, 138, 845, 1979.
51. **Plaut, A. G., Gorbach, S. L., Nahas, L., Weinstein, L., Spanknebel, G., and Levitan, R.,** Studies of intestinal microflora. III. The microbial flora of human small intestinal mucosa and fluids, *Gastroenterology*, 53, 868, 1967.
52. **Lou, M. A., Mandal, A. K., Alexander, J. L., and Thadepalli, H.,** Bacteriology of the human biliary tract and the duodenum, *Arch. Surg.*, 112, 965, 1977.
53. **Bornside, G. H. and Cohn, I., Jr.,** The normal microbial flora, *Am. J. Digest. Dis.*, 10, 845, 1965.
54. **Appelbaum, P. C., Moshal, M. G., Hift, W., and Chatterton, S. A.,** Intestinal bacteria in patients with tropical sprue, *S. Afr. Med. J.*, 57, 1081, 1980.
55. **Bornside, G. H., Welsh, J. S., and Cohn, I., Jr.,** Bacterial flora of the human small intestine, *JAMA*, 196, 109, 1966.
56. **Corrodi, P., Wideman, P., Sutter, V. L., Drenick, E., Passaro, E., Jr., and Finegold, S. M.,** Bacterial flora before and after bypass procedure for morbid obesity , *J. Infect. Dis.*, 137, 1, 1978.
57. **Beeken, W. L. and Kanich, R. E.,** Microbial flora of the upper small bowel in Crohn's disease, *Gastroenterology*, 65, 390, 1973.
58. **Drasar, B. S. and Shiner, M.,** Studies on the intestinal flora, *Gut*, 10, 812, 1969.
59. **Gorbach, S. L., Banwell, J. G., Mitra, R., Chatterjee, B. D., Jacobs, B., and Mazumder, D. N. G.,** Bacterial contamination of the upper small bowel in tropical sprue, *Lancet*, 1, 74, 1969.
60. **Kalser, M. H., Cohen, R., Arteaga, I., Yawn, E., Mayoral, L., Hoffert, W. R., and Frazer, D.,** Normal viral and bacterial flora of the human small and large intestine, *N. Engl. J. Med.*, 274, 500, 1966.

61. **Nichols, R. L., Gorbach, S. L., and Condon, R. E.,** Alteration of intestinal microflora following preoperative mechanical preparation of the colon, *Dis. Colon Rectum*, 14, 123, 1971.
62. **Finegold, S. M., Sutter, V. L., Boyle, J. D., and Shimada, K.,** The normal flora of ileostomy and transverse colostomy effluents, *J. Infect. Dis.*, 122, 367, 1970.
63. **Drasar, B. S. and Hill, M. J.,** *Human Intestinal Flora*, Academic Press, London, 1974, 263.
64. **Peach, S., Lock, M. R., Katz, D., Todd, I. P., and Tabaqchali, S.,** Mucosal-associated bacterial flora of the intestine in patients with Crohn's disease and in a control group, *Gut*, 19, 1034, 1978.
65. **Borriello, S. P.,** Clostridial Flora of the Gastrointestinal Tract in Health and Disease, Ph.D. thesis, Central Public Health Laboratories, London, 1981.
66. **Vargo, D., Moskovitz, M., and Floch, M. H.,** Faecal bacterial flora in cancer of the colon, *Gut*, 21, 701, 1980.
67. **Finegold, S. M., Attebery, H. R., and Sutter, V. L.,** Effect of diet on human fecal flora: comparison of Japanese and American diets, *Am. J. Clin. Nutr.*, 27, 1456, 1974.
68. **Lewis, R. P., Wideman, P., Sutter, V. L., and Finegold, S. M.,** The effect of metronidazole on human fecal flora, in *Metronidazole*, Proc. Int. Metronidazole Conf., Finegold, S. M., Ed., Excerpta Medica, Princeton, N.J., 1977, 307.
69. **Drasar, B. S., Goddard, P., Heaton, S., Peach, S., and West, B.,** Clostridia isolated from faeces, *J. Med. Microbiol.*, 9, 63, 1976.
70. **Finegold, S. M., Sutter, V. L., Sugihara, P. T., Elder, H. A., Lehmann, S. M., and Phillips, R. L.,** Fecal microbial flora in Seventh Day Adventist populations and control subjects, *Am. J. Clin. Nutr.*, 30, 1781, 1977.
71. **Legakis, N. J., Ioannides, H., Tzannetis, S., Golematis, B., and Papavassiliou, J.,** Faecal bacterial flora in patients with colon cancer and control subjects, *Zentralbl. Bakteriol. Hyg. Abt. I: Orig. A*, 251, 54, 1981.
72. **Finegold, S. M., Posnick, D. J., Miller, L. G., and Hewitt, W. L.,** The effect of various antibacterial compounds on the normal human fecal flora, *M. Kröokol Ernährungsforschung*, 10, 316, 1965.
73. **Krook, A.,** Effect of metronidazole and sulfasalazine on the normal human faecal flora, *Scand. J. Gastroenterol.*, 16, 587, 1981.
74. **Gorbach, S. L., Nahas, L., Lerner, P. I., and Weinstein, L.,** Studies of intestinal microflora. I. Effects of diet, age, and periodic sampling on numbers of fecal microorganisms in man, *Gastroenterology*, 53, 845, 1967.
75. **Cohn, I.,** *Intestinal Antisepsis*, Charles C Thomas, Springfield, Ill., 1968, 245.
76. **Finegold, S. M., Flora, D. J., Attebery, H. R., and Sutter, V. L.,** Fecal bacteriology of colonic polyp patients and control patients, *Cancer Res.*, 35, 3407, 1975.
77. **Aries, V. C., Crowther, J. S., Drasar, B. S., Hill, M. J., and Ellis, F. R.,** The effect of a strict vegetarian diet on the faecal flora and faecal steroid concentration, *J. Pathol.*, 103, 54, 1971.
78. **Aries, V., Crowther, J. S., Drasar, B. S., Hill, M. J., and Williams, R. E. O.,** Bacteria and the aetiology of cancer of the large bowel, *Gut*, 10, 334, 1969.
79. **Bornside, G. H. and Cohn, I., Jr.,** Stability of normal human fecal flora during a chemically defined low residue liquid diet, *Ann. Surg.*, 181, 58, 1975.
80. **Hentges, D. J., Maier, B. R., Burton, G. C., Flynn, M. A., and Tsutakawa, R. K.,** Effect of a high-beef diet on the fecal bacterial flora of humans, *Cancer Res.*, 37, 568, 1977.
81. **Levison, M. E. and Kaye, D.,** Fecal flora in man: effect of cathartic, *J. Infect. Dis.*, 119, 591, 1969.
82. **Löhr, W. and Rabfeld, L.,** *Die Bakteriologie der Wurmfortsatzentzündung und der Appendikularen Peritonitis*, Thieme, Stuttgart, 1931, 4.
83. **Weinberg, M., Prévot, A-R., Davesne, J., and Renard, C.,** Recherches sur la bactériologie et la sérothérapie des appendicites aiguës, *Ann. Inst. Pasteur*, Suppl. 42, 1167, 1928.
84. **Jennings, J. E.,** The relation of the Welch bacillus to appendicitis and its complications, *Ann. Surg.*, 93, 828, 1931.
85. **Perrone,** Contribution a l'étude de la bactériologie de l'appendicite, *Ann. Inst. Pasteur*, 19, 367, 1905.
86. **Cazzamali, P. and Miglierina, R.,** La batteriologia delle peritoniti acute, *Arch. Ital. Chir.*, 34, 573, 1933.
87. **Maccabe, A. F. and Orr, J.,** A study of 200 cases of appendicitis with special reference to their bacteriology, *Edinburgh Med. J.*, 59, 100, 1952.
88. **Heyde, M.,** Bakteriologische und experimentelle Untersuchungen zur Aetiologie der Wurmfortsatzentzündung (mit besorderer Berucksichtigung der anaëroben Bakterien), *Beitr. Klin. Chir.*, 76, 1, 1911.
89. **Buttiaux, R. and Tiprez, J.,** Étude de la flore microbienne des appendices dans les appendicites aiguës et chroniques, *C.R. Soc. Biol. (Paris)*, 114, 133, 1933.
90. **Lanz, O. and Tavel, E.,** Bacteriologie de l'appendicite, *Rev. Chir.*, 30, 43, 1904.
91. **Werner, H. and Seeliger, H. P. R.,** Kulturelle Untersuchungen über den Keimgehalt der Appendix unter besonderer Berücksichtigung der Anaerobier, *Zentralbl. Bakteriol. Parasitenkd. Infektionskr. Hyg. Abt. I: Orig.* 188, 345, 1963.

92. **Kahn, M. C.,** Anaerobic spore-bearing bacteria of the human intestine in health and in certain diseases, *J. Infect. Dis.*, 35, 423, 1924.
93. **Nakamura, S., Mikawa, M., Nakashio, S., Takabatake, M., Okado, I., Yamakawa, K., Serikawa, T., Okumura, S., and Nishida, S.,** Isolation of *Clostridium difficile* from the feces and the antibody in sera of young and elderly adults, *Microbiol. Immunol.*, 25, 345, 1981.
94. **Leeming, R. L., Pryce, J. D., and Meynell, M. J.,** *Clostridium welchii* and food poisoning, *Br. Med. J.*, 1, 501, 1961.
95. **Turner, G. C. and Wong, M. M.,** Intestinal excretion of heat resistant *Clostridium welchii* in Hong Kong, *J. Pathol. Bacteriol.*, 82, 529, 1961.
96. **Sutton, R. G. A.,** Enumeration of *Clostridium welchii* in the faeces of varying sections of the human population, *J. Hyg.*, 64, 367, 1966.
97. **Akama, K. and Otani, S.,** *Clostridium perfringens* as the flora in the intestine of healthy persons, *Jpn. J. Med. Sci. Biol.*, 23, 161, 1970.
98. **Sapico, F. L., Emori, H., Smith, L. DS., Bluestone, R., and Finegold, S. M.,** Absence of relationship of fecal *Clostridium perfringens* to rheumatoid arthritis and rheumatoid variants, *J. Infect. Dis.*, 128, 559, 1973.
99. **Smith, H. W. and Crabb, W. E.,** The faecal bacterial flora of animals and man: its development in the young, *J. Pathol. Bacteriol.*, 82, 53, 1961.
100. **Bauer, J. H. and Meyer, K. F.,** Human intestinal carriers of tetanus spores in California, *J. Infect. Dis.*, 38, 295, 1926.
101. **Kerrin, J. C.,** The incidence of *B. tetani* in human faeces, *Br. J. Exp. Pathol.*, 9, 69, 1928.
102. **Tenbroeck, C. and Bauer, J. H.,** The tetanus bacillus as an intestinal saprophyte in man, *J. Exp. Med.*, 36, 261, 1922.
103. **Zubrzycki, L. and Spaulding, E. H.,** Studies on the stability of the normal human fecal flora, *J. Bacteriol.*, 83, 968, 1962.
104. **Mitsuoka, T., Sega, T., and Yamamoto, S.,** Eine verbesserte Methodik der qualitativen und quantativen Analyse der Darmflora von Menschen und Tieren, *Zentralbl. Bakteriol. Parasitenkd. Infektionskr. Hyg. Abt. 1:* Orig., 195.S, 455, 1965.
105. **George, W. L., Sutter, V. L., and Finegold, S. M.,** Toxigenicity and antimicrobial susceptibility of *Clostridium difficile*, a cause of antimicrobial agent-associated colitis, *Curr. Microbiol.*, 1, 55, 1978.
106. **Drasar, B. S., Jenkins, D. J. A., and Cummings, J. H.,** The influence of a diet rich in wheat fiber on the human faecal flora, *J. Med. Microbiol.*, 9, 423, 1976.
107. **Crowther, J. S., Drasar, B. S., Goddard, P., Hill, M. J., and Johnson, K.,** The effect of a chemically defined diet on the faecal flora and faecal steroid concentration, *Gut*, 14, 790, 1973.
108. **Winitz, M., Adams, R., Seedman, D. A., Davis, R. N., Jayko, L. G., and Hamilton, J. A.,** Studies in metabolic nutrition employing chemically defined diets, *Am. J. Clin. Nutr.*, 23, 546, 1970.
109. **Attebery, H. R., Sutter, V. L., and Finegold, S. M.,** Effect of a partially chemically defined diet on normal human fecal flora, *Am. J. Nutr.*, 25, 1391, 1972.
110. **Bishop, R. F. and Anderson, C. M.,** The bacterial flora of the stomach and small intestine in children with intestinal obstruction, *Arch. Dis. Child.*, 35, 487, 1960.
111. **Farrar, W. E., Jr., O'Dell, N. M., Achord, J. L., and Greer, H. A.,** Intestinal microflora and absorption in patients with stagnation-inducing lesions of the small intestine, *Am. J. Digest. Dis.*, 17, 1065, 1972.
112. **Goldstein, F., Wirts, W., Salen, G., and Mandel, R. J.,** Diverticulosis of the small intestine. Clinical, bacteriologic and metabolic observations in a group of seven patients, *Am. J. Digest. Dis.*, 14, 170, 1969.
113. **McKenna, R. D., Beck, I. T., and Epstein, H.,** Clinical significance of small bowel function, *Can. Med. Assoc. J.*, 83, 896, 1960.
114. **Polter, D. E., Boyle, J. D., Miller, L. G., and Finegold, S. M.,** Anaerobic bacteria as cause of the blind loop syndrome, *Gastroenterology*, 54, 1148, 1968.
115. **Mata, L. J., Mejicanos, M. L., and Franklin, F.,** Studies on the indigenous flora of Guatemalan children, *Am. J. Clin. Nutr.*, 25, 1380, 1972.
116. **Heimdahl, A. and Nord, C. E.,** Effect of phenoxymethylpenicillin and clindamycin on the oral, throat and fecal microflora of man, *Scand. J. Infect. Dis.*, 11, 233, 1979.
117. **Finegold, S. M.,** Interaction of antimicrobial therapy and intestinal flora, *Am. J. Clin. Nutr.*, 23, 1466, 1970.
118. **Finegold, S. M., Davis, A., and Miller, L. G.,** Comparative effect of broad-spectrum antibiotics on nonspore-forming anaerobes and normal bowel flora, *Ann. N.Y. Acad. Sci.*, 145, 265, 1967.
119. **Sutter, V. L. and Finegold, S. M.,** The effect of antimicrobial agents on human faecal flora: studies with cephalexin, cyclacillin and clindamycin, in *The Normal Microbial Flora of Man*, Skinner, F. A. and Carr, J. G., Eds., Academic Press, London, 1974, 229.
120. **Heimdahl, A., Nord, C. E., and Weilander, K.,** Effect of bacampicillin on human mouth, throat and colon flora, *Infection*, 7(Suppl. 5), S446, 1979.

121. **Mulligan, M. E., Citron, D. M., McNamara, B. T., and Finegold, S. M.,** The impact of cefoperazone therapy upon the fecal flora, *Antimicrob. Agents Chemother.*, 22, 226, 1982.
122. **Allen, S. D., Siders, J. A., Cromer, M. D., Fischer, J. A., Smith, J. W., and Israel, K. S.,** Effect of LY127935 (6059-S) on human fecal flora, in *Current Chemotherapy and Infectious Disease*, Vol. 1, Nelson, J. D. and Grassi, C., Eds., American Society for Microbiology, Washington, D.C., 1980, 101.
123. **Finegold, S. M., Harada, N. E., and Miller, L. G.,** Lincomycin: activity against anaerobes and effect on normal human fecal flora, in *Antimicrobial Agents and Chemotherapy — 1965*, Hobby, G. L., Ed., American Society for Microbiology, Ann Arbor, Michigan, 1966, 659.
124. **Hewitt, W. L. and Finegold, S. M.,** Laboratory studies with kanamycin, *Ann. N.Y. Acad. Sci.*, 76, 122, 1958.
125. **Finegold, S. M., Sutter, V. L., and Mathisen, G. E.,** Normal indigenous intestinal microflora, in *Human Intestinal Microflora in Health and Disease*, Academic Press, New York, 1983, 3.
126. **Mulligan, M. E., Citron, D., Gabay, E., Kirby, B. D., George, W. L., and Finegold, S. M.,** Alterations in human fecal flora including ingrowth of *Clostridium difficile*, related to cefoxitin therapy, *Antimicrob. Agents Chemother.*, 26, 343, 1984.

Chapter 2

# INFANT BOTULISM

**Stephen S. Arnon**

## TABLE OF CONTENTS

# I. INTRODUCTION

Infant botulism is the newly recognized infectious form of botulism in which ingested spores of *Clostridium botulinum* germinate, multiply, and produce botulinal toxin in the intestinal lumen of the patient.[1,2] Infant botulism should be distinguished from the two other better-known forms of botulism. Food-borne botulism is an intoxication caused by ingestion of improperly preserved food that contains preformed botulinal toxin. Wound botulism, a rarely reported entity, results when *C. botulinum* organisms infect and produce toxin in traumatized tissue.

The neurotoxin produced by *C. botulinum* is the most potent poison known.[3] By extrapolation from studies with other animal species, the lethal dose of this toxin in the bloodstream of man is approximately $10^{-9}$ mg/kg body weight.[4]

Botulinal toxin is a disulfide-bonded, two-chain, simple protein that weighs approximately 150,000 daltons. It irreversibly blocks the release of acetylcholine at peripheral synapses, the most important of which clinically is the neuromuscular junction. Thus, the hallmark of botulism is an acute, flaccid paralysis which invariably begins with the muscles of the head, face, and throat and then extends symmetrically to involve the trunk and extremities. Death results from paralyzed tongue or pharyngeal muscles occluding the airway, or from paralysis of the diaphragm and intercostal muscles.

Recovery from botulism occurs by growth of new nerve endings.[5] With the availability of mechanical respirators and intensive care facilities, the paralyzed patient may be kept alive while this process is completed. These technological developments help explain the decline in the case-fatality rate of botulism from 70% in the 1920s to its present 20% in the U.S.[6]

*C. botulinum* is a spore-forming, obligate anaerobe whose natural habitat is the soil, in which it has been found on every continent where it has been sought. Its spores, carried by dust, are commonly present on fresh fruits, vegetables, and other agricultural products (e.g., honey). The different types of *C. botulinum* can be distinguished serologically by the antigenically distinct toxin molecule each produces. The various toxin types, all of which act by the same mechanism, have arbitrarily been designated by the letters A to G. These different toxin types serve as convenient epidemiological and clinical markers.

Infant botulism is not a new disease, only a newly recognized one. The first clinical cases of botulism in infants were described in 1976,[7] and soon thereafter evidence was obtained that infant botulism, unlike botulism from food poisoning at other ages, is an infectious disease.[2,8] It is now clear in retrospect that a bacteriologically proven case occurred in 1931 and was misdiagnosed.[9] In the mid-1970s the clinical features of infants with botulism were again reported from several centers,[10-12] but only in one instance was infant botulism proposed.[13]

Patients with infant botulism can be divided into three clinical groups: those with (1) mild paralysis, often diagnosed as "failure to thrive", in which hospitalization is not required, (2) moderate to severe paralysis, in which the need for hospitalization is clear, and (3) fulminant paralysis, in which death occurs without warning and without opportunity for hospitalization. As discussed below, infants in the third group are indistinguishable at autopsy from typical cases of sudden infant death syndrome (SIDS, crib death),[14-20] which is the leading post-neonatal cause of infant mortality in the developed countries.

# II. INFANT BOTULISM AS ONE CAUSE OF SUDDEN INFANT DEATH (SIDS)

The possibility that infant botulism might be a cause of SIDS was suggested by (1) the potency and mechanism of action of the toxin, (2) the absence of histological effects from the toxin, and (3) the observation that some patients simply stopped breathing while at home in their mothers' arms. The possibility appeared confirmed when *C. botulinum* was found

**Table 1**
**PERSPECTIVES ON THE SIDS AGE DISTRIBUTION**

"The age distribution of SIDS is unique and striking . . . .Any viable hypothesis for SIDS must account for its characteristic age distribution."

Taken from Schwartz, P. J., *Rev. Perinat. Med.*, 4, 475—524, 1981.

"The SIDS age distribution . . . pattern has been documented time after time and constitutes the single most consistent, provocative and unique characteristic of SIDS yet identified."

Taken from Peterson, D. R., *Epidemiol. Rev.*, 2, 97—112, 1980.

" . . . the peculiar age distribution . . . is completely at variance with all other diseases that cause infants to die. In fact, no other condition that one can think of immediately has such a circumscribed age distribution at any age . . . "

Taken from Froggatt, P., *2nd Int. SIDS Conf.*, University of Washington Press, Seattle, 1970, 77.

"We are aware of no other condition with this unique age distribution."

Taken from Beckwith, J. B., *Curr. Probl. Pediatr.*, 3, 1—34, 1973.

" . . . none of the characteristics of the [SIDS] population are universal, except for the age distribution . . . ."

Taken from Guntheroth, W. G., *Crib Death*, Futura, Mt. Kisco, N.Y., 1982, 47.

in SIDS autopsy specimens from California, Massachusetts, Utah, Washington, and Switzerland.[14-20] In the California study, *C. botulinum* organisms, or both toxin and organisms, were found in 4.7% 10 of 211 SIDS cases and in none of the 69 control infants who died of identifiable causes ($p = 0.02$).[14]

The presence of *C. botulinum* in infants affected by neuromuscular paralysis may be considered diagnostic of infant botulism, as *C. botulinum* is not part of the normal resident human infant intestinal microflora (see Section IV). The exact proportion of SIDS cases that may be attributable to infant botulism is not known. The failure of some studies to find *C. botulinum* in SIDS autopsy specimens may reflect geographical variation in incidence or in methodology.[21,22] Also, as discussed in the next section, substantial logistical and laboratory impediments render difficult the identification of *C. botulinum* and its toxin in autopsy material. However, before turning to this topic, a final point deserves consideration.

In addition to the direct evidence just described linking infant botulism and SIDS, another observation suggests their connection. The age distribution of SIDS had long been considered its unique feature (Table 1), but as more cases of infant botulism were recognized, it became apparent that the infant botulism age distribution matched that of SIDS (Figure 1). Since Figure 1 contains all U.S. cases of infant botulism known as of December 31, 1982, it may be considered as accurate a depiction of its age distribution as is presently epidemiologically obtainable. The similarity of the two age distributions suggests that factors which account for the pathophysiology of infant botulism might likewise be important for some SIDS cases.

## III. METHODOLOGICAL OBSTACLES TO THE IDENTIFICATION OF *CLOSTRIDIUM BOTULINUM* IN SIDS SPECIMENS

The cultural demonstration of *C. botulinum* in . . . (biological samples) . . . by identifying its toxin in the fluid of mixed cultures and its isolation from such cultures is not always a simple task . . . .Primarily, this difficulty arises because no single procedure, no single medium is best for strains of all toxin types and cultural varieties of *C. botulinum*. Consequently, the procedures adopted and the media used represent compromises . . . .It should be added that a certain amount of stubbornness on the part of the investigator is a definite advantage.[23]

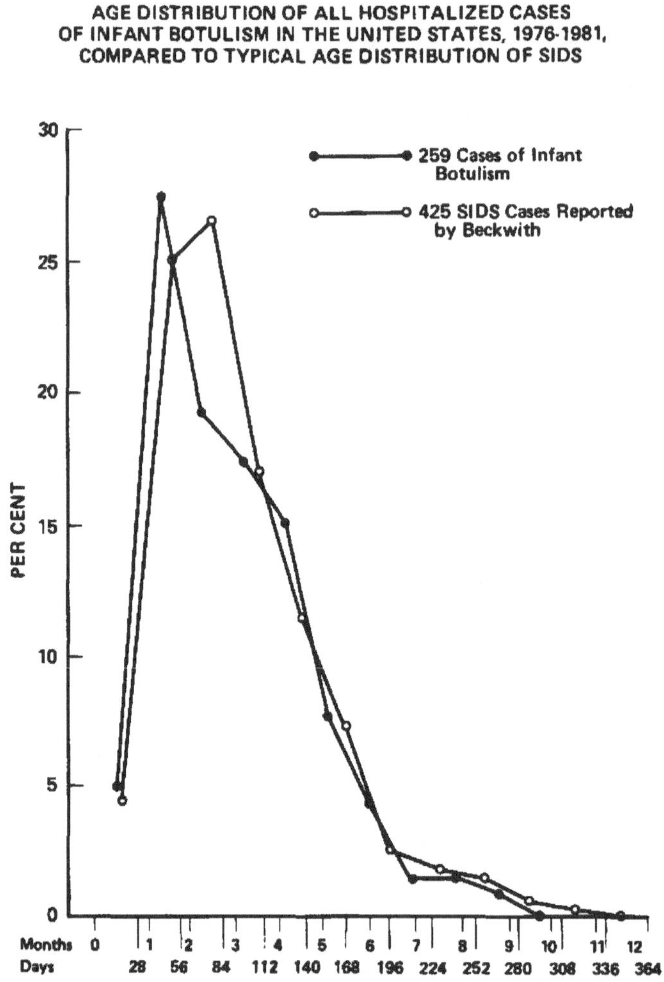


FIGURE 1. Age distribution of all hospitalized cases of infant botulism in the U.S., 1976—1981, compared to a typical age distribution of SIDS reported by Beckwith (Reference 91).

In any attempt to demonstrate *C. botulinum* in SIDS infants, the laboratory difficulties mentioned above are compounded by logistical constraints that affect the biological quality of the autopsy specimens in unknown and unpredictable ways. For instance, at a minimum, several hours elapse between death and autopsy, during which time the intestinal milieu may change appreciably from lack of refrigeration, from autolysis of the epithelium, and from unequal growth rates of different species of intestinal bacteria that may then release proteases and bacteriocins. In addition, after collection it may not be possible to maintain anaerobic and chilled conditions while the specimens are in transit to the laboratory. Attempts to overcome these obstacles by freezing SIDS autopsy specimens may or may not succeed, as the effects of freezing and thawing on the durability of *C. botulinum* vegetative cells and the toxin are not completely known. At the laboratory, the task of identifying *C. botulinum* and its toxin may be complicated by the small size of specimens available and by the use of standard media which permit multiplication of other intestinal bacteria that can inhibit growth and toxin production by *C. botulinum*.[24-26]

## IV. *CLOSTRIDIUM BOTULINUM* AND THE NORMAL INTESTINAL MICROFLORA OF HUMANS

Based on evidence obtained from diverse populations and geographical locales, it appears established that *C. botulinum* is not part of the normal resident intestinal flora of either infants or adults (Table 2).[2,8,14,27-35] For infants specifically, the modern evidence in support of this conclusion may be summarized as follows.

In California since 1976, feces from 395 healthy infants have been studied. Of these, 172 were selected as residence controls (i.e., they lived within a 1.6-km radius of an infant botulism patient), so that they might have a similar exposure to any *C. botulinum* spores present in the general area. All 172 residence control infants had fecal specimens that were negative for *C. botulinum*. In the feces of just 1 of the remaining 223 healthy control infants *C. botulinum* type A organisms, but no toxin, were found. After the fecal specimen that contained *C. botulinum* had been collected, this infant developed constipation (one of the first signs of infant botulism), which was relieved by treatment with suppositories; 3 additional fecal specimens from this infant obtained from 4 to 6 weeks after the initial positive specimen had been passed were negative for *C. botulinum*. Since 1976, further study on 212 California infants with illnesses that prompted their attending physicians to request infant botulism testing have had their feces examined; specimens from all 212 were negative for both *C. botulinum* and its toxin and, based on the clinical and laboratory findings, the infants were judged not to have had infant botulism. Also in this context, it may be noteworthy that the California study of SIDS did not find *C. botulinum* or its toxin in autopsy specimens from 68 control infants who died of identified causes.[14]

In Australia, the question of the possible presence of *C. botulinum* as a member of the normal intestinal flora was approached differently. All clostridia that could be isolated from 98 fecal specimens obtained from 10 healthy infants during their 1st year of life were identified as to species. A total of 12 species were found; none of these was either *C. botulinum* or its atoxigenic counterpart, *C. sporogenes*.[27]

Despite the foregoing evidence, the causative role of *C. botulinum* isolated from SIDS autopsy specimens was once questioned by a report that *C. botulinum* organisms were commonly present, and occasionally the toxin as well, in feces of children living in Utah (including one child as old as 11 years) (Table 2).[17] The high fecal prevalence of *C. botulinum* reported from Utah is at variance with all other studies of the feces of infants and adults (Table 2), and the Utah data would appear to require confirmation.

In contrast to the above findings, in healthy infants and infants ill with other diseases, *C. botulinum* has been found in the feces of each of 157 California patients with gradual onset of infant botulism, both outpatient and hospitalized (Table 2). Furthermore, the feces of 4 of these type B patients contained *C. botulinum* type B in numbers as high as $6.0 \times 10^8$ organisms per gram (which represented 3.3% of the total fecal flora),[36] while in the feces of a type B Australian patient, *C. botulinum* peaked at $8.4 \times 10^6$ organisms per gram.[37] This latter fecal sample simultaneously contained 61,440 mouse $LD_{100}$ of botulinal toxin per gram — by extrapolation, almost 9,000 times the human $LD_{100}$ of toxin if totally absorbed into the circulation.[4] A similar study quantitating excretion by a type A patient identified peak levels of $2.5 \times 10^8$ organisms and 64,000 mouse $LD_{100}$ of toxin per gram of feces.[38]

## V. EPIDEMIOLOGY

With recognition of the spectrum of severity of illness, attention turned to the questions of how and why only some infants became ill, and of those who did, what factors determined their severity of illness? Most current epidemiological information available to answer these

### Table 2
### RESULTS OF STUDIES OF HUMAN FECES DONE WITH METHODS CAPABLE OF DETECTING *CLOSTRIDIUM BOTULINUM* AND ITS TOXIN

| Populations studied | No. of persons | No. (%) positive for *C. botulinum* | No. (%) positive for botulinal toxin | Ref. |
|---|---|---|---|---|
| Healthy California infants, controls for patients with infant botulism[a] | 395 | 1 (0.3)[a] | 0 | 14, 86 |
| California patients whose fecal specimens were submitted to "rule-out infant botulism" | 212 | 0 | 0 | |
| Healthy Australian infants[b] | 10 | 0 | 0 | 27 |
| Healthy British infants | 45 | 0 | 0 | 83 |
| Infants who died, not of SIDS, California | 68 | 0 | 0 | 14 |
| Infants who died, not of SIDS, Washington State | 8 | 0 | 0 | 16 |
| Patients with infant botulism, California | 157 | 157 (100.0) | 155 (98.7)[c] | |
| Utah children[d] | | | | 17 |
| Infant with diarrhea | 1 | 1 (100.0) | 1 (100.0) | |
| Infants and children with atypical illnesses "suggestive of infant botulism" | 9 | 7 (77.8) | 1 (11.1) | |
| Infants and children with "neurological diseases, not botulism" | 45 | 9 (20.0) | 0 | |
| Healthy infants | 32 | 3 (9.4) | 0 | |
| Healthy adult and child family members of patients with infant botulism, California | 23 | 0 | 0 | 2 |
| Healthy adult microbiologists, California | 34 | 0 | 0 | 86 |
| Healthy adult Japanese-Hawaiians | 20 | 0 | Not tested | 28 |
| U.S. patients with colonic polyps | 25 | 0 | Not tested | 29 |
| Healthy U.S. adults | 35 | 0 | Not tested | 29, 30 |
| Healthy Japanese and American adults | 33 | 0[e] | Not tested | 31 |
| Healthy adults in the U.S., Great Britain, Hong Kong, Japan, and Uganda | 175 | 0 | 0 | 32 |
| U.S. patients with pernicious anemia, arthritis, eczema, or intestinal disease | 62 | 0 | Not tested | 30 |
| Healthy adults and children in California | 88 | 0[f] | 0 | 33 |
| Healthy U.S. adults | 10 | 2 (20.0)[f] | 0 | 34 |
| U.S. adults and children without clinical botulism | 92 | 4 (4.3)[g] | 0 | 35 |

[a] Fecal specimens from 15 hospitalized patients were submitted without identification to the Botulism Laboratory of the Centers for Disease Control (CDC), Atlanta, Ga., for confirmatory testing. From each of these 15 specimens, CDC isolated *C. botulinum* of the same single toxin type (8A, 7B) that had originally been identified in California (100% concordance). The single infant with a single-fecal specimen positive for *C. botulinum* was constipated (see text).

[b] Over their 1st year of life, 98 separate fecal specimens were examined, and all were negative for *C. botulinum* organisms and toxin.

[c] Failure to identify botulinal toxin in patients' feces occurred only when the amount submitted was insufficient for testing.

[d] Fecal specimens of 5 patients found to be positive for *C. botulinum* in Utah were sent to the Botulism Laboratory of the CDC. CDC was able to confirm the isolation in only 3 specimens (60% concordance): the first infant was a normal control, the second was diagnosed as having cerebral hypotonia, and the third as having Werdnig-Hoffmann disease. Botulinal toxin was not found in any of the 5 specimens by either laboratory.

[e] The single isolate in this study of a species thought to be *C. botulinum* was later found to be a different *Clostridium* species [Sutter, V. L., personal communication cited in Dowell et al. (Reference 35)].

[f] According to Easton and Meyer,[33] the methods used by Tanner and Dack[34] were inadequate to prevent laboratory-acquired contamination of specimens and consequent false-positive results.

[g] These four asymptomatic persons had consumed food known to contain *C. botulinum* as part of a common-source outbreak of food-borne botulism.

questions, however, is derived from the moderately paralyzed, hospitalized group of infant botulism patients. This group may differ in important but as yet unrecognized ways from patients in the other clinical categories. This limitation, combined with the novelty of the

**Table 3**
**DIAGNOSES CONSIDERED IN 1976 IN PATIENTS WHOSE ILLNESSES WERE SUBSEQUENTLY CONFIRMED AS INFANT BOTULISM**

| Initial impression at time of hospitalization | Common subsequent diagnoses |
|---|---|
| Sepsis | Disorders of amino acid metabolism |
| Viral syndrome | Metabolic encephalopathy |
| Dehydration | Hypothyroidism |
| Pneumonia | Heavy metal poisoning (lead, magnesium, arsenic) |
| Hypotonia of unknown etiology | Myasthenia gravis |
| Failure to thrive | Werdnig-Hoffmann disease (infantile muscular atrophy) |
| | Drug or toxin ingestion |
| | Poliomyelitis |
| | Polyneuropathy secondary to viral infection |
| | Guillain-Barré syndrome or idiopathic infantile polyneuritis |
| | Hirschsprung disease (congenital megacolon) |

Modified with permission from Johnson, R. O., Clay, S. A., and Arnon, S. S., Diagnosis and management of infant botulism, *Am. J. Dis. Child.*, 133, 586, 1979. Copyright 1979, American Medical Association.

infant botulism, renders tentative the conclusions derived from the limited data presently available. Some conclusions may need to be revised substantially as additional investigations further define the epidemiology and pathophysiology of infant botulism.

As judged by the number of hospitalized cases, infant botulism is not a common disease. On the other hand, it is not as uncommon as present figures might suggest, as awareness of its existence still remains limited. In each of the past 8 years, cases of infant botulism have been reported from some U.S. states for the first time. Cases have now been identified in four other countries. As knowledge of infant botulism spreads, the reported incidence should steadily increase.

In considering the epidemiology, it should also be borne in mind that many presently occurring cases may be going unrecognized. This point is best illustrated by considering the various diagnoses given in 1976 to patients with infant botulism, before the true nature of their illness was suspected (Table 3). Today, as then, physicians unfamiliar with infant botulism may have incorrectly diagnosed the disease. The documented temporal and geographical variation in the occurrence of infant botulism needs to be interpreted in light of uneven medical awareness of the existence of the disease. For epidemiological purposes, in California, a case of infant botulism was defined as an illness consistent with the known paralyzing action of botulinal toxin, in which *C. botulinum* was isolated from fecal or autopsy specimens. If the fecal specimen was of adequate size, botulinal toxin was almost always found in it (Table 2).

As of December 31, 1982, a total of 316 hospitalized cases had been recognized worldwide, representing 3 continents and 5 countries — Australia (10 cases), Canada (1 case), Czechoslovakia (1 case), England (1 case), and the U.S. (303 cases);[37-44] 94% of these cases occurred in infants 6 months of age or less. Only 15 SIDS cases and 4 outpatient cases of infant botulism have been documented to date, probably because study of the mild and fulminant ends of the clinical spectrum has been limited.[14-19] For this reason, the present analysis of descriptive epidemiology will be restricted to the hospitalized cases (Tables 4 and 5), which, as previously mentioned, may differ in as yet unrecognized ways from patients in the other parts of the clinical spectrum.

Over two thirds (71%) of the U.S. cases have been identified in just three widely separated states, California, Pennsylvania, and Utah. In the latter two states, the presence of a major pediatric referral center may have contributed in part to increased case recognition in a

**Table 4**
**NUMBERS AND INCIDENCE OF HOSPITALIZED CASES OF INFANT BOTULISM, 1977—1982 CALIFORNIA, PENNSYLVANIA, UTAH, AND THE U.S.**

| | Year hospitalized | | | | | |
|---|---|---|---|---|---|---|
| | **1977** | **1978** | **1979** | **1980** | **1981** | **1982** |
| **Hospitalized Cases** | | | | | | |
| Total U.S. | 43 | 36 | 22 | 65 | 73 | 47 |
| California | 25 | 12 | 9 | 30 | 39 | 29 |
| Pennsylvania | 4 | 4 | 4 | 8 | 11 | 2 |
| Utah | 4 | 7 | 1 | 4 | 1 | 4 |
| **Incidence per 100,000 Live Births** | | | | | | |
| Total U.S. | 1.3 | 1.1 | 0.6 | 1.9 | 2.0 | 1.3 |
| California | 7.2 | 3.4 | 2.5 | 7.8 | 9.2 | 6.8 |
| Pennsylvania | 2.6 | 2.6 | 2.5 | 5.3 | 6.8 | 1.2 |
| Utah | 10.6 | 18.0 | 2.4 | 9.3 | 2.4 | 9.6 |

From Centers for Disease Control, Atlanta, Ga., and National Center for Health Statistics, Rockville, Md. (1981 and 1982 live birth figures, and 1982 case tally, are provisional).

**Table 5**
**TOXIN TYPE, AGE, SEX, AND RACE OF 156 HOSPITALIZED CALIFORNIA CASES OF INFANT BOTULISM, 1976—1982**

| | Hospitalized cases | |
|---|---|---|
| | **No.** | **%** |
| Toxin type | | |
| A | 100 | 64.1 |
| B | 56 | 35.9 |
| Sex | | |
| Male | 79 | 50.6 |
| Female | 77 | 49.4 |
| Race | | |
| White | 118 | 75.6 |
| Hispanic | 28 | 18.0 |
| Asian | 7 | 4.5 |
| American Indian | 2 | 1.3 |
| Black | 1 | 0.6 |
| Age (weeks) | | |
| 2.9—38.0 (range) | | |
| 12.8 (median) | | |
| 13.8 (mean) | | |

circumscribed geographical region.[17,21] The substantial variation in incidence from year to year (Table 4) both within each state and between these three states serves as a reminder of the effect of changes in small numerators on rates and of the difficulities of deriving general conclusions from a limited period of observation.

### A. Toxin Type

With three exceptions, all hospitalized cases of infant botulism have been caused by bacteria that produced either types A or B toxin. The exceptions, all from New Mexico, were one case caused by type F toxin[45] and two cases caused by a toxin apparently with both type B and F antigenicities.[84]

In the U.S., the distribution of hospitalized cases by toxin type has paralleled the known regional soil variation of *C. botulinum*;[46] type A cases predominate from the Rocky Mountains westward, while type B cases predominate from the Mississippi River eastward.[44] This finding suggests that proximity to the telluridic reservoir of *C. botulinum* spores is an important factor in acquisition of disease in some cases.[17,47] However, since *C. botulinum* is so widespread, exposure of most infants to this bacterium is probably a usual, but intermittent, occurrence that only infrequently results in clinically evident infant botulism. For this reason, other environmental and host factors would appear to contribute importantly to susceptibility and resistance to illness.

### B. Age

The age at onset of patients hospitalized with infant botulism is one of the special features of this disease, and one that distinguishes it from every other known human bacterial infection. All other presently recognized bacterial pathogens have the capacity to infect adults and older children as well as infants.*

In infant botulism, 94% of hospitalized cases have been between 2 weeks and 6 months of age, with the median age of onset occurring at 13 weeks (Figure 1). The oldest hospitalized patient was 9 months of age at onset, the youngest 13 days.

### C. Sex

The sex ratio of the hospitalized cases did not differ significantly from the male:female birth ratio (Table 5). Absence of the male preponderance found in many infections of infancy suggests either that females are more susceptible to the less fulminant (hospitalized) form of infant botulism, or alternatively, that males are more vulnerable to the fulminant, SIDS form of the disease and thereby elude hospitalization. Or the explanation for the apparent absence of a male predominance among hospitalized patients may reside with other, as yet unidentified, factors.

### D. Race

Very few nonwhite infants have been hospitalized with infant botulism (Table 5). Meaningful analyses of incidence by race is difficult because the number of cases in each category is small.

### E. Dietary Factors

#### *1. Milk*

Because infants are fed a limited diet consisting mainly of milk, it seemed possible that different milk diets might affect the severity of infant botulism by creating different conditions

* However, the Centers for Disease Control in Atlanta, Ga., have revised their classification scheme for botulism (food-borne, wound, infant, and now, "unclassified") to allow for the possibility that single-case outbreaks of botulism in adults might result from intestinal production of botulinal toxin rather than from food-borne intoxication or from wound infections.[48]

in the infant intestine. Substantially dissimilar intestinal environments could, in turn, influence the ability of *C. botulinum* spores to germinate, multiply, and produce toxin, and thereby influence the severity of this infectious disease.[1,2,4] Because milk is the central element of the infant's diet, the influence of human and formula milk on illness was investigated.[4,49]

Breast-feeding appeared to provide relative protection against SIDS from infant botulism in California, in that all SIDS cases positive at autopsy for *C. botulinum* had been formula-fed, whereas the hospitalized patients were primarily breast-fed. No California infant who died suddenly from infant botulism had been breast-fed within 10 weeks of death. When viewed in the context of age at onset of illness, the disparity in breast-feeding between hospitalized patients and their age-matched controls became even more apparent (Figure 2). After 16 weeks of age, every patient hospitalized with infant botulism was still being nursed. Finally, the feeding experiences of each group were significantly different from each other and from that of age-matched controls.[4,49]

These findings suggested that human milk provided partial protection against fulminant paralysis from infant botulism, and thereby allowed sufficient time for hospitalization to be arranged. Formula milk was associated with the most severe form of the disease, viz., sudden death. Infant botulism may thus be added to the list of other enteric bacterial infections of infancy against which breast-feeding provides relative protection.[50-53]

A possible physiological basis for these epidemiological observations was recently reported. Human milk contains secretory IgA (S-IgA) antibody that can specifically agglutinate vegetative cells of *C. botulinum*. These secretory antibodies are present in the milk both of healthy control mothers whose children never had infant botulism and in the milk of mothers whose babies had infant botulism. However, not every woman's milk contained secretory antibody against all strains of *C. botulinum* tested, and among those women whose milk had agglutinating antibody, substantial variation in titer against a given strain was found.[54] These observations add a further element of biological variability to assessing the possible protective role of breast-feeding in SIDS, in that protection conferred by nursing might be expected to depend in part on the presence, the specificity, and the titer of the antibodies in human milk. For instance, if a woman's milk lacked specific antibody against the particular strain of *C. botulinum* that had colonized her baby's intestine, then the child might die suddenly from infant botulism despite being nursed.

The presence of antibodies in human milk directed against *C. botulinum* demonstrates the effective functioning of the "enteromammary immune system". This recently delineated part of a mother's contribution to her infant's immunological defenses through substances in her milk provides an essential concept for understanding the possible preventive role of breast-feeding in SIDS.

The "enteromammary immune system" describes the remarkable physiological process that connects a lactating woman's intestine to the protective immunological factors in her milk (which include S-IgA antibody, lactoferrin, lysozyme, complement, leukoytes, etc.).[55] In brief, the system functions as follows: lymphocytes in maternal gut-associated lymphoid tissue become sensitized to indigenous or ingested foreign antigens (e.g., foodstuffs, bacteria), and then these cells migrate to the breast where they produce antigen-specific S-IgA antibody. The system performs efficiently and effectively; 3 days after women ingested a nonpathogenic strain of *Escherichia coli*, their milk for the first time contained agglutinating S-IgA antibody against strain-specific *E. coli* antigens;[56] milk from Costa Rican and Pakistani women possessed S-IgA antibody that neutralized the toxins of *E. coli* and *Vibrio cholerae* strains indigenous to those countries.[57-59]

*2. Honey*

Of the several hundred food items fed to California patients who subsequently developed

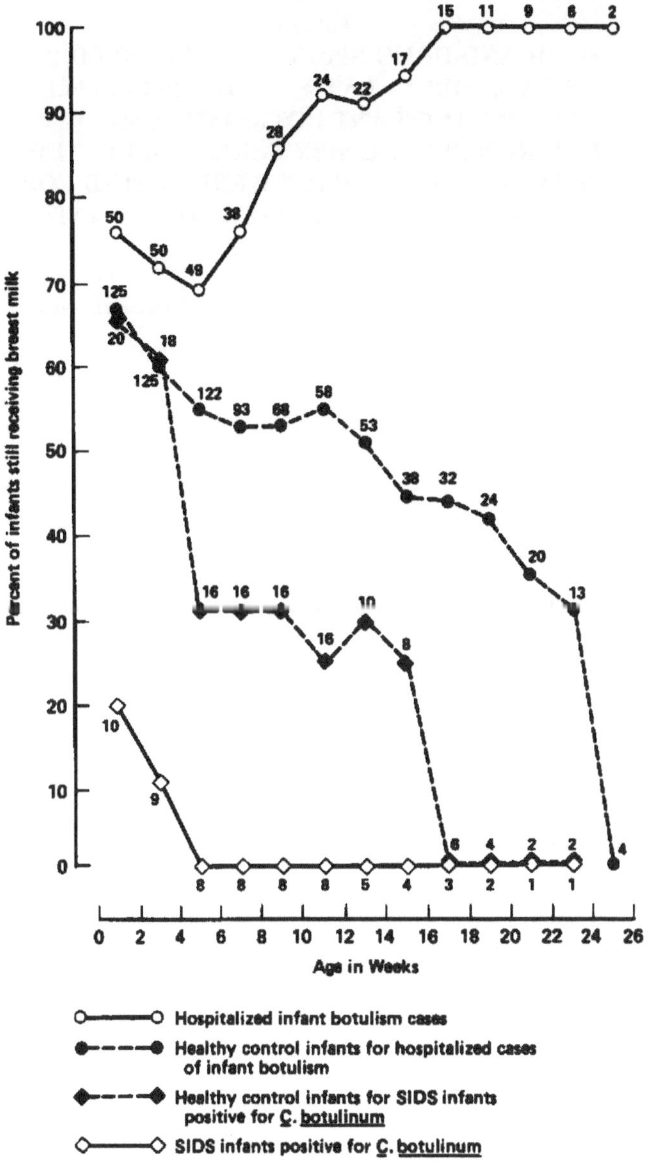


FIGURE 2. The percentage of California infants at each age who were still receiving breast milk in relation to age at onset of illness or of sudden death. The number of infants constituting the denominator of either cases or controls is above each point. This number changes with time because each infant and his or her matched controls were excluded from their respective denominators once illness had begun. (From Arnon, S. S., Damus, K., Thompson, B., Midura, T. F., and Chin, J., *J. Pediatr.*, 100, 568, 1982. By permission of the C. V. Mosby Company, St. Louis.)

infant botulism, only honey was found to contain *C. botulinum* spores (Table 6).[60-62] Furthermore, *C. botulinum* has been repeatedly isolated by different laboratories from honey fed to patients who subsequently became ill, and in every instance the toxin type (A or B) of the *C. botulinum* in the honey matched the toxin type of the infant's illness (Table 7). In California, honey ingestion proved by food ingestion history to be a significant risk factor for illness caused by type B organisms (which are only 1/10 as prevalent as type A organisms in California soils); 44% of type B cases vs. 9% of controls had been exposed to honey before onset of illness, a significant difference ($p < 0.005$). Feeding honey to a baby increased risk of acquiring type B infant botulism almost eightfold.[60]

**Table 6**
**FOOD AND DRUG SPECIMENS INGESTED BY 50 CALIFORNIA PATIENTS HOSPITALIZED WITH INFANT BOTULISM, AND ENVIRONMENTAL SPECIMENS COLLECTED AT THEIR HOMES, THAT WERE TESTED FOR *CLOSTRIDIUM BOTULINUM* TOXIN AND ORGANISMS**

| Type of specimen | No. of specimens tested |
|---|---|
| Food[a] | |
| Honey | 90 |
| Corn syrup (light) | 8 |
| Corn syrup (dark) | 7 |
| Cereal | |
| Rice | 15 |
| Oatmeal | 5 |
| High-protein | 2 |
| Mixed | 6 |
| Barley | 1 |
| Unknown | 1 |
| Commercial formulas | 50 |
| Breast milk | 12 |
| Home-canned foods | |
| Fruits | 21 |
| Vegetables | 8 |
| Commercially canned foods | |
| Fruits | 2 |
| Vegetables | 2 |
| Commercially canned baby foods | |
| Fruits | 8 |
| Juice | 5 |
| Vegetables | 3 |
| Cottage cheese | 1 |
| Oatmeal | 2 |
| Miscellaneous[b] | 54 |
| Drugs | 93 |
| Total food and drugs | 396 |
| Environmental | |
| Dirt | 70 |
| Dust | 46 |
| Water | 13 |
| Miscellaneous | 30 |
| Total environment | 159 |

[a]. The only food items found to contain *C. botulinum* were nine honey specimens (see Tables 7 and 8).

[b] Miscellaneous foods tested included the following: fresh grapes, olive oil, maple syrup, oat cereal (adult type), 10% glucose water, ice cream (vanilla and coconut), sterile water nursette, lollipop, monkey pod powder, dextrose, pearl powder, teething biscuits, gelatin (strawberry), blender washings, sugar, boiled water in bottle, sugar water, manzanilla tea, peppermint leaf tea, rosehips tea, parsley tablets, garlic-onion tablets, cornstarch, honey water, fresh mint leaves, and rice cubes.

Reproduced with permission from Midura, T. F., *Rev. Infect. Dis.*, 1, 652, 1979. The University of Chicago Press, Illinois.

**Table 7**
**CONCORDANCE OF TOXIN TYPE OF PATIENTS' ILLNESSES AND OF *CLOSTRIDIUM BOTULINUM* SPORES IN THEIR HONEY**

| State, year | Toxin type in honey | Toxin type of infant's illness |
|---|---|---|
| California, 1976 | B | B |
| California, 1976 | B | B |
| California, 1977 | B | B |
| Utah, 1977 | B | B |
| Missouri, 1978 | A | A |
| Washington, 1978 | B | B |
| California, 1979 | A | A |
| California, 1979 | A | A |
| Kentucky, 1979 | B | B |
| California, 1982 | A | A |

*Note:* None of the California patients who were fed *C. botulinum*-containing honey lived in households in which an environmental specimen contained *C. botulinum* (see Table 8).

In a separate survey of honey specimens that had no connection to cases of infant botulism, *C. botulinum* types A, B, or both, were isolated from 7.5% of (18 of 241) samples that originated in all regions of the U.S.[63] This figure should be viewed as a minimum estimate of the prevalence of *C. botulinum* in honey, as this and another study[61] demonstrated that the distribution of spores within honey is uneven, presumably because of its high viscosity. Thus, the first aliquot from a given honey sample may be negative for *C. botulinum*, while a subsequent aliquot may be positive.

Although honey represents a significant risk factor for infant botulism and is to date the one identified, avoidable source of *C. botulinum* for infants, honey is *not* the source of spores for most patients. Elimination of honey from the infant diet, although desirable, will not eradicate infant botulism. When honey was first identified as a risk factor, only about one third of patients worldwide had any history of exposure to honey before onset of illness.[60] In California, the figure was lower, roughly one quarter. Since public attention became focused on honey in 1979, the proportion of California patients with a history of preceding honey exposure has steadily decreased to its present level of about 10%.

Because honey is not essential for infant nutrition, several major agencies and organizations have joined in the recommendation that honey not be fed to infants, especially those under 6 months of age. These groups include the American Academy of Pediatrics,[64] the California Department of Health Services,[65] the Centers for Disease Control,[66] the U.S. Food and Drug Administration,[67] the Honey Industry Council of America,[68] and the Sioux Honey Association (the world's largest producer).[69]

### 3. *Other Foods*

As mentioned above, of the over 300 infant foods tested in California, only honey was found to contain *C. botulinum*. No *C. botulinum* was found in 15 samples of corn syrup (either the light or dark variety) tested (Table 6).[61,62] Essentially similar results were recently reported by the U.S. Food and Drug Administration (FDA) in a study of 100 infant foods collected in the Washington, D.C. area.[70] From each of 10 groups of infant foods, 10 samples were examined (dry cereals, dry formula milk, nonfat dry milk, pasteurized whole

**Table 8**
**SOIL AND DUST SAMPLES TESTED FOR *CLOSTRIDIUM BOTULINUM* IN CALIFORNIA, 1976—1978**

| | Homes from which samples were tested | Households from whose samples *C. botulinum* was isolated | |
|---|---|---|---|
| | No. | No. | % |
| Cases | 41[a] | 4 | 9.8 |
| Controls | 19[a] | 2 | 10.5 |

[a] Testing of environmental samples from all households has not been completed.

From Arnon, S. S., Damus, K., and Chin, J., *Epidemiol. Rev.*, 3, 45, 1981. By permission of the Johns Hopkins Press, Baltimore, Md.

cow's milk, canned fruits and fruit juices, honey, corn syrup, granulated cane sugar, and fresh carrots). *C. botulinum* type A was isolated from samples of honey, although the total percentage (2 of 100) was lower than that (8 to 10%) found in other U.S. surveys.

In addition, the FDA survey found *C. botulinum* type B in 8 of 40 samples of retail corn syrup. Its further investigation of 961 retail samples of corn syrups collected nationwide found five (0.5%) that contained *C. botulinum* type B. This finding prompted the FDA to recommend that corn syrup (a nonessential food) not be fed to infants.[70] In early 1984 the principle U.S. manufacturer of corn syrups disclosed that it also had found a "low" level of *C. botulinum* spores in both its light and dark corn syrups, a finding that provided additional support for the recommendation not to feed corn syrups to infants.[87]

### F. Environmental Factors

As was expected from the known ubiquitious distribution of *C. botulinum*, these organisms were occasionally recovered from soil or house (vacuum cleaner) dust collected at the homes of patients and of control infants (Table 8). Although the numbers are small, no difference was found between the proportions of case and control households so exposed, demonstrating that exposure to environmental *C. botulinum* spores potentially exists for many infants. Nonetheless, it is of interest that in each instance, the toxin type of *C. botulinum* isolated from the environmental specimen collected at the patient's home matched the toxin type of his illness (Table 9).

In some cases, disruption of surface soil may contribute to exposure to spores, as suggested by Thompson et al.[17] These investigators reported that 9 of 12 patients hospitalized in Salt Lake City, Utah, lived in newly established residential areas surrounding the city that were described by the parents as "windy, dusty, or having had recent construction". Cases of infant botulism did not occur in the older, established, residential areas. Unfortunately, no soil specimens from the homes in Utah were actually examined for *C. botulinum*, nor was an epidemiologic investigation of control infants undertaken.

In assessing the possible effect of surface soil disruption as a risk factor, it is necessary to exclude other ways in which the patient may have acquired *C. botulinum* spores. It is noteworthy that 10 of the 12 Utah cases (83%) had been fed honey before the onset of illness, the highest rate of exposure to honey yet reported. Moreover, *C. botulinum* of the same toxin type as the patient's illness was isolated from two of the honey specimens fed to two different Utah infants (Table 6). The 2 positive specimens represented 28.6% of the 7 Utah honey specimens available for testing.[17]

**Table 9**
**CONCORDANCE OF TOXIN TYPES OF *CLOSTRIDIUM BOTULINUM* FROM HOSPITALIZED CASES AND FROM ENVIRONMENTAL SPECIMENS TAKEN FROM OR NEAR THEIR HOMES**

| State or country | Environmental sample | Toxin type found in environmental sample | Toxin type of infant's illness | Ref. |
|---|---|---|---|---|
| Australia | Yard soil | A, B | B[a] | 71 |
| | Yard soil | A | A | |
| | Vacuum cleaner dust | A | A | |
| | House water supply (rain water tank) | A | A | |
| California | Vacuum cleaner dust | A | A | 47 |
| | Front yard soil | B | B | |
| | Houseplant soil | B | B | |
| | Front yard soil | B | B | |
| Utah | Filter from window "swamp cooler" | A | A | 72 |

[a] Both *C. botulinum* type A and type B organisms were detected in two separate fecal specimens, but only type B toxin was present in each.

From Arnon, S. S., Damus, K., and Chin, J., Epidemiol. Rev., 3, 45, 1981. By permission of the Johns Hopkins Press, Baltimore, Md.

## VI. ANIMAL MODELS OF PATHOGENESIS

Several animal models have been developed that provide, by analogy, some insight into the pathogenesis of infant botulism. It is historically noteworthy that the capacity of toxin-free *C. botulinum* spores to colonize the gut and cause fatal botulism was first demonstrated in adult guinea pigs by Orr over 50 years ago.[73] Orr had to use large numbers ($10^9$) of spores to produce intestinal infection, presumably because of differences between the adult and the infant gut.

More recently, two other animal models have been developed. Miyazaki and Sakaguchi[74] fed 14-day-old baby chickens toxin-free *C. botulinum* spores and produced fatal botulism. These investigators showed that in this species, the cecum is the site of spore germination and toxin production, as ligation of the cecum at 7 days of age rendered the birds immune to subsequent oral challenge with toxin-free spores.

Using infant mice as a model, Sugiyama and Mills[75] experimentally reproduced the limited age of susceptibility to infant botulism that had been observed clinically in humans.[1,4] They found that *C. botulinum* spores, when injected intragastrically, would germinate in the infant mouse gut only if the animal were between 7 and 13 days of age. However, despite the presence of toxin in the lumen of the colon, the mice remained well.[75] Subsequent studies suggested that the intraluminal toxin was not being absorbed.[76] When germ-free mice were studied, Moberg and Sugiyama[77] found that the minimum age of susceptibility decreased from 7 to 3 days. Adult germ-free mice could be colonized by as few as 10 spores, whereas conventional adult mice could not be colonized even by $10^5$ spores. Also, when adult germ-free mice were placed in the ordinary animal holding room for 3 days, they then became resistant to challenge with $10^5$ spores.[77] Virtually identical results were obtained when the rat was used as the experimental model.[78] Parenthetically, Moberg and Sugiyama[78] found that small amounts ($< 0.5LD_{50}$) of botulinal toxin could pass to infant rats in their mothers' milk.[78] In continuing the preceding work, Burr and Sugiyama[79] treated mice with large oral doses of two broad-spectrum antibiotics (197 mg/kg erythromycin plus 133 mg/kg kanamycin) and found that the number of spores required to establish colonization in the treated mice was at least 50-fold less than that needed to colonize the untreated control animals. This apparent effect of antibiotics in increasing susceptibility to colonization could be abolished by stopping the antibiotics and housing the treated mice with untreated mice for 3 days.[79] These experimental studies, which were extended by use of mice having a limited intestinal flora,[80] suggest that some organisms in the intestinal microflora contribute importantly to the enteric environment that ordinarily keeps *C. botulinum* spores from multiplying in the gut lumen.

## VII. CONCLUSION

In the short span since infant botulism was recognized as a infectious disease, basic information about it has been assembled, but much remains to be learned. Although industrialized countries currently account for all known cases, it is apparent from the ubiquitous distribution of *C. botulinum* that infant botulism must be global in distribution. In comparison to other bacterial infections of infancy, the number of patients hospitalized thus far is small, but the actual morbidity and mortality that results from infant botulism will not be known until awareness of the existence of the disease becomes universal among physicians. Moreover, the mild and fulminant (SIDS) ends of the clinical spectrum have yet to be fully delineated by laboratory techniques more sensitive than now exist. It is already evident that where the disease is recognized, cases of infant botulism outnumber those of food-borne and wound botulism.[81]

The lack of reported cases of infant botulism from the developing countries is most likely temporary. However, it may also reflect indigenous conditions that limit intestinal colonization by *C. botulinum* (e.g., differences in diets, including prevalence and duration of nursing, and in gut flora). Other possible explanations for the lack of reported cases from the developing countries may include the fact that the symptoms of mild to moderate illness (constipation, listlessness, poor feeding, etc.) are so nonspecific that infant botulism is obscured by other causes of infant morbidity and mortality (e.g., malnutrition, dehydration, other infections, etc.) and the fact that laboratories equipped for diagnosing botulism (i.e., those with antitoxins and a mouse colony at hand) are not readily accessible in most developing countries.

In the long run, the major importance of infant botulism may be that it represents the prototype of a class of toxin-mediated intestinal infections by spore-forming bacteria that are able to exploit the unique conditions of the infant gut.[4,20,49,82] The extensive array and potency of toxins produced by some species of *Bacillus* and *Clostridium* suggest that, like *C. botulinum*, they may cause illness (including SIDS) after colonizing the infant intestine. There exist many potential vital target organs for toxins besides the neuromuscular junction (e.g., brainstem, heart, liver). Finally, the discovery of a new toxin type of *C. botulinum* as recently as 1970 (type G) supports the view that " . . . there is no reason to suppose that other serological (toxin) types of *C. botulinum* will not be found".[25] If more are discovered, they may in turn be possible causes of SIDS.

Collectively, this class of toxin-mediated intestinal infections of infancy may account for a larger number of SIDS cases than is presently suspected. For instance, it is unlikely that the virtual identity of the age distributions of infant botulism and SIDS (Figure 1) is mere coincidence. The congruence of these age distribution curves suggests that the two entities share a common underlying intestinal pathophysiology in which human milk, with its content of immunological factors, may play a central defensive role.[20,82] The search for other etiologic agents in this class of diseases, the characterization of their toxin(s) and resultant syndrome(s), and the identification of host susceptibility factors and methods of prevention, collectively constitute a considerable challenge for future research.

# REFERENCES


1. **Arnon, S. S.,** Infant botulism, *Ann. Rev. Med.*, 31, 541, 1980.
2. **Arnon, S. S., Midura, T. F., Clay, S. A., Wood, R. M., and Chin, J.,** Infant botulism: epidemiological, clinical, and laboratory aspects, *JAMA*, 237, 1946, 1977.
3. **Lamanna, C.,** The most poisonous poison, *Science*, 130, 763, 1959.
4. **Arnon, S. S., Damus, K., and Chin, J.,** Infant botulism: epidemiology and relation to sudden infant death syndrome, *Epidemiol. Rev.*, 3, 45, 1981.
5. **Duchen, L. W.,** Motor nerve growth induced by botulinum toxin as regenerative phenomenon, *Proc. R. Soc. Med.*, 65, 196, 1972.
6. **Center for Disease Control,** *Botulism in the United States, 1899—1977, Handbook for Epidemiologists, Clinicians, and Laboratory Workers*, Atlanta, May 1979.
7. **Pickett, J., Berg, B., Chaplin, E., and Brunstetter-Shafer, M.,** Syndrome of botulism in infancy: clinical and electrophysiologic study, *N. Engl. J. Med.*, 295, 770, 1976.
8. **Midura, T. F. and Arnon, S. S.,** Infant botulism: identification of *Clostridium botulinum* and its toxin in faeces, *Lancet*, 2, 934, 1976.
9. **Arnon, S. S., Werner, S. B., Faber, H. K., and Farr, W. H.,** Infant botulism in 1931: discovery of a misclassified case, *Am. J. Dis. Child.*, 133, 580, 1979.
10. **Grover, W. D., Peckham, G. J., and Berman, P. H.,** Recovery following cranial nerve dysfunction and muscle weakness in infancy, *Dev. Med. Child. Neurol.*, 16, 163, 1974.
11. **Center for Disease Control,** Botulism in 1975 — United States, *Morbid Mortal Weekly Rep.*, 25, 75, 1975.
12. **Ramseyer, J. C., Clay, S. A., and Fishman, L. S.,** Electromyographic studies in acute infantile polyneuropathy, *Neurology*, 26, 364, 1976.
13. **Clay, S. A., Ramseyer, J. C., Fishman, L. S., and Sedgwick, R. P.,** Acute infantile motor unit disorder: infantile botulism?, *Arch. Neurol.*, 34, 236, 1977.
14. **Arnon, S. S., Midura, T. F., Damus, K., Wood, R. M., and Chin, J.,** Intestinal infection and toxin production by *Clostridium botulinum* as one cause of sudden infant death syndrome, *Lancet*, 1, 1273, 1978.
15. **Arnon, S. S. and Chin, J.,** The clinical spectrum of infant botulism, *Rev. Infect. Dis.*, 1, 614, 1979.
16. **Peterson, D. R., Eklund, M. W., and Chinn, N. M.,** The sudden infant death syndrome and infant botulism, *Rev. Infect. Dis.*, 1, 630, 1979.
17. **Thompson, J. A., Glasgow, L. A., Warpinski, J. R., and Olson, C.,** Infant botulism: clinical spectrum and epidemiology, *Pediatrics*, 66, 936, 1980.
18. **Sonnabend, O., Sonnabend, W., Heinzle, R., Sigrist, T., Dirnhofer, R., and Krech, U.,** Isolation of *Clostridium botulinum* type G and identification of type G botulinal toxin in humans: report of five sudden unexpected deaths, *J. Infect. Dis.*, 143, 22, 1981.
19. **Sonnabend, O. and Sonnabend, W.,** Different types of *Clostridium botulinum* (A, D, G) found at autopsy in humans. II. Pathological and epidemiological findings in twelve sudden and unexpected deaths, in *Biomedical Aspects of Botulism*, Lewis, G. E., Jr., Ed., Academic Press, New York, 1981, 303.
20. **Arnon, S. S.,** Breast-feeding and toxigenic intestinal infections: missing links in SIDS?, in *Sudden Infant Death Syndrome*, Tildon, J. T., Roeder, L. M., and Steinschneider, A., Eds., Academic Press, New York, 1983, 539.
21. **Brown, L. W.,** Infant botulism, *Adv. Pediatr.*, 28, 141, 1981.
22. **Gurwith, M. J., Langston, C., and Citron, D. M.,** Toxin-producing bacteria in infants: lack of an association with sudden infant death syndrome, *Am. J. Dis. Child.*, 135, 1104, 1981.
23. **Smith, L. D. S.,** *Botulism: The Organism, Its Toxins, The Disease*, Charles C Thomas, Springfield, 1977.
24. **Crisely, F. D. and Helz, G. E.,** Some observations of the effect of filtrates of several representative concomitant bacteria on *Clostridium botulinum* type A, *Can. J. Microbiol.*, 7, 633, 1961.
25. **Smith, L. D. S.,** Inhibition of *Clostridium botulinum* by strains of *C. perfringens* isolated from soil, *Appl. Microbiol.*, 30, 319, 1975.
26. **Jordan, E. D. and Dack, G. M.,** The effect of *Cl. sporogenes* on *Cl. botulinum*, *J. Infect.Dis.*, 35, 576, 1924.
27. **Stark, P. L. and Lee, A.,** Clostridia isolated from the feces of infants during the first year of life, *J. Pediatr.*, 100, 362, 1982.
28. **Moore, W. E. C. and Holdeman, L. V.,** Human fecal flora: the normal flora of 20 Japanese Hawaiians, *Appl. Microbiol.*, 27, 961, 1974.
29. **Finegold, S. M., Flora, D. J., Attebery, H. R., and Sutter, V. L.,** Fecal bacteriology of colonic polyp patients and control patients, *Cancer Res.*, 35, 3407, 1975.
30. **Kahn, M. C.,** Anaerobic spore-bearing bacteria of the human intestine in health and in certain diseases, *J. Infect. Dis.*, 35, 423, 1925.
31. **Finegold, S. M., Attebery, H. R., and Sutter, V. L.,** Effect of diet on human fecal flora: comparison of Japanese and American diets, *Am. J. Clin. Nutr.*, 27, 1456, 1974.

32. **Draser, B. S., Goddard, P., Heaton, S., Peach, S., and West, B.,** Clostridia isolated from faeces, *J. Med. Microbiol.*, 9, 63, 1976.
33. **Easton, E. J. and Meyer, K. F.,** Occurrence of *Bacillus botulinus* in human and animal excreta, *J. Infect. Dis.*, 35, 207, 1924.
34. **Tanner, F. W. and Dack, G. M.,** *Clostridium botulinum*. II. Presence in the human alimentary tract, *JAMA*, 79, 132, 1922.
35. **Dowell, V. R., Jr., McCroskey, L. M., Hatheway, C. L., Lombard, G. L., Hughes, J. M., and Merson, M. H.,** Coproexamination for botulinal toxin and *Clostridium botulinum*, *JAMA*, 238, 1829, 1977.
36. **Wilcke, B. W., Jr., Midura, T. F., and Arnon, S. S.,** Quantitative evidence of intestinal colonization by *Clostridium botulinum* in four cases of infant botulism, *J. Infect. Dis.*, 141, 419, 1980.
37. **Paton, J. C., Lawrence, A. J., and Manson, J. I.,** Quantitation of *Clostridium botulinum* organisms and toxin in the feces of an infant with botulism, *J. Clin. Microbiol.*, 15, 1, 1982.
38. **Paton, J. C., Lawrence, A. J., and Steven, I. M.,** Quantities of *Clostridium botulinum* organisms and toxin in feces and presence of *Clostridium botulinum* toxin the serum of an infant with botulism, *J. Clin. Microbiol.*, 17, 13, 1983.
39. **Shield, L. K., Wilkinson, R. G., Ritchie, M., and Korman, S.,** Infant botulism in Australia — a case report, *Aust. Paediatr. J.*, 17, 59, 1981.
40. **Murrell, W. G., Ouvrier, R. A., Stewart, B. J., and Dorman, D. C.,** Infant botulism in a breast-fed infant from rural New South Wales, *Med. J. Aust.*, 1, 583, 1981.
41. **Douglas, H. M., Manson, J. I., Paton, J. C., Hansman, D. J., Murrell, W. G., and Stewart, B. J.,** Infant botulism (letter), *Med. J. Aust.*, 2, 398, 1980.
42. **McCurdy, D. M., Krishnan, C., and Hauschild, A. H. W.,** Infant botulism in Canada, *J. Can. Med. Assoc.*, 125, 741, 1981.
43. **Neubauer, M. and Milacek, V.,** Infant botulism type B in central Europe, *Zentralbl. Bakteriol. Parasitenkd. Infektionskr. Hyg. Abt. 1: Orig. A*, 250, 540, 1981.
44. **Turner, H. D., Brett, E. M., Gilbert, R. J., Ghosh, A. C., and Liebeschuetz, H. J.,** Infant botulism in England, *Lancet*, 1, 1277, 1978.
45. **Hoffman, R. E., Pincomb, B. J., Skeels, M. R., and Burkhart, M. J.,** Type F infant botulism, *Am. J. Dis. Child.*, 136, 270, 1982.
46. **Smith, L. D. S.,** The occurrence of *Clostridium botulinum* and *Clostridium tetani* in the soil of the United States, *Health Lab. Sci.*, 15, 74, 1978.
47. **Chin, J., Arnon, S. S., and Midura, T. F.,** Food and environmental aspects of infant botulism in California, *Rev. Infect. Dis.*, 1, 693, 1979.
48. **Morris, G. M., Jr. and Hatheway, C. L.,** Botulism in the United States, 1979, *J. Infect. Dis.*, 142, 302, 1980.
49. **Arnon, S. S., Damus, K., Thompson, B., Midura, T. F., and Chin, J.,** Protective role of human milk against sudden death from infant botulism, *J. Pediatr.*, 100, 568, 1982.
50. **Gunn, R. A., Kimball, A. M., Pollard, R. A., Feeley, J. C., and Feldman, R. A.,** Bottlefeeding as a risk factor for cholera in infants, *Lancet*, 2, 730, 1979.
51. **France, G. L., Marmer, D. J., and Steele, R. W.,** Breast-feeding and *Salmonella* infection, *Am. J. Dis. Child.*, 134, 147, 1980.
52. **Larson, S. A. and Homer, D. R.,** Relation of breast versus bottle feeding to hospitalization for gastroenteritis in a middle-class U.S. population, *J. Pediatr.*, 92, 417, 1978.
53. **Bullen, C. L. and Willis, A. T.,** Resistance of the breast-fed infant to gastroenteritis, *Br. Med. J.*, 1, 338, 1971.
54. **Arnold, R. R., Mills, D. C., and Arnon, S. S.,** Natural secretory IgA antibodies to *Clostridium botulinum* in human milk whey, *13th Int. Congr. Microbiol.*, 34(Abstr.), 7, 1982.
55. **Ogra, P. L. and Dayton, D. H., Eds.,** *Immunology of Breast Milk*, Raven Press, New York, 1979.
56. **Goldblum, R. M., Ahlstedt, S., Carlsson, B., Hanson, L. A., Jodal, U., Linin-Janson, G., and Sohl-Akerlund, A.,** Antibody-forming cells in human colostrum after oral immunization, *Nature (London)*, 257, 797, 1975.
57. **Holmgren, J., Hanson, L. A., Carlsson, B., Lindblad, B. S., and Rahimtoola, J.,** Neutralizing antibodies against *Escherichia coli* and *Vibrio cholerae* enterotoxins in human milk from a developing country, *Scand. J. Immunol.*, 5, 867, 1976.
58. **Stoliar, O. A., Pelley, R. P., Kaniecki-Green, E., Klaus, M. H., and Carpenter, C. C. J.,** Secretory IgA against enterotoxins in breast milk, *Lancet*, 1, 1258, 1976.
59. **Simhon, A., Yolken, R. H., and Mata, L.,** S-IgA cholera toxin and rotavirus antibody in human colostrum, *Acta Paediatr. Scand.*, 68, 161, 1979.
60. **Arnon, S. S., Midura, T. F., Damus, K., Thompson, B., Wood, R. M., and Chin, J.,** Honey and other environmental risk factors for infant botulism, *J. Pediatr.*, 94, 331, 1979.

61. **Midura, T. F., Snowden, S., Wood, R. M., and Arnon, S. S.,** Isolation of *Clostridium botulinum* from honey, *J. Clin. Microbiol.*, 9, 282, 1979.
62. **Midura, T. F.,** Laboratory aspects of infant botulism in California, *Rev. Infect. Dis.*, 1, 652, 1979.
63. **Sugiyama, H., Mills, D. C., and Kuo, CL-J.,** Number of *Clostridium botulinum* spores in honey, *J. Food Protect.*, 41, 848, 1978.
64. **Committee on Nutrition and Committee on Infectious Diseases,** American Academy of Pediatrics policy statement: honey and infant botulism, *News Comment*,32, 5, 1981.
65. **Infectious Disease Section, California Department of Health Services,** Honey exposure and infant botulism, *California Morbidity*, No. 27, July 14, 1978.
66. **Center for Disease Control,** Honey exposure and infant botulism, *Morbid Mortal Weekly Rep.*, 27, 249, 255, 1978.
67. **Anon.,** Infant botulism, *FDA Drug Bull.*, 11, 11, 1981.
68. **Rubenstein, R.,** Honey — not the culprit in infant botulism cases, *Gleanings Bee Cult.*, 108, 544, 1980.
69. **Cohn, V.,** Risk of botulism seen in feeding honey to infants under one year, *Washington Post*, p. 1, July 6, 1978.
70. **Kautter, D. A., Lilly, T., Jr., Solomon, H. M., and Lynt, R. K.,** *Clostridium botulinum* spores in infant foods: a survey, *J. Food Protect.*, 45, 1028, 1982.
71. **Murrell, W. G. and Stewart, B. J.,** Botulism in New South Wales, 1980-1981, *Med. J. Aust.*, 1, 13, 1983.
72. **Thompson, J. A., Santos, J. I., and Glasgow, L. A.,** Infantile botulism: spectrum of clinical presentation and epidemiology, *Clin. Res.*, 27, 115A, 1979.
73. **Orr, P. F.,** The pathogenicity of *Bacillus botulinus*, *J. Infect. Dis.*, 30, 118, 1922.
74. **Miyazaki, S. and Sakaguchi, G.,** Experimental botulism in chickens: the cecum as the site of production and absorption of botulinum toxin, *Jpn. J. Med. Sci. Biol.*, 31, 1, 1978.
75. **Sugiyama, H. and Mills, D. C.,** Intraintestinal toxin in infant mice challenged intragastrically with *Clostridium botulinum* spores, *Infect. Immun.*, 21, 59, 1978.
76. **Mills, D. C. and Sugiyama, H.,** Comparative potencies of botulinum toxin for infant and adult mice, *Curr. Microbiol.*, 6, 239, 1981.
77. **Moberg, L. J. and Sugiyama, H.,** Microbial ecologic basis of infant botulism as studied with germfree mice, *Infect. Immun.*, 25, 653, 1979.
78. **Moberg, L. J. and Sugiyama, H.,** The rat as an animal model for infant botulism, *Infect. Immun.*, 29, 819, 1981.
79. **Burr, D. H. and Sugiyama, H.,** Susceptibility to enteric botulinum colonization of antibiotic-treated adult mice, *Infect. Immun.*, 36, 103, 1982.
80. **Wells, C. L., Sugiyama, H., and Bland, S. E.,** Resistance of mice with limited intestinal flora to enteric colonization by *Clostridium botulinum*, *J. Infect. Dis.*, 146, 791, 1982.
81. **Centers for Disease Control,** Botulism — United States, 1979-1980, *Morbid Mortal Weekly Rep.*, 30, 121, 1981.
82. **Arnon, S. S.,** Infant botulism: pathogenesis, clinical aspects and relation to crib death, in *Biomedical Aspects of Botulism*, Lewis, G. E., Jr., Ed., Academic Press, New York, 1981, 331.
83. **Borriello, S. P.,** Clostridial Flora of the Gastrointestinal Tract in Health and Disease, Ph.D. thesis, University of London, 1981.
84. **Hatheway, C. E. and Midura, T. F.,** personal communication.
85. **Schwartz, P. J.,** *Rev. Perinat. Med.*, 4, 475, 1981.
86. **Peterson, D. R.,** *Epidemiol. Rev.*, 2, 97, 1980.
87. **Froggatt, P.,** *2nd Int. SIDS Conf.*, University of Washington Press, Seattle, 1970, 77.
88. **Beckwith, J. B.,** *Curr. Probl. Pediatr.*, 3, 1, 1973.
89. **Guntheroth, W. G.,** *Crib Death*, Futura, Mt. Kisco, N.Y., 1982, 47.
90. California Department of Health Services, Berkeley, 1976—1982, unpublished data.
91. **Beckwith, J. B.,** *Curr. Probl. Pediatr.*, 3, 1, 1973.

Chapter 3

# EVIDENCE FOR CLOSTRIDIAL INVOLVEMENT IN PNEUMATOSIS CYSTOIDES INTESTINALIS

**C. E. Yale and E. Balish**

## TABLE OF CONTENTS

# I. INTRODUCTION AND PURPOSE

Pneumatosis cystoides intestinalis (PCI) is a relatively uncommon disease or condition of previously unknown etiology that is characterized by the presence of gas-filled cysts within the walls of some part of the gastrointestinal (GI) tract. The course of the disease is frequently benign, especially when accidentally discovered in asymptomatic patients. It is, however, sometimes associated with a serious illness, often also of unknown etiology, such as necrotizing enterocolitis (NEC). Recent experiments at the University of Wisconsin Gnotobiotic Laboratory proved that clostridia can produce PCI in animals. These gnotobiotic animal experiments not only support a bacterial etiology of PCI but they also have significant implications in its treatment and the therapy of the conditions or diseases with which it is often associated.

The purposes of this chapter on PCI are

1. To describe the disease in humans.
2. To discuss its etiology with particular emphasis on new evidence which supports the bacterial hypothesis.
3. To briefly outline its treatment, prevention, and prognosis.

# II. DESCRIPTION

The gas-filled cysts of PCI can occur in any part of the GI tract from the esophagus to the rectum. When the cysts are confined to the colon, it is frequently called pneumatosis coli. Similar cysts, which are presumably part of the same pathophysiological process, are frequently found in the mesentery of the involved intestine. Occasionally, they occur in other organs of the abdominal cavity, within the regional lymph nodes, or sometimes are attached to any part of the visceral or parietal peritoneum. PCI is also called pneumatosis intestinalis, intestinal emphysema, cystic lymphopneumatosis, and intestinal or abdominal gas cysts.

## A. Incidence

Du Vernoi[1] first described the disease in humans in 1730 and it was later described in animals, especially swine. In 1952, Koss[2] reported on 213 cases in the literature of the world and in 1974, Shallal et al.[3] found 21 cases in the records of the Mayo Clinic. Although the disease is uncommon, it is probably not as rare as once believed. Numerous cases have been reported in the past 10 years, and it is being recognized and noted more frequently than in the past. For example, Feinberg et al.[4] reported that postoperative abdominal X-rays demonstrated PCI in 24 of 148 patients after an intestinal by-pass for morbid obesity. PCI occurs in both sexes, but it occurs in males about three times as frequently as in females.[2,3] It is seen at all ages, but probably most frequently in infancy and between the ages of 40 and 60 years.

## B. Classification

PCI has been classified in at least two different ways, i.e., primary vs. secondary, as well as adult vs. infantile types.[1,5] The primary, idiopathic type constitutes 15% of the cases and is infrequently associated with other conditions. In general, it is seen more commonly in adults, affects the subserosal tissues of the bowel, is frequently localized to the large intestine (especially the descending colon), produces few if any symptoms, and frequently follows a benign course. By contrast, the secondary type accounts for 85% of the cases and is frequently associated with some other, often serious, condition or disease. In general, secondary PCI is frequently seen in infants, affects the submucosal tissues of the bowel, involves the small intestine, is associated with significant symptoms, and more frequently follows a serious course. PCI has been found in association with several pulmonary conditions such as emphysema, pneumomediastinum, or chronic lung disease. It is, however, uncommon in patients with chronic obstructive pulmonary disease. PCI is more commonly associated with diseases of the GI tract such as peptic ulcer, pyloric obstruction, intestinal obstruction and strangulation, inflammatory bowel disease (including tuberculosis, regional enteritis, and ulcerative colitis), appendicitis, diverticulitis, peritonitis, carcinoma, trauma, NEC, sprue, and a variety of collagen diseases. Occasionally, PCI alone can produce symptoms of a partial bowel obstruction, but usually the symptoms and the clinical course of the patient are more closely linked to the primary disease than to the PCI.

### C. Pathophysiology

Smith and Welter[5] reviewed 33 cases of PCI from the files of the Armed Forces Institute of Pathology and described the pathology in detail. The gas-filled cysts of PCI have been found in every part of the GI tract, from the esophagus to the rectum. They are found in the submucosa and/or subserosa and vary in size from less than a millimeter to several centimeters in diameter. Presumably, as part of the same pathological process, similar cysts may be found in the mesentery adjacent to the bowel, in other intra-abdominal organs, or just beneath or attached to any part of the visceral or parietal peritoneum. The bowel may be extensively involved with thickened, crepitant walls, or it may appear essentially normal except for patches of "soap bubbles" and/or grape-like clusters of cysts attached to its serosa.

Microscopically, the cysts are usually lined by long, flat cells presumed to be histiocytes and separated from one another by connective tissue containing a variable number of inflammatory cells such as leukocytes, lymphocytes, macrophages, plasma cells, and foreign-body giant cells. Occasionally, bacteria are seen in the inflamed connective tissue, but rarely, if ever, in the cysts themselves.[5]

The cysts are usually sterile and analyses of the gas in the cysts have usually shown a significant concentration of hydrogen.[6]

Patients with subserosal intestinal cysts are frequently asymptomatic. These cysts presumably break, reform and break, and reform again and again until scarring stops the process in a given cyst. If enough cysts are involved in this process, the patient can develop a pneumoperitoneum, which has been known to last in a steady state for months.[7] Patients with extensive submucosal cysts are more frequently symptomatic. The number and size of the cysts can progress to the point where the intestinal lumen is partly or completely blocked. It is also postulated that rapid, extensive, submucosal PCI can produce mucosal ischemia, slough of the mucosa, and acute and chronic ulcer formation, which allows the highly toxic intraintestinal contents direct access to the exposed blood vessels, with a resulting fulminating sepsis.

### D. Signs and Symptoms

Patients with primary PCI, especially adults, are frequently asymptomatic.[3,8] In marked contrast, some patients with secondary PCI have severe symptoms and signs of the primary disease that completely overshadow any symptoms from the PCI. Most patients have symptoms and signs that fall between these extremes.

The symptoms due directly to PCI are secondary to a partial bowel obstruction with or without a mucosal ulceration. The patient may complain of nausea and vomiting, increased abdominal girth, crampy abdominal pain, an alteration of bowel function such as constipation or diarrhea, excessive flatulence, mucus or blood in the stools, rectal urgency, or fecal incontinence. The symptoms frequently mimic those of a malabsorption or an irritable bowel syndrome. In the absence of a fulminating primary disease, symptoms and signs of a severe GI hemorrhage are unusual and a GI perforation with peritonitis and sepsis is rare.

The vital signs are usually normal in the absence of a serious primary disease. On examination, the abdomen may be distended and tympanitic (especially with a pneumoperitoneum), the involved bowel may be palpable and tender, and obstructive sounds may be heard on auscultation. Rectal examination may show a rectal mass and mucus or blood in the stool.

### E. Diagnosis

The physical exam can be highly suggestive, but it is infrequently diagnostic. Studies on the patient's blood and urine are rarely helpful. Flat and upright X-rays of the patient's abdomen are frequently diagnostic, especially if the large bowel is involved and a pneumoperitoneum is present.[7,8] A barium examination of the upper and lower GI tract may be helpful, and endoscopy, especially of the colon and rectum, may be diagnostic. Recently, Gillon et al.[9] showed that the hydrogen content of the end-expiratory air was elevated in patients with PCI.

### F. Differential Diagnosis

Primary PCI must be differentiated from irritable bowel syndrome, malabsorption disease such as sprue, inflammatory bowel disease, fat necrosis and sclerosing lipogranulomatosis, pneumoperitoneum (e.g., from a perforated peptic ulcer or a ruptured diverticulum), and partial or complete bowel obstruction from a variety of causes. PCI has been mistaken for benign or malignant bowel tumors such as enterogenous cyst, submucosal lymphoid hyperplasia, lipoma, multiple polyposis, villous adenoma, lymphoma, and carcinoma.

Secondary PCI is associated with a long list of GI and pulmonary conditions, as previously noted. Frequently, it is very important to recognize and treat the primary disease. For example, the knowledge that PCI is often associated with acute NEC can help in making the diagnosis of this primary disease, in which prompt diagnosis and treatment are so critical.[10]

## III. ETIOLOGY

Bacterial, biochemical, mechanical, neoplastic, and nutritional theories have been used to explain the origin of PCI. The biochemical, nutritional, and neoplastic theories were adequately summarized by Koss in 1952,[2] and since they are rarely supported today, they will not be discussed in further detail.

### A. Mechanical Hypothesis

The mechanical hypothesis has been popular during the past decade. According to this theory, gas is forced into the intestinal cysts in one of two ways. In the first, pulmonary alveoli rupture and air escapes from the chest by dissecting along the aorta and mesenteric blood vessels to the bowel. In the second, gas in the lumen of the intestine enters the submucosa through a mucosal tear or ulcer.

In support of the first of the two mechanical theories are the facts that PCI is seen with some frequency in patients with chronic pulmonary disease,[11] and traumatic rupture of the alveoli can lead to air that escapes the thorax and follows the blood vessels.[7] Against this theory are the facts that PCI occurs in many patients without pulmonary disease, the gas in the cysts contain $H_2$ in much greater concentrations than found in air,[6,12] and the gas cysts do not rapidly resorb as does the air forced into tissues following trauma.

The second mechanical theory may be more plausible, since it suggests that this gas originates within the lumen of the bowel due to bacterial fermentation, contains $H_2$, and is forced through a mucosal defect into the submucosal tissues by any condition that causes an increase in intra-abdominal pressure. In support of this theory are the facts that PCI is seen in association with many GI diseases in which there is ulceration and/or partial bowel obstruction. Against this theory is the fact that it does not adequately explain the etiology of PCI in patients who do not have a preexisting pulmonary or GI condition. It seems best not to invoke the theory of multiple etiologies unless necessary.

**B. Bacterial Hypothesis**

This theory suggests that the gas in the cysts is formed by gas-producing bacteria within the lymphatics or tissues of the GI tract. It is supported by the facts that:

1. PCI, simulating that in the human, can be produced in experimental animals by gas-forming bacteria.[13-16]
2. The same species of organisms are frequently found in the intestine of patients with PCI.[17-18]
3. The intestinal bacteria can translocate across the intestinal mucosa.[19,20]
4. The cysts contain $H_2$ which is very likely produced by bacterial fermentation.[6,12]
5. End-expiratory breath samples of patients with PCI have abnormally high levels of $H_2$.[9]
6. Methods which suppress the gas-forming anaerobes, such as oxygen therapy,[12] an elemental diet,[21] and antibiotics[22,23] are known to speed the resolution of PCI.

Against this bacterial theory have been the findings that the cysts are usually sterile, and very few, if any, bacteria are found in the tissues adjacent to the cysts.

Recent studies in experimental animals clearly lend support to the bacterial hypothesis. Stone et al.[15] produced PCI in guinea pigs by injecting a pure culture of either *E. coli*, *E. aerogenes*, or *C. perfringens* into the bowel wall. We first noted an increased incidence of PCI in germ-free rats with a surgically created closed-loop obstruction that had been inoculated with *C. perfringens*.[13,14] PCI has never been seen pre- or postoperatively in any germ-free rat. In our laboratory, PCI was also produced in germ-free rats without bowel obstruction by injecting *C. perfringens* into the wall of the distal ileum or cecum.[14] We later showed that the incidence of PCI depended on the inoculum size and that a simple intraperitoneal injection of $10^7$ viable *C. perfringens* produced PCI in 100% of our germ-free rats. To avoid the possibility of trauma to the bowel or of an intramural or intraluminal injection, the *C. perfringens* were placed in the peritoneal cavity through a small midline incision. Similar experiments were done in an attempt to produce PCI in germ-free rats monocontaminated with a variety of other common intestinal organisms, including *E. coli* and *E. aerogenes*, without success. To date, only *C. tertium* and *C. perfringens* produced PCI[24] in 10 species of clostridia in germ-free rats. The PCI produced by the intraperitoneal injection of *C. perfringens* in germ-free rats was similar to that seen in humans. Submucosal, subserosal, peritoneal, and wound PCI were seen individually and in all combinations in these animals, from PCI limited to the omental peritoneum to extensive involvement of the bowel, the peritoneum, and the wound. The development and spontaneous disappearance of PCI was demonstrated by the finding of PCI in 50% of the animals 2 weeks after injection, all the animals at 4 and 6 weeks after injection, 83% at 8 and 10 weeks, and varying but smaller amounts up to 26 weeks after injection (Figure 1). Using aseptic techniques, the peritoneal cavities and cyst walls were sampled, and then cultured under strict anaerobic conditions. All cultures were sterile. This is consistent with the findings in humans. We know that *C. perfringens* caused PCI in our animals, and, therefore, we do not believe that "sterile cysts" should be used as an argument against the bacterial hypothesis of PCI in humans.

We believe that the bacterial hypothesis can now explain the occurrence of all cases of PCI. In one manner or another, and perhaps enhanced by the transient intra-abdominal pressures known to be associated with chronic pulmonary diseases, gas-forming bacteria, especially the clostridia, gain entrance to the tissues of the bowel wall or peritoneal cavity,

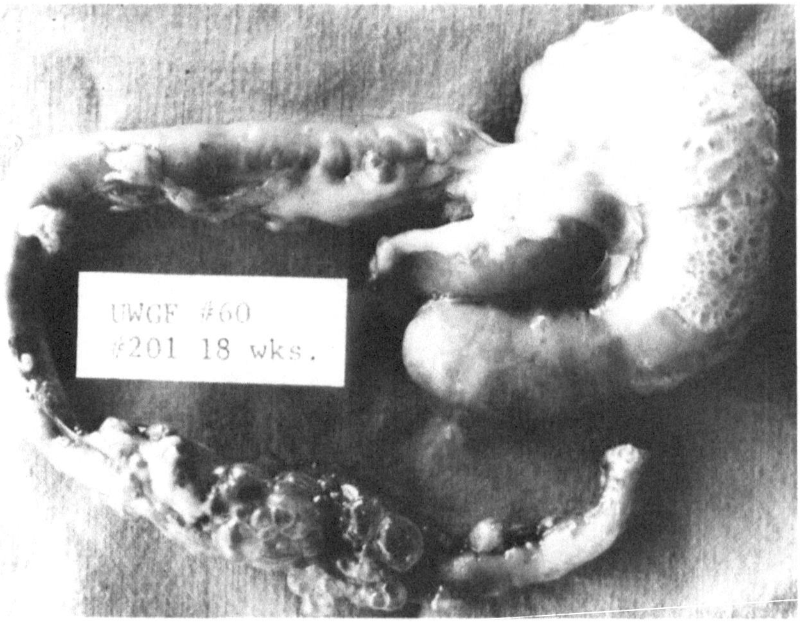


FIGURE 1. Pneumatosis cystoides intestinalis involving the cecum and colon of the germ-free rat 18 weeks after an intraperitoneal monocontamination with *C. perfringens*.

find a suitable substrate for fermentation, and produce gas in the infected tissues. Gas cysts, containing a concentration of $H_2$ greater than in air, are formed. These cysts persist as long as the infecting bacterium and a suitable substrate for fermentation are available in sufficient concentration in the intestinal tract for significant gas production.

Proof that PCI can be caused by the gas-forming anaerobic clostridia has significant implications in establishing a plan for treating a devastating disease such as NEC which is known to be associated with PCI.[10,17,18] PCI is, however, a relatively rare and often benign condition. The real impact of establishing the bacterial theory of PCI is not in the number of lives that are saved, but in the discovery that simple, almost innocuous, contamination of the peritoneal cavity of an animal with clostridia can produce profound changes in the structure of the involved segments of bowel and yet leave behind very little, if any, of the usual evidence of a present or past infection. One cannot help but postulate that a similar microorganism-induced process accounts for the pathological changes seen in other major segmental inflammatory bowel diseases such as regional enteritis and ulcerative colitis — both of which are of a much greater health and economic significance to the world than PCI.

## IV. TREATMENT, PREVENTION, AND PROGNOSIS

Asymptomatic primary PCI may need no treatment. The cysts may spontaneously disappear or remain unchanged for long periods of time. If the disease is persistent and extensive, or if it becomes symptomatic, primary PCI should be treated. Secondary PCI may require treatment, but only in association with proper treatment of the primary disease.

During the past decade, significant progress has been made in the nonsurgical treatment of PCI, primarily because of a better understanding of its etiology. Now that most, if not all, of the cases of PCI can be explained by the bacterial hypothesis, there is a more rational approach to its treatment. Symptomatic treatment has always been used, but recently investigators have reported that an elemental diet,[21] antibiotics,[22,23] and oxygen by tent, mask, or hyperbaric chamber[12,25-29] have all been effective in treating PCI.

In 1973, Forgacs et al.,[12] first reported on the successful treatment of PCI with a high concentration of oxygen. Their patient was placed in a head tent containing 70 to 75% oxygen for a period of 6 days, during which time the patient's significant rectal involvement regressed and his chronic diarrhea disappeared. The rate at which gases diffuse from the cysts is related to the difference between the partial pressure of gases in the cyst and the surrounding tissues, and the venous capillary blood. Breathing oxygen moves nitrogen out of the tissues. As much of the oxygen is utilized in the tissues, the partial pressure of gases in the venous capillary blood is markedly reduced, and the gases move out of the cysts. Oxygen treatment may also work because it increases the amount of oxygen in the tissues and decreases the number of anaerobic gas-forming organisms such as clostridia, which are known to produce PCI in experimental animals.[14,24] Several additional reports have confirmed the temporary and sometimes permanent efficacy of treating PCI with oxygen breathing.[25-29]

Van der Linden and Marsell[21] recently reported on a patient with pneumatosis coli and severe diarrhea whom they treated with an elemental diet, resulting in regression of the cysts. They found abnormally high end-expiratory $H_2$ values in patients with PCI, and believe that this relates to the amount of $H_2$ formed in the colon, presumably secondary to bacterial metabolism. Since gas in the cysts is known to contain a significant amount of hydrogen,[6,12] they postulated that feeding an elemental diet, which is totally absorbed in the small intestine, denies the colonic flora sufficient substrate to continue to produce enough $H_2$ to perpetuate the PCI. These findings also correlate with the observation that patients who have had an intestinal by-pass are prone to develop excessive flatulence and PCI[4,22] — presumably in this case because of the arrival in the colon of increased amounts of carbohydrates available for bacteriologic fermentation.

In 1980, Ellis[23] reported on the dramatic relief of symptoms following treatment with metronidazole in 2 patients having symptomatic pneumatosis coli. Metronidazole is believed to be effective because it inhibits the anaerobic gas-producing bacteria.

Resection of a segment of bowel massively involved with PCI has been an effective and necessary treatment in the past. Surgery has its place in treating the primary disease associated with PCI, whether it is the prompt resection of necrotic bowel in a patient with NEC or the substitution of a gastric by-pass for an intestinal by-pass to maintain weight control in patients with morbid obesity without the incapacitating side effects of the intestinal by-pass. As the etiology of PCI is better understood and the medical treatments become more effective, the surgical treatment of PCI will decrease.

The **prevention** of PCI depends upon our understanding of the etiology of the disease. Control of the amount of substrate furnished to the intestinal microorganisms and/or the number of gas-forming organisms in the GI tract should decrease the incidence of secondary PCI in those conditions or diseases known to be associated with PCI. Primary PCI will probably continue to occur, since the precipitating causes are not recognized in advance and the disease is frequently asymptomatic.

The **prognosis** of primary PCI is excellent, especially with a better understanding of the cause of the disease and the introduction of more effective medical methods of treatment. The prognosis of secondary PCI depends upon early diagnosis and more effective treatment of the primary disease.

## REFERENCES


1. **Ecker, J. A., Williams, R. G., and Glay, K. L.**, Pneumatosis cystoides intestinalis — Bullous emphysema of the intestine, *Am. J. Gastroenterol.*, 56(2), 125, 1971.
2. **Koss, L. G.**, Abdominal gas cysts (pneumatosis cystoides intestinorum hominis), *AMA Arch. Pathol.*, 53, 523, 1952.
3. **Shallal, J. A., VanHeerden, J. A., Bartholomew, L. G., and Cain, J. C.**, Pneumatosis cystoides intestinalis, *Mayo Clin. Proc.*, 49, 180, 1974.
4. **Feinberg, S. B., Schwartz, M. Z., Clifford, S., Buchwald, H., and Varco, R. L.**, Significance of pneumatosis cystoides intestinalis after jejunoileal bypass, *Am. J. Surg.*, 133, 149, 1977.
5. **Smith, B. H. and Welter, L. H.**, Pneumatosis intestinalis, *Am. J. Clin. Pathol.*, 48(5), 455, 1967.
6. **Hughes, D. T. D., Gordon, K. C. D., Swann, J. C., and Bolt, G. L.**, Pneumatosis cystoides intestinalis, *Gut*, 7, 553, 1966.
7. **Nelson, S. W.**, Extraluminal gas collections due to diseases of the gastrointestinal tract, *Am. J. Roentgenol.*, 115, 225, 1972.
8. **Meyers, M. A., Ghahremani, G. G., Clements, J. L., Jr., and Goodman, K.**, Pneumatosis intestinalis, *Gastrointest. Radiol.*, 2, 91, 1977.
9. **Gillon, V., Tadesse, K., Logan, R. F. A., Holt, S., and Sircus, W.**, Breath hydrogen in pneumatosis cystoides intestinalis, *Gut*, 20, 1008, 1979.
10. **Stone, H. H., Webb, H. W., and Kovalchik, M. T.**, Pneumatosis intestinalis of infancy, *Surg. Gynecol. Obstet.*, 130, 806, 1970.
11. **Gruenberg, J. C., Grodsinsky, C., and Ponka, J. L.**, Pneumatosis intestinalis: a clinical classification, *Dis. Colon Rectum*, 22(1), 5, 1979.
12. **Forgacs, P., Wright, P. H., and Wyatt, A. P.**, Treatment of intestinal gas cysts by oxygen breathing, *Lancet*, 1, 579, 1973.
13. **Yale, C. E.**, Etiology of pneumatosis cystoides intestinalis, *Surg. Clin. North Am.*, 55, 1297, 1975.
14. **Yale, C. E., Balish, E., and Wu, V. P.**, Etiology of pneumatosis cystoides intestinalis, *Arch. Surg.*, 109, 89, 1974.
15. **Stone, H. H., Allen, W. B., and Smith, R. B.**, III, Infantile pneumatosis intestinalis, *J. Surg. Res.*, 8, 301, 1968.
16. **Meyer, R. C. and Simon, J.**, Intestinal emphysema (pneumatosis cystoides intestinalis) in a gnotobiotic pig, *Can. J. Comp. Med.*, 41, 302, 1977.
17. **Kosloske, A. M. and Ulrich, J. A.**, A bacteriologic basis for clinical presentations of necrotizing enterocolitis, *J. Pediatr. Surg.*, 15, 558, 1980.
18. **Kosloske, A. M., Ulrich, J. A., and Hoffman, H.**, Fulminant necrotizing enterocolitis associated with clostridia, *Lancet*, 2, 1014, 1978.
19. **Berg, R. D.**, Mechanisms confining indigenous bacteria to the gastrointestinal tract, *Am. J. Clin. Nutr.*, 33, 2472, 1980.
20. **Berg, R. D.**, Promotion of the translocation of enteric bacteria from the gastrointestinal tract of mice by oral treatment with penicillin, clindamycin or metronidazole, *Infect. Immun.*, 33, 854, 1981.
21. **Van der Linden, W. and Marsell, R.**, Pneumatosis cytoides coli associated with high $H_2$ excretion. Treatment with an elemental diet, *Scand. J. Gastroenterol.*, 14, 173, 1979.
22. **Drenick, E. J., Ament, M. E., Finegold, S. M., Corrodi, P., and Passarg, E.**, Bypass enteropathy. Intestinal and systemic manifestations following small-bowel bypass, *JAMA*, 236, 269, 1976.
23. **Ellis, B. W.**, Symptomatic treatment of primary pneumatosis coli with metronidazole, *Br. Med. J.*, 1, 763, 1980.
24. **Yale, C. E. and Balish, E.**, Pneumatosis cystoides intestinalis, *Dis. Colon Rectum*, 19(2), 107, 1976.
25. **Hoflin, F. and Van der Linden, W.**, Pneumatosis cystoides intestinalis treated by oxygen breathing, *Scand. J. Gastroenterol.*, 9, 427, 1974.
26. **Van der Linden, W. and Hoflin, F.**, Pneumatosis cystoides coli recurs after oxygen treatment. A clue to pathogenesis?, *Eur. Surg. Res.*, 10, 225, 1978.
27. **Holt, S., Gilmour, H. M., Buist, T. A. S., Marwick, K., and Heading, R. C.**, High flow oxygen therapy for pneumatosis coli, *Gut*, 20, 493, 1979.
28. **Simon, N. M., Nyman, K. E., Divertie, M. B., Rovelstad, R. A., and King, J. E.**, Pneumatosis cystoides intestinalis: treatment with oxygen via close-fitting mask, *JAMA*, 231, 13544, 1975.
29. **Masterson, J. S. T., Fratkin, L. B., Osler, T. R., and Trapp, W. G.**, Treatment of pneumatosis cystoides intestinalis with hyperbaric oxygen, *Ann. Surg.*, 187(3), 245, 1978.

Chapter 4

# THE ROLE OF CLOSTRIDIA IN THE PATHOGENESIS OF NEONATAL NECROTIZING ENTEROCOLITIS

**R. M. Kliegman**

## TABLE OF CONTENTS

## I. INTRODUCTION

Neonatal necrotizing enterocolitis (NEC) is a disease of unknown etiology. Although recent years have witnessed a resurgence of interest in this syndrome, NEC has probably occurred among newborn infants as early as the latter part of the 19th century. This serious neonatal gastrointestinal (GI) illness has many similarities to enteric infectious diseases of newborn and adult animals. NEC has been compared to severe enterotoxemias among human adults such as pseudomembranous colitis (PMC) and enteritis necroticans or Pig-Bel. Multiple epidemiological findings have suggested that NEC is similar to these GI infectious diseases among both animals and older human subjects. The presence of bacteremia, neutropenia, patients clustered in place and time, and so on have also implicated an infectious agent or its toxin in the pathogenesis of this disease. There are, however, confusing observations, perhaps related to the staging of the disease or the question of primary bacterial invasiveness as opposed to secondary bacterial effects (upon a previously damaged intestine) which suggest a more opportunistic rather than pathogenetic role for infectious agents in the disease process. Alternately, identification of a primary pathogen or toxin may be delayed or masked by the events which ensue after the initial GI injury. These secondary events, such as GI perforation and peritonitis take time to develop and may permit the normal intestinal flora to proliferate and interfere with the recovery of the initiating agent. Be that as it may, this chapter will review the epidemiology of NEC with particular reference to the role of specific clostridial organisms in the pathogenesis of this syndrome.

## II. PATIENT CHARACTERISTICS

Neonatal NEC is predominantly a disease of premature infants occurring during their hospitalization in a neonatal intensive care nursery (NICU).[1-4] Although premature infants have been discharged from the unit only to return with NEC, and although term infants have developed NEC in well baby nurseries, over 90% of affected patients are inhospital premature infants. The mean birth weight of affected NICU patients is no different than that for the overall NICU population (1200 to 1500 g) and depends on the nature of high-risk deliveries and infant transport services.[5-8] At Rainbow Babies and Childrens Hospital, Cleveland, Ohio, the mean birth weight of affected patients was 1460 g and the average gestational age was 31 weeks.[4-5] The range, however, was 26 to 42 weeks, as 9 of 123 patients with NEC (or 7.3%) were full-term infants. There were no sex or racial factors related to this disease, nor were there any differences in the incidence between inborn patients and those infants transferred to our unit from the community.

The annual incidence of NEC at our hospital varies between 1.1 to 7.7% of all patients admitted to the unit. This data encompasses all infants including those expiring prior to the initiation of enteric alimentation (an apparent prerequisite for NEC). When one excludes infants who had never been enterally fed, the annual incidence increases (as it does when one examines the incidence in low birth weight infants, <1500 g), and may then approach 15%. Annual incidence data needs to be tempered by the presence or absence of clustered cases. Over the course of a 9-year study of NEC, there were 3 years with 6 or fewer patients with NEC, while there were 2 years where 6 cases of NEC occurred in just 1 month![5] Although seasonal (summer and fall) epidemics of NEC have been reported, this has not been consistently observed among all investigations. In addition, within individual cities, or even within individual training programs which service more than one neonatal unit, the occurrence of NEC may be quite variable.[9-12] In one service responsible for two city hospitals, NEC is very common in one unit while it is infrequently reported in the second unit, although house staff and attending physicians rotate between the two units. The usual circumstances related to the patterns of disease incidence, however, are a low occurrence rate or endemic incidence which may or may not be augmented by incidental epidemics of this disease scattered between months or years of a few isolated cases. Infant characteristics during epidemic periods may be different than those during endemic epochs as the epidemic patients may have higher Apgar scores, be older in age at the onset of illness, usually fed a longer time, and have a lower incidence of the respiratory distress syndrome (RDS). Although epidemics have been thought to be related to ICU crowding, being over census, to a sub-optimal nurse-to-patient ratio, or to the absolute number of "very sick" premature infants, these hypotheses have not been proven. On occasion, however, epidemics of NEC have been reported to occur during periods of GI illness among the working personnel in the NICU. Thus, the incidence and nature of the affected population may be altered by epidemics or clustered episodes of NEC. The resulting variability in the incidence of the disease may produce confounding factors in any prospective epidemiological or therapeutic study of this disease.[13]

NEC usually occurs within the 1st or 2nd week of life and almost universally after enteric alimentation has begun. Of the 123 patients investigated by the author, the mean age of onset was the 12th day of life, the median was the 7th day, while the most frequent was the 3rd day of life.[5] NEC may be seen on the 1st day or as late as 3 months after birth. It has recently been suggested that GI maturity may alter the susceptibility to NEC as the postconceptional age (gestational and neonatal age) of those infants affected later in the newborn period may be equivalent to those affected in the first few days of life. In other words, the less mature a preterm infant is at birth, the longer the period of time (until GI maturity occurs) when this infant is at risk to develop NEC.

## III. DISEASE CHARACTERISTICS

When the disease does present, it may demonstrate within a wide spectrum of signs, symptoms, severity, and mortality. Before continuing, however, specific diagnostic criteria must be defined to avoid including patients who do not have NEC. The hallmark for the diagnosis of NEC without prior histopathological examination of affected tissue is the roentgenographic demonstration of pneumatosis intestinalis.[5,11] Abnormal GI gas production, as evident by gas in the hepatic-portal venous system or intramural intestine, must be present to diagnose NEC without tissue samples (see below). The clinical criteria for the presumptive diagnosis of NEC are often nonspecific. These include, in order of frequency, abdominal distention, ileus, apnea and bradycardia, formula retention in the stomach, hematochezia, abdominal tenderness, emesis, and diarrhea. Further signs and symptoms include a "septic-like appearance", shock, acidosis, discolored anterior abdominal wall, and the formation of a right lower-quadrant mass (abscess). Many of the early signs and symptoms are easily confused with the more common event of feeding problems which are, in part, due to immature GI function, such as diminished motility. In addition, the early signs of bacterial or viral sepsis are often confused with those of NEC. Only until specific signs of intra-abdominal pathology are present, usually heralding the onset of peritoneal irritation or frank peritonitis, may the diagnosis of NEC be suspected. At this time, the diagnosis may be confirmed by abdominal roentgenographic examination demonstrating abnormal GI gas within the venous system of the liver or intestinal mucosa, or the presence of GI perforation and free intra-abdominal air. Many infants are suspected of having NEC but never demonstrate these abnormal gas patterns or further evidence of intra-abdominal pathology. Although they may respond to restriction of enteric feedings and systemic antibiotics, these infants probably did not have NEC and may have had the frequent problem of feeding (formula) intolerances noted among premature infants. As no specific diagnostic criteria are available to determine those with NEC (in the absence of pneumatosis), these patients should not be included in therapeutic or epidemiological studies. These "suspect" cases are a heterogeneous group, formed predominantly by infants without NEC.

FIGURE 1. Histopathology of NEC. Upper area of small bowel demonstrating complete loss of mucosal architecture with coagulation necrosis. Lower portion demonstrates intact morphology; however, superficial layer of villi demonstrate necrosis. Overall, lower portion has hemorrhage and inflammatory response. (Magnification × 100.)

Although NEC may occur in the absence of pneumatosis intestinalis, the diagnosis must then be confirmed by histopathological tissue analysis (Figure 1). Pathologically, this disease most often affects the terminal ileum and ascending colon, but the entire GI system may be involved (for some reason the duodenum is usually spared).[11,14,15] Early pathological changes include gaseous blebs in the submucosa and subserosa in addition to mucosal edema. A bland (coagulation) autolysis may follow only to be replaced by mucosal hemorrhage and further necrosis which progresses to transmural gangrene. A diphtheritic membrane (pseudomembrane) may occasionally be seen lining the ulcerated mucosa, and is formed by fibrinous exudate, leukocytes, and necrotic mucosal debris. Bacteria are also occasionally noted in this debris or within the gas blebs; less often they are noted within the wall of the bowel. An acute inflammatory response may also be noted, but this is a less frequent observation, while arterial vessel thrombosis is an unusual finding. The progression from mucosal edema and hemorrhage to the coagulation type of transmural necrosis with subsequent viscus perforation and peritonitis are apparently the usual events in the most severe cases. As stated earlier, the multiple stages of pathology may not depict the inciting event because secondary factors may become involved once transmural necrosis and peritonitis develop. Early stages of the pathological sequence in NEC are analogous to that observed in PMC, and may also resemble the pathology seen in the clostridia-related enterotoxemias in both man and animals (see below).

As discussed in the beginning of this section, NEC may present within a wide spectrum of manifestations.[16-19] It may present as a mild illness in previously healthy near-term infants with bright red blood in the rectum and localized rectal pneumatosis intestinalis. In contrast, it may present rapidly (within hours) in a premature infant manifesting shock, acidosis, disseminated intravascular coagulation (DIC), bacteremia, and peritonitis. The former patient has a very low morbidity with no mortality, while the latter, more severely affected patient has a mortality of 50 to 80%.[6] Factors related to this high mortality are those patients with shock, bacteremia, DIC, low birth weight, and peritonitis with ascites.[6,17] Prior investigations of neonatal anaerobic sepsis have suggested that those due to clostridial organisms have a higher mortality than nonclostridial bacterial anaerobic septicemias, while an additional report had stated that infants with NEC associated with *Clostridium perfringens* had a more fulminant disease with a higher mortality. The latter investigation has not, however, been confirmed by additional studies.

## IV. MILK FEEDINGS AS A FACTOR

In most series of patients with NEC, over 95% of the confirmed cases had been fed by the enteric route, hours to days or weeks prior to the onset of the illness.[5,11,20,21] Three important aspects of enteric feedings may be related to NEC.[22-25] The first may be related to the obligate increased GI (mucosal) blood flow required during enteric alimentation. Under circumstances associated with diminished systemic perfusion (RDS, umbilical catheterization), a relative mucosal ischemia may develop when enteric feedings are introduced. (For further discussion of intestinal ischemia as a factor in NEC, see below). Two additional aspects of milk feeding and relative GI mucosal ischemic or direct injury are the volume of formula fed and the osmotic strength of the solution.[24-28] While hyperosmotic formula (600 to 800 mOsm/ℓ) have firmly been incriminated in producing mucosal injury in animal models and probably among preterm infants, the relationship to the volume of isotonic milk feedings and NEC is less clear.[24] No controlled study has implicated the absolute volume of formula nor the rate of change of milk feedings in the pathogenesis of NEC.[27] In one controlled investigation, there was no effect of rapid vs. slow increments in enteric feeding regimens and the occurrence of NEC.[25,26] Nevertheless, the essential role of enteric alimentation has recently been highlighted serendipitously in a controlled study investigating growth and nutrition among infants fed by total parenteral alimentation, or by enteric feedings. Only infants fed by the enteric route developed NEC.

A second hypothesis related to enteric milk alimentation has proposed that immunological factors present in fresh breast milk should be protective against mucosal damage from enteric organisms.[22,23] This is analogous to the proposed protective effects of immunological factors in milk during periods of diarrhea associated with bacterial and viral enteric pathogens during infancy in rural, urban, and even suburban communities. Human breast milk contains a multitude of protective immunological factors which may be effective in preventing neonatal enteric infections. Secretory IgA directed against potential pathogens in the maternal flora, in addition to breast milk macrophages and lymphocytes, are important factors.[29-31] Other nonspecific factors include lysozyme, lactoperoxidase, complement, bifidus growth factor, and smaller quantities of IgG and IgM. Animal[32-37] models of NEC among newborn rats have demonstrated a protective effect of expressed rat breast milk compared to artificial formula feeding.[23] When further examined, this protection was most beneficial when the cells of the rat milk were present and viable. However, human neonates have developed NEC after receiving pasteurized or refrigerated human milk as their only enteric milk feeding.[38] In the latter report, milk cellular viability was not significantly altered by the 48 hr of refrigeration. The incidence of NEC was not altered nor the severity affected by human milk feedings. Since these earlier reports, additional patients have been reported who were fed only human milk which was either fresh, refrigerated, pasteurized, or frozen prior to feeding. Thus, the importance of the immunocompetent cells of the milk and other immunoprotective factors appear not to directly influence the occurrence of NEC among high-risk infants. However, the only study reporting incidence may have been confounded by three important variables: (1) potential bacterial contamination of milk, (2) donor rather than maternal milk feeding, and (3) immunological stimulation by neonatal flora of maternal immunocompetent cells due to separation of mother and child. Until these three variables are controlled in an appropriate study, the final relationship between breast milk and NEC will not be determined.

The third hypothesis related to NEC and milk feeding is that formula may serve as a substrate to support abnormal enteric bacterial growth.[39-42] Quantitative stool bacterial counts are depressed among infants maintained by intravenous feedings alone. When enteric alimentation is started, stool colony forming units of organisms increase. In addition, neonates, especially premature infants, malabsorb both fat and lactose. The latter substrate may enter the colon and subsequently support bacterial growth in the large bowel. The bacterial fermentation products of lactose include short-chain organic acids and hydrogen gas. The former may contribute to systemic acidemia among lactose-malabsorbing premature infants. Hydrogen gas produced by bacteria in the colon may be absorbed across the colonic venous plexus and eventually be excreted in exhaled air in a quantitative relationship to its enteric production rate. As hydrogen gas is only produced by bacteria within the bowel, it is a good marker for enteric bacterial fermentation of malabsorbed carbohydrates. Prior studies have documented large quantities of lactose appearing in preterm infants' stools, in addition to acidic compounds which may potentially result in diarrhea. Of greater importance is the fact that the luminal gas and the gaseous composition within the pneumatosis intestinalis is predominantly hydrogen among patients with NEC.[39] In addition, preliminary studies have suggested that breath hydrogen excretion, which is high among fed premature infants, may be further increased among patients with NEC.[42] Patients maintained without enteric feedings or germ-free animals excrete no hydrogen gas. Thus, it would appear that malabsorbed carbohydrates enter the colon where they must be fermented to gaseous products and volatile organic acids. Hydrogen gas will be produced in direct proportion to the bacterial count and the load of malabsorbed lactose.

Additional evidence for the importance of carbohydrate malabsorption and subsequent fermentation is the data relating positive reducing substances in the stool to the onset of NEC.[25] These reducing substances increase in content in stool prior to the onset of NEC.

Whether gaseous distention itself alters intestinal perfusion or is simply a marker for pathological bacterial proliferation remains to be proven. This relationship to the onset of feeding is reminiscent of certain clostridial enterotoxemias which require enteric feeding to be reproduced in an animal model or a specific alteration in diet among humans to stimulate toxin production and disease (see below). The combination of carbohydrate fermentation together with local or total intestinal ileus may result in sufficient stasis, bacterial overgrowth, anaerobiosis, and altered redox state to alter the bacterial microenvironment and result in enterotoxin production.

## V. ISCHEMIA AS A PURPORTED RISK FACTOR

Past hypotheses related to the etiology of NEC all evoked GI ischemia (hypoxia) as an important component for the pathogenesis of NEC.[43-58] This was, in part, due to the similarity of the histopathology of NEC to ischemic gastroenteritis or to GI infarction among adults. Additional supportive evidence was related to observations of a dive-like reflex in fetuses and newborn animals where systemic hypoxic-ischemic insults resulted in shunting of blood away from the intestines.[52,55,56] In these studies, an even greater shunting of blood was diverted from the GI mucosa compared to other areas of the intestine or colon. The potential ischemic mucosal injury may result in a breakdown of the protective glycocalyx layer and mucosal cell barrier and render the submucosa susceptible to bacterial invasion or toxic effects. The role of bacteria appears important in this ischemic enteritis theory, as germ-free animals do not develop histopathological evidence of ischemia as do animals with bacterial contamination. In addition, those contaminated animals have a much faster demise than do the germ-free animals.[57,58]

**Table 1**
**NEONATAL NECROTIZING ENTEROCOLITIS (NEC)**

Past theories
- Fetal hypoxia producing selective GI mucusal ischemia
- Bacterial influences
- Formula feeding as substrate for abnormal bacterial gas production
- Absence of immunoprotective components of breast milk

Present discrepancies
- Controlled studies demonstrate equivalent risk factors in NEC and non-NEC patients
- Low-risk patients develop NEC
- Epidemics of NEC
- NEC weeks to months after birth
- NEC in patients NPO
- NEC in patients fed human milk

Systemic and/or selective GI ischemia may occur in the perinatal period following asphyxia neonatorium, the respiratory distress syndrome (RDS), placement of umbilical arterial catheters, exchange transfusion, neonatal shock-like states, congestive heart failure, etc. Past uncontrolled descriptive reports of patients with NEC demonstrated a high incidence of these common neonatal problems which may also potentially result in hypoxia or ischemia. Low Apgar scores, RDS, umbilical vessel catheterization, patent ductus arteriosus, and exchange transfusion were often observed as the antecedent events in the perinatal history of patients with NEC. Because of the presence of many of these historical events, their role as hypoxic-ischemic initiators of GI injury was strengthened. However, as noted in Table 1, there were disconcerting observations which did not agree with the importance of these purported risk factors. NEC was observed to occur in patients without any known identifiable risk factors.[18,43,48] In addition, infants developed NEC weeks to months after a brief period of fetal hypoxia.[4] This long delay seemed inappropriate as one would expect the purported injury to be maximal at the time of the hypoxic insult.

Furthermore, the incidence of these risk factors among patients with NEC and appropriately matched control patients from the same unit during the same time period was not different.[20,21,43,49,53,54] This data has suggested that these risk factors are common to the entire low birth weight population of infants and are not specific antecedent events in the pathogenesis of NEC.

Additional evidence which questioned the importance of these risk factors is the occurrence of waves of epidemics or clustered cases of NEC. These epidemics often affected ''no-risk'' or low-risk patients as frequently as those with other neonatal problems such as RDS.[48]

## VI. INFECTIOUS ETIOLOGY OF NEC

In both past and present theories of NEC, bacterial pathogens have always been proposed to be one component of the disease.[59-77] Whether these are primary pathogens directly involved in GI injury or secondarily invade mucosal tissues damaged by other mechanisms has not been resolved. Indeed, one bacterium may initiate the mucosal injury to only later be overgrown by opportunistic local flora during the more terminal stages of the disease. Prior reports have observed that 30 to 40% of infants with NEC have positive blood cultures at the immediate onset of signs and symptoms of GI or systemic disturbances. *Escherichia coli*, *Klebsiella*, and other enterobacteria were the predominant organisms recovered from these nonepidemic cases of NEC (Table 2).[48,59,63-69,71,74-77] More recently, clostridial organisms and viral agents have been implicated in isolated and clustered cases of NEC, while fungi have rarely been isolated from infants with NEC.[61,65,66,68,69] The high incidence of

**Table 2**
**ORGANISMS ASSOCIATED WITH NEC**

| Organism |
|---|
| *E. coli* |
| *Klebsiella pneumoniae* |
| *Enterobacter cloacae* |
| *Pseudomonas* |
| *Salmonella* |
| *C. butyricum* |
| *C. perfringens* |
| *Clostridium* (other) |
| Echo virus |
| Coxackie virus |
| Coronavirus |
| Enterovirus |
| Aspergillosis |
| *Candida albicans* |
| *Torulopsis glabrata* |

**Table 3**
**INFECTIOUS DISEASE ASPECTS OF NEC**

| |
|---|
| Clustering of patients in "epidemics" |
| Single organism recovered at individual centers |
| Controlled by infectious disease control measures |
| Purported risk factors less relevant (disease in low-risk infants and controlled studies) |
| Association with alterations of stool flora |
| Associated with omphalitis |
| ? Seasonal incidence |
| Associated with GI illness in nursery personnel |
| Similarities to PMC |
| Similarities to enterotoxemias of human adults |
| Similarities to enterotoxemias of newborn mammals |
| Intra-abdominal abscess formation |

positive blood cultures, together with the frequent clinical appearance of a "septic-like" state, has suggested that a primary infectious etiology is functional in the pathogenesis of NEC. Additional epidemiological observations relating NEC to an infectious disease are noted in Table 3.

Although NEC epidemics have been reported in individual nurseries where no single agent had been recovered, the incidence of disease abruptly declined with the institution of specific infectious disease control measures (Table 4).[60,62,76] These earlier reports did not have available the more sophisticated infectious disease identification methods which must be employed today (Table 5). Nevertheless, specific agents have been identified during clustered cases of NEC in individual nurseries. These organisms include *E. coli, Klebsiella,*[64] *C. butyricum, C. perfringens,*[39] *C. difficile, Enterobacter cloacae, Salmonella, Pseudomonas,*[63,67] enteroviruses,[66,68] and recently, coronavirus.[48] Although not direct evidence as a primary initiating etiological event, the isolation of these organisms or their products among patients with NEC during an epidemic strongly suggests the essential role of these infectious agents. Of these organisms, clostridial species have recently been implicated in the pathogenesis of NEC. Because of the similarities of NEC to clostridial enterotoxemias of animals and adult man, and to the pathology of PMC, together with recent isolates of

**Table 4**
**INFECTIOUS DISEASE CONTROL MEASURES FOR EPIDEMICS OF NEC**

*Strict handwashing* with products effective against Gram-negatives (e.g., Betadine)
*Long sleeve gowns and gloves*: rewash, regown, and reglove with each patient
*Isolation of confirmed cases*: use single room for NEC patients
*Cohort nursing*: separate nurses for NEC patients and those colonized with specific organism; nurses colonized to work with colonized patients
*No personnel with GI symptoms* should work in the nursery
*Separate diaper and laundry pails*
*Monitor suspect cases*: use stool-reducing substances and hematest; institute modified control on patients with positive reducing substances or hematest

**Table 5**
**INFECTIOUS DISEASE APPROACH TO NEC**

Organism isolation (aerobic, anaerobic) (virus)
- Blood
- Stool
- Peritoneum
- Wounds
- Other[a]

Further organism identification
- Fermentation profiles G.L.C. of isolate
- Electron microscopy of stool (virus)
- Tissue Gram-stain, histopathology, and immunofluorescence
- Systemic antibody response (acute and convalescence serum)

Toxin production
- Cytopathic and enterotoxic effects of filtrates of stool/cultured organisms
- Neutralized by identifiable antitoxin
- Protected by toxoid/antitoxin therapy
- Identification of naturally acquired antitoxin

Reproduce effects of organism or toxin in animal model

[a] Urine, cerebral spinal fluid, bowel.

various clostridial organisms and/or toxins from tissues or other biological samples, research related to these organisms and NEC have generated new and exciting epidemiological data. Before examining these investigations in more detail, we will review normal and abnormal neonatal colonization and earlier observations of clostridial disease in animals, infants, and adult man.

## VII. NEONATAL BACTERIAL COLONIZATION

Because an organism is recovered from a cutaneous, enteric, or other mucosal site does not imply that it is pathogenic. Colonization of an infant's stool, for example, does not suggest invasiveness or pathogenicity. Likewise, the presence of a known pathogen in an infant's stool, skin, or nasopharynx does not mean the infant is infected. Known pathogens or their toxins have been isolated from the stool of normal infants in the complete absence of disease, while these organisms or toxins are usually only present in adults during disease states.[78-86]

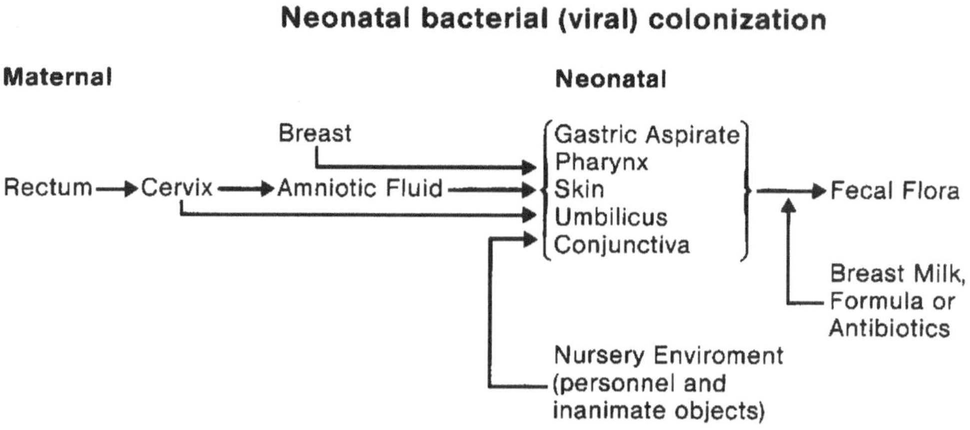


FIGURE 2. Routes and influences of neonatal enteric colonization.

Infants born following a period of prolonged rupture of the amniotic membranes (greater than 12 to 24 hr) may have colonization initiated *in utero*.[79] A very small percentage of these infants may develop invasive bacterial disease prior to birth. This usually occurs as an ascending infection from contamination of amniotic fluid and membranes by the maternal fecal or cervical flora. Fetal aspiration pneumonitis or funicitis follows colonization and may result in fetal septicemia. This is the proposed pathogenesis of the early onset of group B streptococcal disease (or *E. coli* sepsis). Rarely do organisms cross the placenta as in *Listeria monocytogenes* fetal sepsis. Nevertheless, the vast majority of infants born following prolonged rupture of amniotic membranes are not infected but may be colonized at the time of birth.

Infants born by vaginal delivery without prolonged rupture of the membranes may become colonized with maternal or environmental organisms as outlined in Figure 2. This may be modified by maternal antibiotics, Cesarean section, premature birth, neonatal intensive care, and breast or formula feeding.[79,80,85] The normal term infant develops the appropriate combination of aerobic and anaerobic fecal flora to establish colonic bacterial microenvironments. Within these niches "normal flora" bacteria or their products may autoregulate potential bacterial pathogens as has been described among adult colonic bacteria (bacterial cross-interference effects).

The source of the initial bacterial flora of the normal neonate originates from the mother's rectal, or more directly, in her cervical bacteria. There is a high correlation between maternal cervical (rectal) organisms and those later recovered in neonatal gastric aspirates.[79] Additional organisms colonize the infant through the conjunctiva, umbilicus, and pharynx. These organisms may originate from environmental objects, hands of caretakers (nurses, doctors, technicians), and be modulated by breast milk or formula feedings, in addition to the intensive care practices of systemic broad spectrum antibiotics and invasive placement of endotracheal, nasogastric, or umbilical central catheters.[85]

Meconium is usually sterile during the first few hours after birth, but may demonstrate bacterial organisms within 24 hr of life.[80] Gram-negative facilitative organisms (*E. coli*) are the first bacteria isolated from infants' meconium and stool during the 1st day of life.[82,83,86] Breast-fed infants develop a fecal flora containing a large number of bifidobacteria, in part due to the bifidus factor in breast milk and to the chemical composition of this milk producing an acidic colonic environment. In contrast, the number of organisms' characteristics are similar to that in adult flora among formula-fed infants where anaerobic bacteria outnumber aerobic bacteria in the colon (total count $10^{11}$ organisms per gram of stool). Bacterial counts are lower in the ileum ($10^8$) and stomach or jejunum ($10^5$) where aerobic organisms equal or outnumber strict anaerobic bacteria, respectively. Clostridial organisms colonize the newborn fecal flora as early as the 1st day of life.[82,83,86] These organisms are as not consistently recovered as *C. perfringens* and may be seen in only 33 to 50% of stool specimens from neonates. Bacteroides are also present in neonatal fecal cultures and are more often recovered from term infants delivered by vaginal rather than cesarean delivery.[79]

Both fecal and conjunctival colonization may occur with clostridial organisms. In addition to *C. perfringens*, normal term or preterm infants (who are asymptomatic) may be colonized with *C. butyricum*, *C. difficile*, or *C. paraputrificum*.

The presence of *C. butyricum* and/or *C. difficile* in stool is extremely unusual among healthy adults. In contrast to its absence among adult control patients, both *C. difficile* and its cytopathic toxin (even in high titers $1{:}10^4$) have been isolated from essentially asymptomatic-well premature and term neonates and in growing infants up to 5-months old (see below). This apparent paradox of *C. difficile* cytopathic toxin in healthy infants' stool may relate to local endogenous bacterial interference, absence of mucosal receptors, or other as yet unidentifiable factors found in the less mature human intestinal mucosa. Developmental studies of this organism and toxin among young animals are urgently indicated to determine this resistance.

The local colonic population of bacterial growth and invasiveness potential are controlled by many factors.[78] Transient emergence or colonization by known pathogens (toxin producers or invasive bacteria), breast-milk feeding, invasive intensive care practices (e.g., broad spectrum antibiotics), and feeding tube placement may all upset the local bacterial homeostasis in the lumen of the premature infant's colon. Abnormal aerobic colonization among preterm infants in a NICU has been related to broad-spectrum antibiotic therapy and prolonged duration of hospitalization.[80,85] Colonization was delayed following antibiotics, cesarean section, and being kept NPO. In addition, abnormal colonization by aerobic Gram-negative organisms frequently anteceded the occurrence of invasive bacterial disease such as sepsis or pneumonia.[85] Infants with normal flora were never noted to develop invasive disease while those with high titers of abnormal or pathogenic floral organisms acquired nosocomial infections with these bacteria. Although no work to date has investigated these relationships among the anaerobic fecal flora, it may be proposed that alterations in the flora of the colonic microenvironment could result in either toxin production or local tissue invasion by the previously considered "normal flora". In this regard, specific bacterial interference for the growth of *C. difficile* has been recently reported. Lactobacillus, bifidobacteria, and group D enterococci all individually inhibited the in vitro growth of *C. difficile*. Thus, potential alterations of fecal flora may permit growth of normal organisms if competing bacterial survival is diminished.

Certainly, clostridial organisms may become pathogens in infancy as evident by neonatal tetany and infantile botulism. *C. perfringens*, after colonizing the newborn conjunctiva, may result in conjunctivitis as this organism is resistant to the effects of silver nitrate. Additional clostridial infections include omphalitis and subsequent anterior abdominal wall infection with or without myonecrosis and gas gangrene formation. Scalp abscess or cellulitis secondary to fetal scalp monitoring, aspiration pneumonia, and sepsis are often related to anaerobic bacteria (more often bacteroides than clostridia). Nevertheless, neonatal clostridial sepsis is thought to be the most lethal of the anaerobic organisms producing bacteremia. These anaerobic infections are associated with prolonged rupture of membranes, prematurity, amnionitis, RDS, and fetal distress. On occasion, classical clostridial infection with shock, sepsis, and frank hemolysis and hemoglobinuria have been reported. A previous report of clostridial infections involved an infant who, following resection of a tumor, later developed a sigmoid infarct and sepsis. Another report is of a 6-day-old with gangrene of the jejunum and colon. Both these prior case reports were very suggestive that clostridia were functioning as enteric neonatal pathogens. Prior patient reports of omphalitis without myonecrosis and our recent observation of a high incidence of omphalitis prior to the onset of NEC also suggests the umbilicus as either a portal of entry or a preliminary infectious site for clostridial enteric disease (see below).

## VIII. ENTERIC CLOSTRIDIAL DISEASE OF ANIMALS

Evidence has been presented that the newborn human's feces or colon become rapidly colonized with both Gram-negative facultative anaerobic organisms and obligate anaerobes within the 1st hours to days of life. Of the latter organisms, various species of clostridia are present in neonatal stool within the same time reference as the usual neonatal enteric flora. As suggested above, neonatal clostridial infections have also been documented and include sepsis, hemolytic-icteric bacteremia, tetanus, botulism, omphalitis, and other local tissue infections. Both systemic toxin effects and local toxin-induced tissue necrosis have been involved in the broad spectrum of neonatal clostridial diseases.[87-97] Although enterotoxemias among newborn animals have been known in the veterinary literature for many years, there is a paucity of data in the pediatric literature related to this topic. The role of clostridia in animal enterotoxemias and enteritis will be briefly reviewed prior to discussion of human necrotizing enteritis-like syndromes (for a full review see Chapter 10).

Clostridial disease may be initiated by inoculation of the offending agent into a tissue, or by altering the local microbiological balance (microenvironment) of a specific bacterial flora. In the presence of an altered oxidation reduction state (which may not necessarily require complete anaerobiosis) local organisms may proliferate and product toxin. In the GI tract, tissue damage may result following a break in the protective mucosal barrier after trauma, surgery, gastroenteritis of nonclostridial origin, or cancer and chemotherapy.[89,96] These conditions may result in local factors which permit clostridial organisms to become pathogenic.[84] In addition, colonic flora may undergo periodic alterations or intermittently acquire a new organism within a previously stable eco-system.[84] The appearance of this new member may unbalance the bacterial homeostasis and result in GI injury. The new organism may need to have the ability to bind to enterocytes in order to initiate the sequence of proliferation and toxin production. It could be hypothesized that neonatal clostridial enterotoxemias may be related to an immature GI immunological defense mechanism, diminished gastroenteric nonspecific protection due to malnutrition, or to dietary changes which dramatically alter the microenvironment of colonic flora.

Toxin production is the rule among clostridial organisms and although most species produce more than one specific toxin, there is considerable overlap between different clostridia and the toxins they produce.

*C. perfringens* may produce 12 toxins which includes an enterotoxin, lecithinase, collagenase, hyaluronidase, etc. *C. novyi* and *C. histolyticum* may also produce similar toxins as *C. perfringens*.[89,90,96] The cytotoxin of *C. sordellii* and *C. difficile* are similar and are both neutralized by *C. sordellii* antitoxin. It would appear that many clostridial toxins produce extensive tissue necrosis and perpetuate the alterations of redox state required for clostridial proliferation.

In vitro animal models of enteritis have demonstrated the important role of clostridial organisms in the tissue-destructive effects on the bowel during periods of oxygen deprivation. Germ-free animals demonstrate little histological evidence of ischemic enteritis as well as demonstrating improved survival when compared to "normal flora" animals or those monocontaminated with *C. perfringens*.[57,58] In addition, in a similar animal model, monocontamination with *C. perfringens* resulted in pneumatosis intestinalis while other germ-free animals or those monocontaminated with other organisms were unsuccessful in producing pathological enteric gas (see Chapter 3). Furthermore, germ-free neonatal animals monocontaminated with clostridia develop an enteritis similar to NEC, while more mature adult animals do not and remain well. It would seem that neonatal colonization by one bacteria may result in an overgrowth phenomenon unchecked by competing flora leading to disease. This data emphasizes the importance of local bacterial interference. *E. coli* production of "colicins" or *Pseudomonas* release of pyocins or the pH alterations and $H_2O_2$ production of the lactobacillus may control the potential pathogens of the local flora. The establishment and maintenance of a stable, nonpathogenic, microbiological colonic environment may be an essential component of enteric disease prevention. Such a role may be relevant for *C. difficile* and its interference by colonic lactobacillus, enterococci, and bifidobacteria as discussed above in the section on enteric flora colonization. Alterations such as dietary changes, broad-spectrum systemic antibiotics, and malnutrition may adversely affect the interspecies bacterial balances in a stable enteric bacteriological mileu. The resultant perturbations may selectively favor an organism's growth, producing excessive proliferation and potential disease states.

Immature GI function related to mucosal integrity or immunological protection may be pathogenic in the clostridial-produced enterocolitis of newborn animals. Epidemics of a necrotizing enteritis are common among newborn piglets and are due to *C. perfringens* type C exotoxin. These epidemics may be prevented by maternal vaccination resulting in colostral antitoxin production. In addition, type-specific antitoxin administered to the newborn piglet will avoid the enteritis.[92] *C. perfringens* type B enterotoxemias are similar to those of the piglet disease, affect both calves and piglets, and are also responsible for lamb dysentary. *C. perfringens* type A may also cause enterotoxemia among unweaned lambs and newborn alpacas as well as necrotic enteritis of fowl. In addition, *C. perfringens* type D may be responsible for a fatal enterotoxemia among sheep, goats, and calves.[87] Similar episodes of enterotoxemia-produced necrotizing gastroenteritis may be observed among older, more mature animals, and are usually associated with dietary changes.[87,94,96] Enhanced consumption of high carbohydrate diets may result in malabsorption with excessive substrate provision to colonic (or rumen) bacteria. Together with ileus, bloating, distention, and stasis, an altered redox state may develop and augment clostridial proliferation.[96] This is the proposed pathogenesis of pulpy kidney disease, an enteritis seen in adult sheep after the springtime increment of grass consumption and is due to a toxin of *C. perfringens* type D. Alternately, dietary factors which favor enteric toxin stability may produce disease. Intragastric inoculation of *C. perfringens* type C into guinea pigs results in disease only if sweet potato is also fed.[91] It appears that sweet potato has a trypsin inhibitor which prevents trypic digestion of this protease-sensitive toxin. In the absence of the action of trypsin, enterotoxin production exceeds its destruction and severe enteritis may develop. The important condition of ileus may also result from a direct toxin effect. Following GI stasis, the bacteria may bind to the enterocyte and further proliferate. The essential role of GI stasis, which augments bacterial growth, and may also alter local pH and redox states in favor of clostridial toxin production, appears critical for the development of these disease states.

## IX. ENTERIC CLOSTRIDIAL DISEASE OF ADULT HUMANS

Although *C. perfringens* type-A food poisoning may be a common enterotoxemia of man, there are other more important, GI diseases produced by clostridial species which have relevance to neonatal NEC. These life-threatening enterotoxemias produce a potentially fatal necrotizing enteritis following colonic toxin production. Clostridial intestinal gas gangrene, seen concomitantly with ischemic enterocolitis, may also be associated with trauma, leukemia, and neutropenia, or may be idiopathic; it is similar in pathology to NEC.[95] Darmbrand was a necrotizing enteritis observed for a brief period of time in post-World War II Germany and was caused by *C. perfringens* type-F enterotoxin.[87] This enteritis was seen in epidemics following specific diet changes or periods of poor nutritional intake. Histologically, one notes intestinal gas production, edema, necrosis, and a paucity of inflammatory cells. The disease often progresses from transmural necrosis and hemorrhage to GI perforation and peritonitis, and later to a post-inflammatory intestinal stenosis or stricture. Both the pathology and disease sequellae of Darmbrand (enteritis necroticans) are very similar to that of NEC. Enteric feeding of the presumptive toxin to guinea pigs results in an identical illness in the animal world.

An additional well-investigated human enterotoxemia of humans is Pig-Bel.[91-94] This disease affects children more than adults, may be fatal, and is due to *C. perfringens* type-C enterotoxin (see Chapter 5). Pig-Bel occurs predominantly in the New Guinea Highlands following the dietary changes which occur during the festive pig roasts. The sudden alteration of diet, combined with malnutrition, may result in a relative deficiency of pancreatic trypsin producing local toxin production (while toxin catabolism is inhibited). As in the other severe human clostridial enterotoxemias, the histopathology of this often fatal disease is similar to neonatal NEC, demonstrating edema, hemorrhage, and submucosal and subserosal gaseous blebs, together with patchy areas of GI gangrene. As with NEC, the original interpretation of the pathology of this disease had suggested that ischemia produced Pig-Bel. The disease course is similar to NEC in that there may be subclinical states, fulminent shock-like states, or, more often, a patient presents with ileus only to later progress to intestinal perforation and peritonitis. Complications later in the disease are similar to those seen in NEC and include malabsorption and stricture formation. Most patients who survive, however, do so with an intact GI system. Treatment includes antitoxin to type-C alpha and beta toxins and surgery. Recent evidence has suggested that this disease entity is not only isolated to the New Guinea Highlanders as it has also been observed in Southeast Asia and Africa. Although clostridial organisms may be present in the stool of asymptomatic patients, it would appear that additional factors are needed to initiate the disease process such as low protease (trypsin) content of pancreatic secretion or the presence of trypsin inhibitors present in the diet. Sweet potato or soy protease inhibitors are capable of reproducing the disease in an animal model. Protein deficiency malnutrition classically may lower pancreatic trypsin secretion and predispose to a Pig-Bel type enteritis. Appropriate substrates are also needed for the organisms to proliferate. More recently, prevention of this disease has proven effective by immunizing children with *C. perfringens* beta toxoid. Similar preventive measures were successful in sheep, cattle and piglet clostridial enteritis using analogous immunization regimens.

Pseudomembranous colitis (PMC) is a disease with many similarities to neonatal NEC.[87,88,98-117] Clinically, PMC usually presents in adults following a period of antibiotic therapy, although cases have occurred without prior antibiotics.[98-100,112] GI surgery, intestinal obstruction, hypotension, uremia, and other factors have been related to nonantibiotic-associated PMC. PMC becomes manifested as fever, diarrhea, and melena. The pathology demonstrates pseudomembrane formation of a grey or white to yellow plaque over a denuded intestinal mucosa. There is mucosal edema, contact bleeding, and hyperemia. The membrane is composed of a fibrin-leukocytic exudate with mucus and may at times resemble the observations noted during adult-onset ischemic GI necrosis. Although these features are similar to the pathology of NEC, there is no pneumatosis observed in PMC and transmural necrosis is infrequently observed. In addition, PMC usually affects the colon while NEC may affect both the large and small bowel. Similar to NEC, PMC has been reported in epidemics and horizontal spread among hospitalized patients with altered intestinal bacterial flora or normal siblings have been observed.

Further implications of PMC as an infectious disease have been reported as *C. difficile* and its toxin appear to be the primary offending agents in this disease in the adult and older child.[100,106,117] This organism is not usually recovered from normal adult's stool nor is its toxin readily demonstrable. *C. difficile* may be demonstrated on inanimate objects in hospitals such as toilets, floors, and bedding.[103] It is not usually seen in the air or food samples. It was isolated from the hands of hospital personnel and the urogenital tract of patients in a venereal disease clinic. Between 3 to 10% of normal people may have this organism as an evanescent member of the fecal flora. Antibiotic-pretreated animals may acquire lethal cecitis by cross-contamination in the animal facility, suggesting potential airborne spread, although inanimate objects or hand carriage may not have been ruled out as the source of infection.[99] These observations may explain the clustering of patients and also emphasizes the need for enteric precautions in the management of patients infected with *C. difficile*.[103,104,109,111]

*C. difficile* cytopathic toxin has been reported to be similar to heat-stable *E. coli* enterotoxin as both toxins stimulate enteric guanylate cyclase activity.[115,116] *C. difficile* cytopathic toxin may also reciprocally inhibit adenylate cyclase action. An additional laboratory has reported that the cytopathic toxin alters intracellular microtubular action stability, thus disrupting cell architecture.[115]

PMC can be reproduced in antibiotic-treated animals spontaneously or by inoculation of *C. difficile* or its toxin. Antitoxin protects these animals. Therapy for PMC is directed at eliminating the organism with the antibiotic Vancomycin. Thus, among adults, PMC is an additional specific enterotoxemia due to a clostridial organism with many features which resemble those observed during neonatal NEC (for a full description of PMC see Chapter 7).

## X. CLOSTRIDIAL IMPLICATIONS FOR NEC

With all the documentation of the role of clostridial enterotoxemias in newborn animals and adult humans, together with the recent evidence of the pathogenesis of PMC as another clostridial enterotoxemia, pediatric workers began to investigate the role of these organisms in the pathogenesis of NEC. Specifically, three organisms have been implicated: *C. perfringens, C. butyricum*, and *C. difficile*.[118-149]

The theoretical basis for the role of clostridia in the pathogenesis of NEC is much greater than the present evidence. If NEC is related to GI ischemia, the perfect milieu for anaerobic bacterial proliferation may be established and may result in toxin production. The essential component of enteric feeding, combined with immature GI function (digestive, immunologic — see below) may also be analogous to the dietary changes which are known to induce clostridial enterotoxemias among animals and man (both mature and immature of these species). The potential combination of GI ischemia, stasis, and carbohydrate loading of the colonic clostridial flora may produce a sufficiently altered redox state to result in toxin stability and enhanced clostridial growth. Abnormal enteric gas production under experimental conditions is also best suited to the clostridial organisms.

Very early reports indicated that clostridial intestinal infections were possible among newborn infants.[119,126-128] These included an infant with GI necrosis and foci of clostridia within the colon and a 6-day-old infant with gangrene of the jejunum and colon who also demonstrated *C. perfringens* and gas in the intravascular system. Additionally, these organisms have long been implicated as a pathogen in neonatal omphalitis.[5,128] This infection of the umbilical stump and anterior abdominal wall has recently been associated to commonly occur either immediately prior to or during the course of NEC.

Engel[39] had reported three patients with NEC who had evidence of clostridial involvement. These infants presented with NEC as a clustered event and had *C. perfringens* isolated from their blood. In addition, the histopathology described from these infants' GI system was similar to that of the piglet enteritis caused by *C. perfringens*. Volsted-Pederson[147] reported on seven patients with NEC where *C. perfringens* type-A was implicated. This organism was cultured in one infant, while six of seven infants who underwent operative therapy for NEC had these organisms demonstrated on microscopic examination of the involved bowel. The authors remarked on the histopathological similarity of NEC to adult-onset GI gas gangrene.

Further evidence of *C. perfringens* as an agent for NEC was reported from an investigation of the relationship between bacterial neuraminidase production and red blood cell antigens.[141] Clostridia-producing neuraminidase as an exotoxin may expose an unusual red cell antigen, the Thomsen-Friedenreich antigen (TFA). TFA-activated red cells may agglutinate if exposed to agglutinins and result in altered red cell survival. Of four patients with NEC, all had TFA activation, while clostridial species were isolated from two patients. In an additional study, 9 of 26 patients with NEC had demonstrated TFA exposure; 2 of these 9 had added evidence of clostridial involvement (*C. perfringens* and *C. butyricum*).

Kosloske and colleagues[136] had reported a series of infants with NEC and evidence of *C. perfringens* involvement in their disease. Of 12 patients with NEC, 7 had evidence for *C. perfringens* infection. This data included positive identification of the organisms from blood or peritoneal cultures (via paracentesis) or microscopic appearance of bacteria in the affected bowel wall. Patients with NEC and evidence for clostridial infection were felt to have a more fulminant course than patients without evidence of this bacteria.[136]

We have observed that *C. perfringens* is the fifth most common pathogen recovered from blood cultures of patients with NEC at our hospital and that it is the sixth most common from the data available in the literature.[48,135] Many of these reports occurred prior to the establishment of excellence in anaerobic microbiological techniques and must therefore be an underestimation of the clostridial recovery rate. Peritoneal cultures from intraoperative or paracentesis samples demonstrate clostridial organisms at the same frequency or greater than that reported from blood cultures. Furthermore, we have reported seven patients with NEC who had evidence for *C. perfringens* involvement in their disease process (three additional patients with *C. perfringens* and NEC also occurred after this was published).[135] In contrast to prior reports, these infants were indistinguishable from patients without evidence of clostridia in terms of demography, nature of illness, and mortality. Of these seven patients, five had *C. perfringens* isolated from peritoneal cultures, one had umbilical exudates, and one had blood together with stool cultures. In addition, histopathological evidence of these organisms was noted in the affected bowel wall at autopsy or examination of resected tissue in three patients (Figure 3). It is conceivable that these organisms play no role in the primary pathogenesis of NEC and merely gain access to the blood, bowel, or peritoneum as secondary invaders following intestinal injury or bowel compromise. However, the histopathological evidence from at least four centers, together with pure growth of *C. perfringens* from certain patients with NEC, suggests a primary role in the pathogenesis of NEC.

*C. butyricum* has recently been implicated in NEC by Howard and co-workers.[131] They reported on 12 infants with NEC occurring over a 6-week period of time; 5 patients were involved in a 10-day epoch. These infants were also low-risk neonates (seven were full-term). Possible evidence of *C. butyricum* from gas chromatography analysis of fermentation products in blood specimens was demonstrated in nine of ten. Blood cultured over an extended period also recovered this organisms in seven of nine infants. The organism was cultured from two stool specimens. Further investigation revealed that *C. butyricum* was present in the breast milk pump trap and the distal tubing components of the unit.[126] In addition, the bacteria was isolated from the floor of the special care nursery and a 1% hypochlorite solution. Two disconcerting points of information were that *C. butyricum* usually colonizes extra-alimentary sites such as the eye, umbilicus, skin, and inanimate objects, and that prior to this report, *C. butyricum* had not usually been considered to be toxigenic. However, a more recent report has demonstrated toxin production from *C. butyricum* isolated from a patient with NEC.[146] This toxin-producing organism was resistant to penicillin and aminoglycosides, but was sensitive to Vancomycin, Chloramphenicol, and Clindamycin. The sites of isolation of this organism were the peritoneum and the cerebrospinal fluid. The toxin produced cytopathic effects in the Walker cell line and in human fibroblasts. Antiserum to *C. sordelli, C. perfringens, C. septicum,* or *C. histolyticum* toxins did not neutralize the effect of this new toxin.

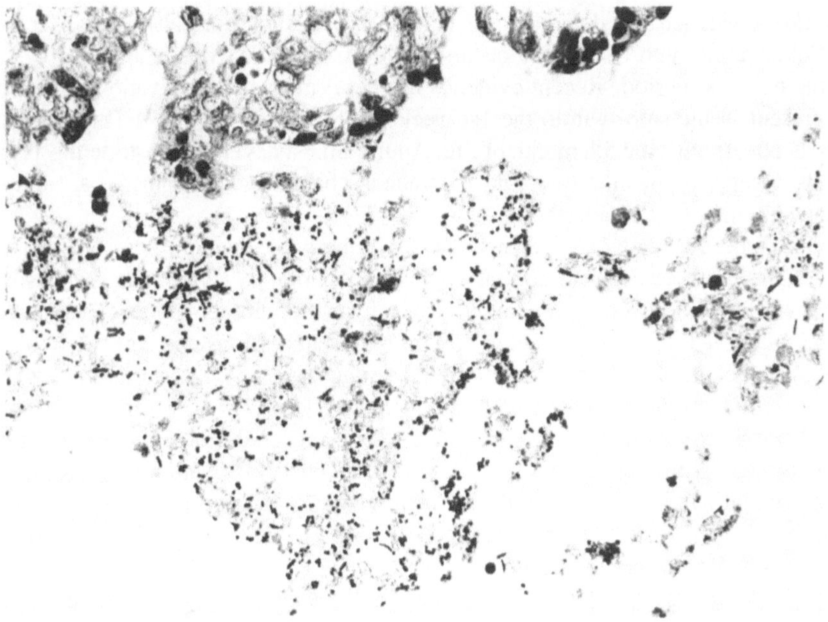

FIGURE 3. Large Gram-positive rod-like organisms in luminal exudate of small bowel in patient with NEC who grew *C. perfringens*. (Magnification × 400.)

These reports resulted in further analyses related to the role of *C. butyricum* in neonatal colonization and pathogenesis of NEC. In one report of 10 infants who were 2 to 30 days old and who had NEC, 6 of 10 had *C. butyricum* isolated while 7 of 19 without NEC also had this organism present in stool. An additional report of three patients with NEC associated with *C. butyricum* and no cytopathic toxin also described 5 of 17 control patients had *C. butyricum* as the only clostridial organism.[129] Another investigation reported on six infants with NEC, five of whom had *C. butyricum* in the stool, while seven of eight controls also had evidence of this bacteria.[149] These reports, together with the near consistent previous inability to isolate enterotoxins from *C. butyricum* suggests that this organism may not be pathogenic but merely a member of normal neonatal (or high-risk premature) fecal flora. What is not evident however, is the multitude of potential possibilities relative to initiating events which may stimulate a member of the fecal flora to becoming pathogenic. Altered defense mechanisms, broad spectrum antibiotics, or dietary factors such as volume and carbohydrate nature of the formula feedings must all be considered as essential confounding variables. As noted in Pig-Bel and neonatal animal clostridial enterotoxemias, immunological defenses and dietary perturbations may be responsible for a member of the anaerobic colonic bacterial flora becoming pathogenic and producing enterotoxin. Thus, although *C. butyricum* may colonize the normal neonatal stool, microenvironmental intestinal circumstances may have altered this organism's position from one of "normal flora" to one of "pathogen".

*C. difficile* is a third clostridial organism investigated as a pathogen in NEC. The toxin of this bacteria is involved in the pathogenesis of PMC among adults and older infants. The similarities of PMC and NEC discussed above suggest that the role of *C. difficile* in NEC be critically investigated.

One reported outbreak of NEC, where 11 patients had documented disease while another 9 were suspicious of NEC, compared the recovery of *C. difficile* toxin in the stool in affected patients to 13 normal age-matched control infants. Clostridia were recovered from 8 of 11 patients with NEC but were of multiple species. There was evidence of *C. difficile* toxin in the stool of three patients while no toxin was present in the stool of controls; suspect cases had toxin in two patients. No *C. difficile* was isolated from these particular stool samples.[121]

Much older evidence has demonstrated that *C. difficile* is a normal neonatal fecal organism.[81] Colonization with *C. difficile* occurs in the first few days of life and may be carried beyond the neonatal period. Recent evidence has confirmed this observation as *C. difficile* may be present in the stool within the 1st week of life.[124,133,140,142,144,148] This colonization may persist at least until the 5th month of life. Adults almost never demonstrate this organism in stool specimens, suggesting that a developmental change occurs in the fecal flora somewhere between infancy and adulthood.

Of greater interest, and one which creates a remarkable paradox, is the recovery of cytopathic *C. difficile* toxin from normal asymptomatic infants. Cytopathic titers as high as $1 \times 10^4$ may be demonstrated in healthy term infants. In another report of intensive care patients, *C. difficile* toxin was present in 55% of NICU admissions and correlated to vaginal delivery and breast milk feedings (another report suggested that in term infants, toxin was less in breast-fed infants). Furthermore, toxin recovery was suppressed by concomitant systemic therapy with ampicillin and gentamicin. Toxin appeared 4 days after these antibiotics were discontinued. Another recent report described recovery of *C. difficile* organisms from nursery personnel hands (but not stool) and inanimate objects, together with many infant's stool specimens with (59%) or without (51%) NEC. Disease state did not correlate to cytopathic toxin titer in this report.

A very recent epidemic of NEC has again and more convincingly implicated *C. difficile* in the pathogenesis of NEC.[130] During a 2-month period, 13 infants developed NEC. Of 13 infants, 12 (92%) had *C. difficile* cytopathic toxin recovered from the stool while 61% of stool specimens demonstrated growth of this organism. Control infants demonstrated toxin in only 11%, while none yielded the organism. Oral Vancomycin therapy (given to cases and close contacts) and anti-infectious disease control measures were two factors thought to terminate the epidemic.

Further evidence which needs to be considered related to *C. difficile* and NEC should include the investigation of both the cytopathic toxin and the recently discovered enterotoxin.[107,113,114,116] Both are readily separatable, are immunologically distinct, and although they have similar molecular weights, differ in many biological properties. Although the enterotoxin has minor cytopathic effects, it differs in its susceptibility to pH and and has a markedly positive effect in the rabbit ileal loop assay. Further data related to NEC and *C. difficile* which needs to be determined are specific strains of *C. difficile*, quantitative stool colony counts, and individual toxin potency. Almost all strains of this organism produce toxin; however, the potency may vary among different strains.

Thus, although it would appear that *C. difficile* may be a normal enteric (stool) organism among premature and term infants, it may, under circumstances, be pathogenic. It may produce cytopathic toxin without any clinically detectable adverse GI pathology in newborns. The significance of toxin strength and concomitant enterotoxin production is unknown. Therefore, for both NEC, PMC, and other reported enterocolitis (like enteritis occurring during infancy), cautious interpretation related to the pathogenicity of *C. difficile* is warranted until further information is available. As evident from these past reports, a compatible control group always needs to be included to define pathogenicity, as exemplified by the discovery of high titers of cytopathic toxin among many asymptomatic healthy neonates and infants. What determines colonization and what stimulates disease conditions is the key to the understanding of both NEC and PMC. Most probably, systemic or local enteric factors alter GI function and subsequently perturb the colonic bacteria's microenvironment.[110] These factors (none of which are presently identifiable) also need to be considered in the pathogenesis of NEC.

## XI. ADDITIONAL POTENTIAL FACTORS AND NEC

These potential factors are outlined in Tables 6 and 7. The incidence of NEC increases

**Table 6**
**IMMATURITY OF GI DEFENSES (NONSPECIFIC AND SPECIFIC)**

Delayed-onset local secretory IGA production
↓ Trypsin activity?
↓ Bile salt pool
↓ Motility
↓ Gastric acidity
↓ Glycocalyx (function and/or structure)?
Delayed gut closure
↑ Bacterial binding to epithelial cells
Absence of colostral protective mechanisms

**Table 7**
**ADDITIONAL VARIABLES IN NEC**

1. Role of neonatal feeding practices
2. Immaturity of GI system
   a. ↓ Mucosal barrier to antigens
   b. ↓ Mucosal immunity
   c. ↓ Bile salts pools
   d. ? Pancreatic enzymes
3. Broad spectrum antibiotic's effect on flora
4. Systemic illness
5. Iatrogenic toxins (plasticizers, etc.)
6. Alterations of GI bacterial flora ecosystems by combinations of 1 to 5.

with decreasing gestational age. NEC is predominantly a disease of hospitalized premature infants, although an occasional full-term infant in a neonatal intensive care unit becomes ill with NEC.[150-156] More rarely, an infant is discharged to home and readmitted 1 to 4 weeks later with NEC. Overall, it is a disease of the newborn period. GI function and structure are not completely developed in the full-term infant.[150,151,155] Premature infants have an even greater restriction of GI function and morphology. These developmental aspects of newborn GI immaturity may be divided into functional deficits related to protective immunological factors and those which relate to nutrient digestion.[156]

In the absence of breast-milk feeding, neonates have diminished GI secretory IgA levels. Salivary and fecal secretory IgA is absent in the 1st week of life only to appear between 10 to 14 days after birth. By 20 days of age, 90% of fed infants develop these secretory local antibodies. If enteric feedings are not offered, the appearance of fecal IgA may be delayed. In addition, the mechanical barrier to the entry of macromolecules (toxins, antigens, etc.) into the mucosa is more patent among immature animals compared to the complete antigenic impermeability (gut closure) in the normal adult GI system. This barrier against bacteria, their products, and possible immune complexes may protect the enterocytes from further inflammatory or toxin-associated injury. In the absence of sIgA from breast milk, the neonate's immature GI immunoprotection may predispose to disease entities such as NEC (Table 6).[154-156]

Furthermore, newborn intensive care practitioners may adversely alter the development of a normal colonic microbiological flora. The normal interaction of these organisms controls the local microenvironment and prevents single bacterial species from excessive proliferation with subsequent toxin production and from becoming locally tissue-invasive. In the presence of broad-spectrum systemic antibiotics or the strict "aseptic" care of premature infants, mono- or oligocolonization of the colonic flora may ensue. The usual interaction of checks and balances (interference) between a much more varied flora may be absent and disease may be manifested more readily.[154]

Neonatal feeding practices may also introduce iatrogenic factors such as the altered GI colonization noted in a nasojejunally alimented infant. Nasogastric tubes may leach out abnormal chemicals which may be local or systemic toxins. These may include the plasticizer employed in the manufacturing of polyvinylchloride tubes.[150]

Unphysiological feeding practices may also stress an immature GI system. Excessive osmolality or nutrient intake may overwhelm the poorly developed digestive system and result in malabsorption, gas formation, distention, and local ischemia or bacterial toxin production. Hypertonic formula feedings may directly alter mucosal function and result in GI injury.[150]

Direct mucosal injury or excessive substrate provision (nutrients which are malabsorbed) for bacterial growth, toxin production, or subsequent invasive behavior may go unchecked by a poorly developed local immune system. Bacterial toxin production may also remain unneutralized, while diminished proteolytic activity (trypsin) may prolong the half-life of locally produced enterotoxins. Stasis and ileus in sick neonates, and the usually physiological hypomotility of premature infants may also enhance bacterial overgrowth, tissue adherence, and toxin production.

One of the many additional factors to be considered in the balance between colonic bacterial colonization and these organism's potential evaluation toward the pathogenesis of NEC are the local nonspecific defense mechanisms in the GI tract. Bile salts are a potential candidate, as they have been demonstrated to neutralize endotoxin.[152,155] They may also be responsible for the observed survival of animals fed supralethal doses of endotoxin. In addition, bile salts may potentiate growth of certain colonic flora while inhibiting others. Newborn neonates who are full term have diminished bile salt pools compared to older children and adults, while premature infants demonstrate an even smaller bile acid pool size. At the onset of enteric alimentation, fecal endotoxin levels increase markedly and may be related to enhanced bacterial growth and/or augmented toxin production. Furthermore, the diminished bile acid pool size coupled with bacterial growth and/or toxin production may produce an excess of the biological effect of the toxin as it is not neutralized by bile salt action. Recently, an animal model of NEC among newborn rats has demonstrated a preventative and therapeutic role of bile salt administration. Formula-fed newborn rats developed NEC while those fed formula supplemented with the bile salt deoxycholic acid (0.1%) avoided the disease.[152] These observations stress the importance of nonspecific antibacterial defense mechanisms in determining the local checks and balances between the bacteria of the microflora in the colon. Alterations of local bile salt concentrations, intestinal organism proliferation, or toxin production which has its toxic effect unchecked may result in GI disease rather than a stable colonic ecosystem.

Each of these confounding variables needs to be considered as an important component for the pathogenesis of NEC. Immature GI absorptive and protective function together with subtle, as yet unidentified alterations within the microenvironment of the colonic flora may combine in the final common pathway which becomes manifested as NEC.

## XII. APPROACH TO THE PATIENT WITH NEC

Regardless of one's personal hypothesis related to the pathogenesis of NEC, all investigators presently treat patients as if NEC is an infectious disease.[153] In addition to therapy for an infectious disease, the GI aspects of NEC need to be considered, such as bowel decompression and discontinuing enteric alimentation. Furthermore, the development of associated complications such as septic and/or hypovolemic shock, disseminated intravascular coagulation and hemorrhage, in addition to intestinal perforation, needs to be carefully monitored and treated appropriately. Table 8 provides a systemic problem-oriented approach to patients with documented NEC. This documentation may only be confirmed by histopathological evidence or the roentgenological demonstration of pneumatosis intestinalis or hepatic portal gas. Infants without this documentation but with minor signs and symptoms need their care to be individualized, while those with serious manifestation shock, peritonitis, DIC) should be considered as having NEC. Until further diagnostic criteria are developed, these severely affected infants without evidence for pathological gas production should be considered and treated as NEC. In the event of a clustering of patients, the usual GI infectious disease prevention measures should be employed (Table 4). These have been helpful in both our nursery and other hospitals who have used this approach to terminate epidemics of this GI disease.

## Table 8
## APPROACH TO THE PATIENT WITH NEC

| | |
|---|---|
| Cultures: | Blood, CSF, urine, stool, and paracentesis if indicated |
| Antibiotics: | Based on nursery's organisms and sensitivities; begin anaerobic coverage if peritonitis is present |
| NPO: | With GI decompression |
| Shock: | maintain BP and urine output<br>Fluids: fresh/frozen plasma, whole blood<br>Catecholamines: dopamine, dobutamine<br>Steroids |
| DIC: | Exchange transfusion? |
| X-rays: | Frequent abdominal X-rays to detect ascites or perforation |
| Supportive care: | Elevate head of the bed<br>Maintain adequate hct, pH, $pO_2$, $pCO_2$<br>Mechanical ventilation for apnea — shock |
| Epidemics: | Institute infectious disease control measures |
| Surgery: | As indicated |
| Nutrition: | Once stabilized begin intravenous alimentation |

# REFERENCES


1. **Frantz, I. D., L'Heureux, P., Engel, R. R. et al.,** Necrotizing enterocolitis, *J. Pediatr.*, 86, 259, 1975.
2. **Hopkins, G. B., Gould, V. E., Stevenson, J. K., et al.,** Necrotizing enterocolitis in premature infants, *Am. J. Dis. Child.*, 120, 229, 1970.
3. **Karayalcin, G., Kim, K. Y., and Acs, H.,** Necrotizing enterocolitis following exchange transfusion, *N.Y. State J. Med.*, March 1976.
4. **Kliegman, R. M. and Fanaroff, A. A.,** Neonatal necrotizing enterocolitis, a nine year experience. I. Epidemiology and uncommon observations, *Am J. Dis. Child.*, 131, 603, 1981.
5. **Kliegman, R. M. and Fanaroff, A. A.,** Neonatal necrotizing enterocolitis, a nine year experience. II. Outcome assessment. *Am. J. Dis. Child.*, 131, 608, 1981.
6. **Lake, A. and Walker, W.,** Neonatal necrotizing enterocolitis: a disease of altered host defense, *Clin. Gastroenterol.*, 6, 463, 1977.
7. **Mizrahi, A., Barlow, O., Berdon, W. et al.,** Necrotizing enterocolitis in premature infants, *J. Pediatr.*, 66, 697, 1965.
8. **Polin, R., Pollack, P. F., Barlow, B. et al.,** Necrotizing enterocolitis in term infants, *J. Pediatr.*, 89, 460, 1976.
9. **Reid, W. D. and Shannon, M. P.,** Necrotizing enterocolitis — a medical approach to treatment, *Can. Med. Assoc. J.*, 108, 573, 1973.
10. **Roback, S. A., Foker, J., Frantz, I. F. et al.,** Necrotizing enterocolitis, *Arch. Surg.*, 109, 314, 1974.
11. **Santulli, T. V., Schullinger, J. N., Heird, W. C., et al.,** Acute necrotizing enterocolitis in infancy: a review of 64 cases, *Pediatrics*, 55, 376, 1975.
12. **Stevenson, J. K., Graham, C. B., Oliver, T. K., Jr. et al.,** Neonatal necrotizing enterocolitis. A report of 21 cases with 14 survivors, *Am. J. Surg.*, 118, 260, 1969.
13. **Touloukian, R. J.,** Neonatal necrotizing enterocolitis: an update on etiology, diagnosis, and treatment, *Surg. Clin. North Am.*, 56, 281, 1976.
14. **Benirschke, K.,** Rep. 68th Ross Conf. on Pediatric Research, Ross Laboratories, Columbus, Ohio, 1974.
15. **Hathaway, W.,** Rep. 68th Ross Conf. Pediatric Research, Ross Laboratories, Columbus, Ohio, 1974.
16. **Leonidas, J. C. and Hall, R. T.,** Neonatal pneumatosis coli: a mild form of neonatal necrotizing enterocolitis, *J. Pediatr.*, 89, 456, 1976.
17. **Leonidas, J. C., Drasna, I. H., Fox, H. A. et al.,** Peritoneal fluid in necrotizing enterocolitis: a radiologic sign of clinical deterioration, *J. Pediatr.*, 82, 672, 1973.

18. **O'Neill, J. A., Jr., Stahlman, M. T., and Meng, H. C.,** Necrotizing enterocolitis in the newborn: operative indications, *Ann. Surg.* 182, 274, 1975.
19. **Richmond, J. A. and Mikity, V.,** Benign form of necrotizing enterocolitis, AJR, 123, 301, 1975.
20. **Yu, V. Y. H., Tudehope, D. I., and Gill, G. J.,** Neonatal necrotizing enterocolitis. I. Clinical aspects, *Med. J. Aust.*, 1, 685, 1977.
21. **Yu, V. Y. H., Tudehope, D. I., and Gill, G. J.,** Neonatal necrotizing enterocolitis. II. Perinatal risk factors, *Med. J. Aust.*, 1, 688, 1977.
22. **Ahlstedt, S., Carlsson, B., Fallstrom, S. P. et al.,** Antibodies in human serum and milk induced by enterobacteria and food proteins, in *Immunology of the Gut*, Ciba Foundation Symp. 46, Elsevier/North Holland, Amsterdam, 1977.
23. **Barlow, B., Santulli, T. V., Heird, W. C. et al.,** An experimental study of acute neonatal enterocolitis. The importance of breast milk, *J. Pediatr. Surg.*, 9, 587, 1974.
24. **Book, S., Herbert, J. J., and Jung, A. L.,** Necrotizing enterocolitis in infants fed an elemental formula, *Pediatr. Res.*, 8, 379, 1974.
25. **Book, L. S., Herbst, J. J., and Jung, A. L.,** Carbohydrate malabsorption in necrotizing enterocolitis, *Pediatrics*, 57, 201, 1976.
26. **Book, L. S., Berbst, J. J., and Jung, A** Comparison of fast- and slow-feeding schedules to the development of necrotizing enterocolitis, *J. Pediatr.*, 89, 463, 1976.
27. **Brown, E. G. and Sweet, A. Y.,** Preventing necrotizing enterocolitis in neonates, *JAMA*, 240, 2452, 1978.
28. **Eidelman, A. and Inwood, R.,** Necrotizing enterocolitis and enteral feedings, *Am. J. Dis. Child.*, 134, 545, 1980.
29. **Goldman, A. S.,** Human milk, leukocytes, and immunity, *J. Pediatr.*, 90, 167, 1977.
30. **Gotoff, S. P.,** Neonatal immunity, *J. Pediatr.*, 85, 149, 1974.
31. **Head, J.,** Immunobiology of lactation, *Semin. Perinatol.*, 1, 195, 1977.
32. **Larsen, S. A., Jr. and Homer, D. R.,** Relation of breast versus bottle feeding to hospitalization for gastroenteritis in a middle-class U.S. population, *J. Pediatr.*, 91, 417, 1978.
33. **Pitt, J.,** Breast milk leukocytes, *Pediatrics*, 58, 769, 1976.
34. **Pitt, J., Barlow, B., and Heird, W. C.,** Protection against experimental necrotizing enterocolitis by maternal milk. I. Role of milk leukocytes, *Pediatr. Res.*, 11, 906, 1977.
35. **Stoliar, O. A., Pelley, R. P., Kaniecki-Green, E. et al.,** Secretory IgA against enterotoxins in breast milk, *Lancet*, 1, 1258, 1976.
36. **Svirsky-Gross, S.,** Pathogenic strains of coli (0,111) among prematures and the use of human milk in controlling the outbreak of diarrhea, *Ann. Pediatr.*, 190, 109, 1958.
37. **Winberg, J. and Wessner, G.,** Does breast milk protect against septicaemia in the newborn?, *Lancet*, 1, 1091, 1971.
38. **Kliegman, R. M., Pittard, W. R., Fanaroff, A. A.,** Necrotizing enterocolitis in neonates fed human milk, *J. Pediatr.*, 95, 450, 1979.
39. **Engel, R.,** Rep. 68th Ross Conf. on Pediatric Research, Ross Laboratories, Columbus, Ohio, 1974, 66.
40. **Engel, R. R., Virnig, N. L., Hunt, C. E. et al.,** Origin of mural gas in necrotizing enterocolitis, *Pediatr. Res.*, 7, 292, 1973.
41. **Gall, L S.,** The role of intestinal flora in gas formation, *Ann. N.Y. Acad. Sci.*, 150, 27, 1968.
42. **Kirschner, B. S., Lahr, C., Lahr, D. et al.,** Detection of increased breath hydrogen in infants with necrotizing enterocolitis, *Gastroenterology*, 72, 1080, 1977.
43. **Bunton, G. L., Durbin, G. M., McIntosh, N. et al.,** Necrotizing enterocolitis. Controlled study of 3 years experience in a neonatal intensive care unit, *Arch. Dis. Child.*, 52, 772, 1977.
44. **Coello-Ramirez, P., Gutierres-Topete, G., and Lifshitz, F.,** Pneumatosis intestinalis, *Am. J. Dis. Child.*, 120, 3, 1970.
45. **Corkery, J. J., Dubowitz, V., Lister, J. et al.,** Colonic perforation after exchange transfusion, *Br. Med. J.*, 4, 345, 1968.
46. **Guinan, M., Schaberg, D., Bruhn, F. W., Richardson, C. J. et al.,** Epidemic occurrence of neonatal necrotizing enterocolitis, *Am. J. Dis. Child.*, 133, 594, 1979.
47. **Hakanson, D. O. and Oh, W.,** Necrotizing enterocolitis and hyperviscosity in the newborn infant, *J. Pediatr.*, 90, 458, 1977.
48. **Kliegman, R. M.,** Neonatal necrotizing enterocolitis: implications for an infectious disease, *Pediatr. Clin. North Am.*, 26, 327, 1979.
49. **Kliegman, R. M., Hack, M., Jones, P. et al.,** Epidemiologic study of necrotizing enterocolitis among low birth weight infants, *J. Pediatr.*, 100, 440, 1982.
50. **Leake, R. D., Thanopoulos, B., and Nieberg, R.,** Hyperviscosity syndrome-associated with necrotizing enterocolitis, *Am. J. Dis. Child.*, 129, 1192, 1975.
51. **L'E Orme, R. and Eades, S. M.,** Perforation of the bowel in the newborn as a complication of exchange transfusion, *Br. Med. J.*, 4, 349, 1968.

52. **Lloyd, J. R.,** The etiology of gastrointestinal perforations in the newborn, *J. Pediatr. Surg.*, 49, 77, 1969.
53. **Ryder, R. W., Shelton, J. D., Guinan, M. E.,** and **the Committee on Necrotizing Enterocolitis,** Necrotizing enterocolitis: a prospective multicenter investigation, *Am. J. Epidemiol.*, 112, 113, 1980.
54. **Stoll, B. J., Kanto, W. P., Jr., Glass, R. I. et al.,** Epidemiology of necrotizing enterocolitis: a case control study, *J. Pediatr.*, 96, 447, 1980.
55. **Touloukian, R. J., Posch, J. N., and Spencer, R.,** The pathogenesis of ischemic gastroenterocolitis of the neonate: selective gut mucosal ischemia in asphyxiated neonatal piglets, *J. Pediatr. Surg.*, 7, 194, 1972.
56. **Touloukian, R. J., Kadar, A., and Spencer, R. P.,** The gastrointestinal complications of neonatal umbilical venous exchange transfusion: a clinical and experimental study, *Pediatrics*, 51, 36, 1973.
57. **Yale, C. E. and Balish, E.,** The importance of clostridia in experimental intestinal strangulation, *Gastroenterology*, 71, 793, 1976.
58. **Yale, C. E., Blaish, E., and Wu, J. P.,** The bacterial etiology of pneumatosis cystoides intestinalis, *Arch. Surg.*, 109, 89, 1974.
59. **Bell, M., Shackelford, P., Feigin, R., et al.,** Epidemiologic and bacteriologic evaluation of neonatal necrotizing enterocolitis, *J. Pediatr. Surg.*, 14, 1, 1979.
60. **Book, L. S., Overall, J. C., Jr., Herbst, J. J. et al.,** Clustering of necrotizing enterocolitis. Interruption by infection-control measures, *N. Engl. J. Med.*, 297, 984, 1977.
61. **Chany, C., Moscovici, O., Lebon, P. et al.,** Association of coronovirus infection with neonatal necrotizing enterocolitis, *Pediatrics*, 69, 209, 1982.
62. **Guinan, M., Schaberg, D., Bruhn, F. W., et al.,** Epidemic occurrence of neonatal necrotizing enterocolitis, *Am. J. Dis. Child.*, 133, 594, 1979.
63. **Henderson, A., Maclaurin, J., and Scott, J. M.,** *Pseudomonas* in a Glasgow baby unit, *Lancet*, 2, 316, 1969.
64. **Hill, H. R., Hunt, C. E., and Matsen, J. M.,** Nosocomial colonization with *Klebsiella*, type 26, in a neonatal intensive care unit associated with an outbreak of sepsis, meningitis, and necrotizing enterocolitis, *J. Pediatr.*, 85, 415, 1974.
65. **Izant, R.,** Surgical complications of neonatal necrotizing enterocolitis, *Arch. Surg.*, 107, 223, 1973.
66. **Johnson, F. E., Crnic, D. M., Simmons, M. A. et al.,** Association of fatal coxsackie $B_2$ viral infection and necrotizing enterocolitis, *Arch. Dis. Child.*, 52, 802, 1977.
67. **Krasna, I., Kurgan, A., and Noy, C.,** *Pseudomonas* septicemia: necrotizing bowel lesions (NEC) and skin lesions in a 5 month old child, *J. Pediatr. Surg.*, 14, 481, 1979.
68. **Lake, A. M., Lauer, B. A., and Clark, J. C.,** Enterovirus infections in neonates, *J. Pediatr.*, 89, 787, 1976.
69. **Mangurten, R. and Fernandez, B.,** Neonatal Aspergillosis accompanying fulminant necrotizing enterocolitis, *Arch. Dis. Child*, 55, 559, 1980.
70. **Moomjian, A. S., Packman, G. J., Fox, W. W. et al.,** Necrotizing enterocolitis-endemic vs. epidemic form, *Pediatr. Res.* 12, 530, 1978.
71. **Powell, J., Bureau, M., Park, C., et al.,** Necrotizing enterocolitis epidemic following an outbreak of *Enterobacter cloacae* type 3305573 in a neonatal intensive care unit, *Am. J. Dis. Child.*, 134, 1152, 1980.
72. **Sawyer, R. B., Sawyer, K. C., and List, J. E.,** Infectious emphysema of the gastrointestinal tract in the adult, *Am. J. Surg.*, 120, 579, 1970.
73. **Shapiro, M. and Stein, H.,** Necrotizing enterocolitis and exchange transfusion, *S. Afr. Med. J.*, 47, 1236, 1973.
74. **Speer, M. E., Taber, L. H., Yow, M. D., et al.,** Fulminant neonatal sepsis and necrotizing enterocolitis associated with a "nonenteropathogenic" strain of *Escherichia coli*, *J. Pediatr.*, 89, 91, 1976.
75. **Stein, H., Beck, J., Solomon, A., et al.,** Gastroenteritis with necrotizing enterocolitis in premature babies, *Br. Med. J.*, 2, 616, 1972.
76. **Virnig, N. L. and Reynolds, J. W.,** Epidemiological aspects of neonatal necrotizing enterocolitis, *Am. J. Dis. Child.*, 128, 186, 1974.
77. **Waldhausen, J. A., Herendeen, T., and King, H.,** Necrotizing colitis of the newborn: common cause of perforation of the colon, *Pediatr. Surg.*, 54, 365, 1963.
78. **Berg, R. and Savage, D.,** Immunological responses and microorganisms indigenous to the gastrointestinal tract, *Am. J. Clin. Nutr.*, 25, 1364, 1972.
79. **Brook, I., Barrett, C., Brinkman, C. et al.,** Aerobic and anaerobic bacterial flora of the maternal cervix, and newborn gastric fluid and conjunctiva: a prospective study, *Pediatrics*, 63, 451, 1979.
80. **Goldmann, D., Leclair, J., and Macone, A.,** Bacterial colonization of neonates admitted to an intensive care environment, *J. Pediatr.*, 93, 288, 1978.
81. **Hall, I. and O'Toole, E.,** Intestinal flora in newborn infants, *Am. J. Dis. Child.*, 49, 390, 1935.
82. **Mata, L. and Arrutia, J.,** Intestinal colonization of breast-fed children in a rural area of low socioeconomic level, *Ann. N.Y. Acad. Sci.*, 176, 93, 1971.

83. **Mata, L., Mejicgnos, M., and Jimenez, F.,** Studies on the indigenous gastro-intestinal flora of Guatemalan children, *Am. J. Clin. Nutr.*, 25, 1380, 1972.
84. **Savage, D.,** Associations and physiological interactions of indigenous microorganisms and gastrointestinal epithelia, *Am. J. Clin. Nutr.*, 25, 1372, 1972.
85. **Sprunt, K., Leidy, G., and Redman, W.,** Abnormal colonization of neonates in an intensive care unit: means of identifying neonates at risk of infection, *Pediatr. Res.*, 12, 998, 1978.
86. **Yow, M.,** Personal communication cited by McCracken, G. and Nelson, J., in *Antimicrobial Therapy for Newborns,* Grune & Stratton, New York, 1977, 132.
87. Editorial, Clostridia as intestinal pathogens, *Lancet,* 1, 1113, 1977.
88. Editorial, Pseudomembranous colitis, *Lancet,* 1, 1080, 1978. (See subsequent section)
89. **Finegold, S. M., Bartlett, J. G., Chow, A. W. et al.,** Management of anaerobic infections, *Ann. Intern. Med.*, 83, 375, 1975.
90. **Finegold, S. M.,** in *Anaerobic Bacteria and Human Disease,* Academic Press, New York, 1977.
91. **Lawrence, G. and Cooke, R.,** Experimental Pig-Bel: the production and pathology of necrotizing enteritis due to *Clostridium welchii* type C in the guinea pig, *J. Exp. Pathol.*, 61, 261, 1980.
92. **Lawrence, G., Shann, F., Freestone, D. et al.,** Prevention of necrotizing enteritis in Papua New Guinea by active immunization, *Lancet,* 1, 227, 1979.
93. **Lawrence, G. and Walker, P. D.,** Pathogenesis of enteritis necroticans in Papua New Guinea, *Lancet,* 1, 125, 1976.
94. **Murrell, T. G. C., Roth, L., Egerton, J. et al.,** Pig-bel: enteritis necroticans, *Lancet,* 1, 217, 1966.
95. **Sawyer, R. B., Sawyer, K. C., and List, J. E.,** Infectious emphysema of the gastrointestinal tract in the adult, *Am. J. Surg.*, 120, 579, 1970.
96. **Stern, M. and Batty, I.,** in *Pathogenic Clostridia,* Butterworths, London, 1975.
97. **Zeissler, J., and Rassfeld-Sternberg, L.,** Enteritis necroticans due to *Clostridium welchii* type F, *Br. Med. J.*, 267, 1949.
98. **Adler, S., Chandrika, T., and Berman, M.,** *Clostridium difficile* associated with pseudomembranous colitis. Occurrence in a 12 week old infant without prior antibiotic therapy, *Am. J. Dis. Child.*, 135, 820, 1981.
99. **Bartlett, J., Onderdonk, A., Cisneros, R. et al.,** Clindamycin-associated colitis due to a toxin-producing species of *Clostridium* in hamsters, *J. Infect. Dis.*, 136, 701, 1977.
100. **Bartlett, J., Chang, T., Gurwith, M. et al.,** Antibiotic-associated pseudomembranous colitis due to toxin producing Clostridia, *N. Engl. J. Med.*, 298, 531, 1978.
101. **Burdon, D., Thompson, H., Candy, D. et al.,** Enterotoxin(s) of *Clostridium difficile, Lancet,* ii, 258, 1981.
102. **Donta, S., Stuppy, M., and Myers, M.,** Neonatal antibiotic-associated colitis, *Am. J. Dis. Child.*, 135, 181, 1981.
103. **Fekety, R., Kim, K., Brown, D. et al.,** Epidemiology of antibiotic-associated colitis: isolation of *Clostridium difficile* from the hospital environment, *Ann. Int. Med.*, 70, 906, 1981.
104. **Greenfield, C., Szawathowski, M., Noone, P. et al.,** Is pseudomembranous colitis infectious?, *Lancet,* 1, 371, 1981.
105. **Hyams, J., Berman, M., and Helgason, H.,** Nonantibiotic-associated enterocolitis caused by *Clostridium difficile* in an infant, *J. Pediatr.*, 99, 750, 1981.
106. **Larson, H., Price, A., Honour, P., Borriello, S. P.,** *Clostridium difficile* and the aetiology of pseudomembranous colitis, *Lancet,* 1, 1063, 1978.
107. **Libby, J., Sullivan, N., Tassell, R. et al.,** Relationship of two toxins of *Clostridium difficile,* Abstr. #B44, Presented at the ASM Annual Meeting, Dallas, Texas, March 3, 1981, 22.
108. **Rifkin, G., Fekety, F., Silva, J. et al.,** Antibiotic-induced colitis. Implication of a toxin neutralized by *Clostridium sordellii* antitoxin, *Lancet,* ii, 1103, 1977.
109. **Rogers, T., Petrou, M., Lucas, C. et al.,** Spread of *Clostridium difficile* among patients receiving non-absorbable antibiotics for gut decontamination, *Br. Med. J.*, 283, 408, 1981.
110. **Rolfe, R., Helebian, S., and Finegold, S. M.,** Bacterial interference between *Clostridium difficile* and normal fecal flora, *J. Infect. Dis.*, 143, 470, 1981.
111. **Schussheim, A. and Goldstein, E.,** Antibiotic-associated pseudomembranous colitis in siblings, *Pediatrics,* 66, 932, 1980.
112. **Silva, J. and Fekety, R.,** Clostridia and antimicrobial enterocolitis, *Ann. Rev. Med.*, 32, 327, 1981.
113. **Taylor, N., Thorne, G., and Bartlett, J.,** Separation of an enterotoxin from the cytotoxin of *Clostridium difficile, Clin. Res.*, 28, 285, 1980.
114. **Taylor, N., Taylor, R., and Bartlett, J.,** *Clostridium difficile* toxins: localization and binding studies, Abstr. #207, presented at the ASM Interscience Conference on Antimicrobial Agents and Chemotherapy, New Orleans, October 23, 1980.
115. **Thelestam, M. and Bronnegard, M.,** Interaction of cytopathogenic toxin from *Clostridium difficile* with cells in tissue culture, *Scand. J. Infect. Dis. Suppl.*, 22, 16, 1980.

116. **Vesely, D., Straub, K., Nolan, C. et al.,** Purified *Clostridium difficile* cytotoxin stimulates guanylate cyclase activity and inhibits adenylate cyclase activity, *Infect. Immun.*, 33, 285, 1981.
117. **Viscidi, R. and Bartlett, J.,** Antibiotic-associated pseudomembranous colitis in children, *Pediatrics*, 67, 381, 1981.
118. **Bignold, L. and Harvey, H.,** Necrotizing enterocolitis associated with invasion by *Clostridium septicum* complicating cyclic neutropenia, *Aust. N.Z. J. Med.*, 9, 426, 1979.
119. **Bogdan, J. and Rapkin, R.,** Clostridia infection in the newborn, *Pediatrics*, 58, 120, 1976.
120. **Bradley, J., Szawatkowski, M., Noone, P., et al.,** Letter reply, *Lancet*, 1, 389, 1978.
121. **Cashore, W., Peter, G., Lauermann, M. et al.,** Clostridia colonization and clostridial toxin in neonatal necrotizing enterocolitis, *J. Pediatr.*, 98, 308, 1981.
122. **Chaney, N.,** *Clostridium* infection in mother and infant, *Am. J. Dis. Child.*, 134, 1175, 1980.
123. **Cooperstock, M.,** *Clostridium difficile*, enterocolitis and Hirschsprungs disease, *Lancet*, 1, 800, 1982.
124. **Donta, S. and Myers, M.,** *Clostridium difficile* toxin in asymptomatic neonates, *J. Pediatr.*, 100, 431, 1982.
125. **Epidemiological Research Lab.,** *Clostridium welchii* from blood culture, *Br. Med. J.*, 845, 1976.
126. **Flynn, D., Howard, F., Bradly, J. et al.,** Necrotizing enterocolitis, *Lancet*, 1, 545, 1977.
127. **Freedman, S. and Hollander, M.,** *Clostridium perfringens* septicemia as a postoperative complication of the newborn infant, *J. Pediatr.*, 71, 576, 1962.
128. **Gormley, D.,** Neonatal anaerobic (clostridial) cellulitis and omphalitis, *Arch. Dermatol.*, 113, 683, 1977.
129. **Gothefors, P. and Blenkharn, I.,** *Clostridium butyricum* and necrotizing enterocolitis, *Lancet*, 1, 52, 1978.
130. **Han, V., Sayed, H., Chance, G. et al.,** Outbreak of *C. difficile* necrotizing enterocolitis (NEC): a case for oral Vancomycin therapy? *Pediatr. Res.*, 16, 290A, 1982.
131. **Howard, F., Bradley, J., Flynn, D., et al.,** Outbreak of necrotizing enterocolitis caused by *Clostridium butyricum*, *Lancet*, ii, 1099, 1977.
132. **Isenberg, A.,** *Clostridium welchii* infection, *Arch. Surg.*, 92, 727, 1966.
133. **Kelsey, M. and Vince, A.,** Clostridia in neonatal faeces, *Lancet*, ii, 100, 1979.
134. **Kindley, A., Roberts, P., and Tullock, W.,** Neonatal necrotizing enterocolitis, *Lancet*, 1, 649, 1977.
135. **Kliegman, R. M., Fanaroff, A. A., Izant, R. et al.,** Clostridia as pathogens in neonatal necrotizing enterocolitis, *J. Pediatr.*, 95, 287, 1979.
136. **Kosloske, A., Ulrich, J., and Hoffman, H.,** Fulminant necrotizing enterocolitis associated with Clostridia, *Lancet*, ii, 1014, 1978.
137. **Laverdiere, M., Robert, A., Chicoine, R. et al.,** Clostridia in necrotizing enterocolitis, *Lancet*, ii, 377, 1978.
138. **Loc, C., Larroche, J., and Beaudoin, P.,** Infection à *Clostridium perfringens* et entérocolite ulceró-nécrosante, *Arch. Fr. Pediatr.*, 37, 233, 1980.
139. **Mitchell, R., Etches, P., and Day, D.,** Non-toxigenic Clostridia in babies, *J. Clin. Pathol.*, 34, 217, 1981.
140. **Rietra, P., Slateros, K., Zane, H. et al.,** Clostridial toxin in feces of healthy infants, *Lancet*, ii, 319, 1978.
141. **Seger, R., Joller, P., Bird, G. et al.,** Necrotizing enterocolitis and neuraminidase-producing bacteria, *Helv. Paediatr. Acta*, 34, 121, 1980.
142. **Sherertz, R. and Sarubbi,** The prevalence of *Clostridium difficile* and toxin in a nursery population: a comparison between patients with necrotizing enterocolitis and an asymptomatic group, *J. Pediatr.*, 100, 435, 1982.
143. **Sonnabend, O., Sonnabend, W., Rauh, G. et al.,** Clostridien-infektionen mit und ohne manifeste gas-gangran, *Schweiz. Med. Wochenschr.*, 107, 34, 1977.
144. **Stark, P. and Lee, A.,** Clostridia isolated from the feces of infants during the first year of life, *J. Pediatr.*, 100, 362, 1982.
145. **Stoll, B., Nahmias, A., Wickliffe, C. et al.,** Bacterial toxin and neonatal necrotizing enterocolitis, *J. Pediatr.*, 96, 114, 1980.
146. **Sturm, R., Staneck, J., Stauffer, L. et al.,** Neonatal necrotizing enterocolitis associated with penicillin-resistant toxigenic *Clostridium butyricum*, *Pediatrics*, 66, 928, 1980.
147. **Volsted-Pederson, P., Halveg, A., Hansen, F. et al.,** Necrotizing enterocolitis of the newborn — Is it gas-gangrene of the bowel?, *Lancet*, ii, 715, 1976.
148. **Welch, D. and Marks, M.,** Is *Clostricium difficile* pathogenic in infants?, *J. Pediatr.*, 100, 393, 1982.
149. **Smith, M. F., Borriello, S. P., Clayden, G. S., and Casewell, M. W.,** Clinical and bacteriological findings in necrotizing enterocolitis: a controlled study, *J. Infect.*, 2, 23, 1980.
150. **Brown, E. and Sweet, A., Eds.,** *Neonatal Necrotizing Enterocolitis*, Grune & Stratton, New York, 1980.
151. **CIBA Foundation,** *Development of Mammalian Absorptive Processes*, No. 70, Excerpta Medica, Amsterdam, 1979.

152. **Diaz, J., Samson, H., Kessler, D. et al.,** Experimental necrotizing enterocolitis: the possible role of bile salts in its etiology and treatment, *Pediatr. Res.* 14, 595, 1980.
153. **Fanaroff, A. A. and Kliegman, R. M.,** Nectrotizing enterocolitis, in *Care of the High Risk Neonate*, Klaus, M. and Fanaroff, A. A., Eds., W. B. Saunders, Philadelphia, 1979, 136.
154. **Lawrence, G., Bates, J., and Gaul, A.,** Pathogenesis of neonatal necrotizing enterocolitis, *Lancet*, ii, 137, 1982.
155. **Lebenthal, E., Ed.,** *Textbook of Gastroenterology and Nutrition in Infancy*, Vols. 1 & 2, Raven Press, New York, 1981.
156. **Ross Laboratories,** The Mucosal Immune System in Health and Disease, Rep. 81st Ross Conf. Pediatric Research, Ross Laboratories, Columbus, Ohio, 1981.

Chapter 5

# PIG-BEL

**P. D. Walker**

## TABLE OF CONTENTS

## I. THE DISEASE

Enteritis necroticans or Pig-Bel is a necrotizing hemorrhagic jejunitis affecting the young adult population of the Highlands of Papua New Guinea.[1,2] It is a major cause of death and suffering in Highland children and the most common cause of hospital death in children after the 1st year.[3] Although the disease is well recognized in this area, there is also considerable evidence that it occurs in other parts of Southeast Asia, particularly in Indonesia,[4] Malaysia,[5] and the southern parts of China.[6] A disease with symptoms similar to Pig-Bel has also been described in Uganda, Bangladesh, and the British Solomon Island Protectorates.[7-9] The disease certainly bears a remarkable similarity to "Darmbrand" which occurred in Germany shortly after World War II.[10,11] More recently, Murrell[12] has argued convincingly that the disease is routed in antiquity. The unhygienic environmental conditions prevailing in the Highlands of Papua New Guinea today mimic those in Europe prior to the mid-19th century. In particular, the intake of protein was low for a significant proportion of the population during a bad harvest. Early literature as far back as Hippocrates associates diarrheal disease outbreaks with hot summers and drought conditions in human settlements of ancient times. Two main diarrheal conditions were described; colic, or severe abdominal pain, from which recovery often took place and Ileus or Ileac Passion from which patients seldom recovered. A medieval writer in the reign of Queen Elizabeth I gives a classical description of enteritis necroticans in children and was the first to relate the diet to meat.[13]

The disease presents itself in a number of forms: acute Pig-Bel, a fulminating disease where the patient may die within 24 hr and a more mild subacute form difficult to distinguish from gastroenteritis from which recovery usually takes place. A smaller group of patients present months or years after an acute episode with either subacute intestinal obstruction and evidence of malnutrition, some otherwise unexplained malabsorption, or develop sudden intestinal obstruction. This disease is customarily referred to as chronic Pig-Bel.

## II. ACUTE PIG-BEL

The typical symptoms are abdominal pain, usually abdominal distention, vomiting, and passage of bright blood. In most cases, pig or other meat has been eaten about 48 hr before the onset, but the time interval can vary from less than 24 hr to 1 week. Meat is not an essential ingredient and other foodstuffs such as raw peanuts have been involved. Pain is persistent, often with superimposed colic. Vomiting continues intermittently but by 24 to 48 hr most patients have developed constipation and there is no further adverse blood loss. The vomit is a foul-smelling, murky fluid flecked with particles of shed intestinal mucosa. Tenderness is most severe in the upper abdomen.[14]

In an acute toxic case abdominal signs may be overshadowed by severe shock, thus complicating diagnosis. The shock syndrome developing later in the disease is related to fluid and electrolyte loss and perhaps septicemia more than the original clostridial toxins.[14]

Conservative treatment of acute disease has as its basis the rapid replacement of estimated fluid losses, together with the provision of electrolytes, protein, and calories. Large deficits of potassium are also encountered. Additional measures include intestinal decompression, administration of penicillin, and Chloramphenicol intravenously. Vitamin supplements are recommended.[3]

The use of surgery in Pig-Bel is a matter of some debate, some surgeons undertaking it in all but mild cases. The following policy is based on recommendations by Sheppard,[14] a specialist surgeon at Goroka Hospital.

Operation has been undertaken when:

1. Toxicity persists in a patient in spite of resuscitation and adequate treatment.

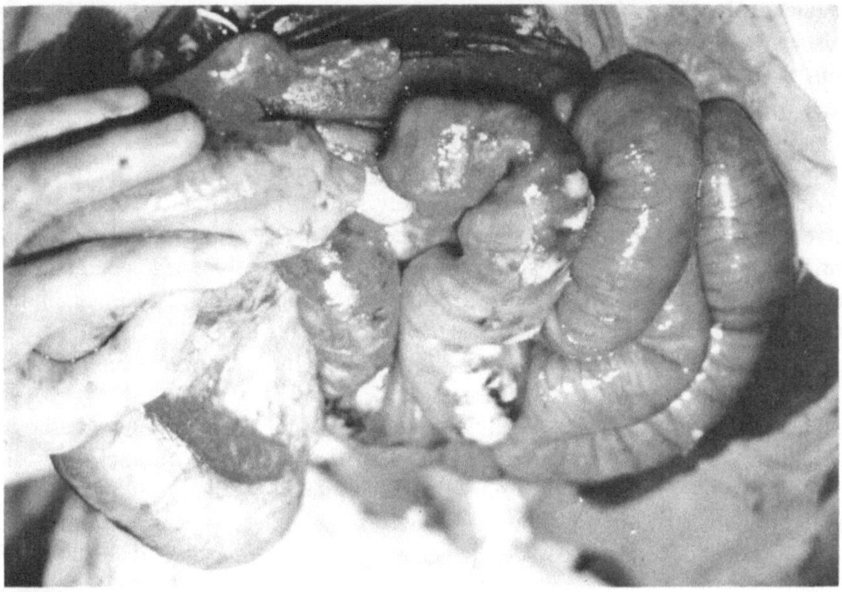

FIGURE 1. Dilated loops of the jejunum as seen at operation.

2. Persistence of intestinal obstruction as shown by large volume of aspirate and failure of distention to resolve, together with continuing pain.
3. Suspect perforation.
4. Recurrent severe bleeding.

The first two indications mean waiting no more than 2 to 3 days providing the situation remains stable or improved. Any deterioration requires surgery. The reasons for this are that many patients are poor operative risks and some with severe clinical disease do recover without surgery. Decisions relating to the amount of the intestine to be removed are often difficult in early stages. A waiting policy is not recommended for patients admitted at a late stage of acute disease as operations tend to become more technically difficult after the 1st week and can be very troublesome when 10 to 12 days have elapsed since onset. Operation is contraindicated in mild cases where diagnosis is doubtful and conservative measures likely to succeed. Surgery is not indicated in very severe cases where resuscitative measures cannot reverse the trend to shock.

At operation the surgeon is confronted with loops of dilated small intestine (Figure 1). The jejunum is the usual primary site of disease, although in less typical cases, the ileum appears to have been the chief target. The area affected is likely to be about 50 to 200 cm in length, although the whole jejunum and ileum can be involved in extreme cases. Macroscopic changes in the wall vary from edema and slight discoloration to red and plum-colored areas with necrotic patches scattered throughout the worst sections. Gangrenous segments may be present in late stages. Abscesses are found within the mass of adherent small bowel loops, the whole usually wrapped in omentum and transverse colon. The mesentery is very edematous and large, fleshy lymph nodes are a constant feature. A few lines of gas bubbles may extend across the bowel wall. Typically, some 50 to 200 cm of jejunum will require removal; ideally, the anastomosis is of near normal tissue.[14]

## III. CHRONIC PIG-BEL

Most of the patients are malnourished and require intravenous feeding before surgery can be undertaken. The operative procedure may be to do no more than simply remove constricting bands. However, most bands are associated with one or more strictures requiring resection.[14]

## IV. PATHOLOGY

The pathology of Pig-Bel has been extensively studied by Cooke.[15] The small intestine is involved in a segmental fashion; two infected segments can be separated by a normal segment. Examination of the mucosal surface also demonstrates the segmental nature of the disease. Areas of redness on the serosal surface correspond to areas of thick, light-green membrane covering the mucosal surface. The mucosal folds of small intestine between the areas of membrane affected are thickened. Multiple tiny green spots are seen on the surface of the mucosal folds.

A longitudinal section shows thickening of the mucusal folds due to accumulation of clear edema fluid in the submucosa. The green spots on the surface of the thickened folds are areas of necrosis which may involve only mucosal epithelium or as full thickness necrosis through the wall. As the disease progresses, these areas coalesce to form a confluent green membrane. As necrosis advances, the bowel wall becomes thin and friable and liable to perforation.

A microscopic section through the thickened mucosal folds shows intact epithelium with accumulation of fluid in the submucosa. There is virtually no inflammatory cell infiltration. The area of small, green dots shows infarcation of the mucosal epithelium, edema, and heavy infiltration of polymorphs extending from the mucosal surface right through the serosa. There are variable amounts of hemorrhagic infarcation and necrosis of the wall.

A section through the confluent green membrane shows that, although the whole surface is apparently involved, there are segments of intact epithelium, submucosa, and underlying muscle adjacent to the fully affected segments. The junction between normal and affected bowel is clearly demarcated by a line of polymorphs extending from the mucosa to the serosa. This resembles the edge of an acute infarcation in solid organs.

The necrotic mucosal epithelium contains numerous bacteria including large Gram-positive rods. Underneath the epithelium is a well-defined junction zone beneath which is a heavy infiltration of polymorphs. Along this junction zone thrombi can be observed in the smaller blood vessels. Edema and infiltration of polymorphs are seen in the submucosa, and at the junction of normal and abnormal bowel there is marked endothelial proliferation within veins which may be associated with thrombosis of the smaller blood vessels.

In the muscle layer infiltration of polymorphs and variable necrosis of the muscle fibers is seen. The serosa shows marked edema, vascular dilatation, and peritonitis. Large gas cysts may be seen in any or all layers of the bowel wall.[15]

## V. EPIDEMIOLOGY

The earliest cases of Pig-Bel were recorded in the Bena-Bena Valley of New Guinea in 1960,[16,17] although Roth had recorded the condition in his hospital notes as "gangrenous gastroenteritis" prior to these reports.[12] During the subsequent years, intensive studies were undertaken to determine the epidemiology of the disease. The following is a summary of the essential features of this study.[18,19]

Case records were kept of patients with known and suspected disease during the period January 1961 to November 1964. A total of 210 cases were reviewed. Bacteriological investigations were undertaken and antitoxin titers determined in 24 proven cases; serial samples were taken from 21 cases with the disease. Sera from 38 exposed relatives and 216 randomized individuals were also matched against 42 control European sera taken from New Guinea and England. The epidemiology was studied on six different occasions during this period, the first three coinciding with pig feasting. Studies were also made on slaughtering methods, preparation of meat, cooking methods, and pork distribution to determine sources of contamination. Samples were also taken of pig feces.

It was found that the prevalence of the disease varied considerably; initial data indicated a range of from 4 to 22 per 10,000 of the population. In the Upper Chimbu region of New Guinea the prevalence was as high as 48.3 per 10,000 of the population and this was associated with a high prevalence of beta-antitoxin titers.

Mortality was approximately 13.8 per 10,000 compared with a case mortality rate of 36.0%. If mild forms of the disease were excluded, the mortality rate rose to 40.8% and was highest in the acute toxic group (84.4%). Deaths due to enteritis necroticans varied between 2.1 and 3.5% at local hospitals during the period. Rates were again higher in the 1 to 12 age group (16.1%).

High and low prevalence areas were identified. The prevalence of Pig-Bel rose as did other diarrheal diseases when pig feasting activities commenced and the spread of cases was shown to be along the directional movements of pork. Hospital records indicated that peaks of Pig-Bel occurred mostly in the dry season between the months of April and September when major epidemics coincided with the larger pig-killing ceremonies held at this time of the year because of the prevailing climate and good harvest.

A comparison was made of race, age, and sex. Prevalence was greater in the range 2 to 10 years (52.3%) of which the largest segment (29%) occurred in the 6 to 10 age group. In general, the ratio of male-to-female was 2.2:1 and only in the 16 to 20-year-old age group were there more females than males.

Pig feasting figured centrally in all social events, including ceremonies, celebrations, and sacrifices. Ownership of pigs is therefore the basis of a man's social status and prestige as well as a source of food. Within this system large pig kills occur in cycles when numbers have built up. Considerable preparations are made in advance of pig feasts and singing and dancing precede the event.

Pigs were clubbed to death and the hair singed over an open fire before the carcasses were lifted onto leaf mats of bananas, bread fruit, and tree fern leaves. Appropriate incisions were made and the intestines removed. Women took the abdominal viscera to a nearby stream for washing after which the bowels were wrapped in leaves ready for cooking. While the washing proceeded, the men continued the final filleting of the carcass. Earth pits were dug and lined with banana leaves, tree fern fronds, and leaves of a variety of bread fruit. The meat, sweet potato or bananas, and chopped greens were added with preheated stones and included with the meal. Layer by layer the oven was filled up until the large leaves were earthed over a final insulating layer formed by the pigs' quarters and flanks. Water was poured in and the oven sealed by another layer of leaves and mounded with dirt so that the food was cooked under steam pressure in its own moisture (Figures 2 to 7). Thermocouples placed in the center of the ovens showed that the cooking had a mean temperature of 78°C. The results indicated that large chunks of meat were not thoroughly cooked and after cooking the meat cooled rapidly. Further handling of the meat in the relatively dirty environment resulted in further contamination and in transit, a half-side of pork might change hands several times and take several days to reach its destination. From the foregoing it can be seen that not only during cooking, but also in the subsequent handling there is ample opportunity for contamination.

Dietary histories of 140 patients showed that 133 had a prior pork meal. From 38 patients having surgical resections, 17 *Clostridium welchii* type C strains were isolated and 9 from suspected mild cases and contacts of proven cases. No *C. welchii* type C was isolated from the feces of normal individuals.

Out of 24 individuals with Pig-Bel, 23 had detectable antitoxin, 10 of which had very high levels (> 5 units/mℓ). Significant levels of beta antitoxin was found in 71.3% of the

FIGURE 2. Pigs clubbed to death prior to preparation of carcasses.

normal population. There was a significant difference in level between high and incident areas and detectable levels in relatives of persons with known Pig-Bel were significantly higher than in the normal population. Highest levels were found in the age group greater than 40 (90.3%). The percentage of individuals showing titers in the European population was the lowest (9.52%).

Before considering other aspects of the pathogenicity of Pig-Bel, it is pertinent to consider Darmbrand and enterotoxemias of animals in which *C. perfringens* type-C has been implicated.

## VI. DARMBRAND

During and shortly after World War II, large numbers of people were affected with a similar disease in Norway and Germany, especially Hamburg,[20] Lubeck,[10,11] and Keele. The disease reached epidemic proportions but spontaneously resolved about 1950. The disease was referred to as necrotizing enteritis,[21,22] Darmbrand (burnt bowel),[10,11] necrotizing jejunitis,[23] jejunitis actua philgmonosa,[24] and enteritis gravis.[25] There was considerable dispute regarding the etiology. Three hypotheses were postulated. The first group believed that an infectious agent was involved. *C. perfringens* type F (*C. welchii* F) had been isolated from cases by Zeissler and Rassfeld-Sternberg,[26-28] although a neurotropic virus was postulated by others.[29,30] A second school of thought felt that bacterial invasion was secondary, resulting from disturbances in the circulation caused by eating a rich diet.[31] A third group postulated that the necrosis had an allergic basis.[32,33] This was based on experiments in which guinea

FIGURE 3. Singeing of pigs to remove hair prior to filleting.

pigs previously sensitized to meat protein by feeding were challenged with large meat meals following an interval of a meat-free diet. Following feeding, death occurred by anaphalactic shock accompanied by edema, esinophilic infiltration, and fibrinoid swelling in the terminal mucosal vessels of the upper intestine. It was suggested that the animals had developed a sensitivity to meat protein and that the early lesions of Darmbrand were produced similarly as a result of dietary change from carbohydrate to protein.[33]

The incidence of the disease ranged from 0.04 to 3.59 per 10,000 of the population[25] and exceptionally up to 16.[34] Mortality varied from 22 to 60%.[22,25,31-34] These figures compared with an overall mortality of 36% in New Guinea.[18] A notable difference in the New Guinea disease was the age distribution. In Darmbrand, more cases occurred in the age group 40 to 60 compared to 0 to 10 in New Guinea.

Because of the history of an antecedent diet rich in protein (usually of improperly cooked or canned meat preparation) in persons with enteritis necroticans, it was felt the change from the previously meager rations of carbohydrate to an influx of richer foods was an underlying factor in the epidemics.[35,36]

## VII. VETERINARY DISEASES

*C. perfringens* type C (*C. welchii* C) produces a number of diseases in animals including sheep, calves, lambs, and piglets. The disease most analogous to Pig-Bel is the disease in weaned sheep known as "struck" which is associated with Romney Marsh in England.

FIGURE 4. Filleting of the carcasses.

However, the majority of detailed observations have been made with *C. perfringens* type C infections in neonates and these will also be described as they provide valuable information in understanding the etiology and pathology of Pig-Bel.

*C. perfringens* type C infections causing "struck" in sheep were described in the early 1930s.[37] The disease is associated with Romney Marsh (England) during certain periods of the year, especially late winter or early spring and appears to be associated with certain pastures and dietary changes. Changes in the small intestine ranged from moderate to very acute congestion of part or several parts of its length, frequently accompanied by ulceration of the mucous membrane. These changes were generally present in the jejunum, varying from small areas to extensive lesions and could be numerous, involving many feet of bowel. Histologically, the primary lesions consisted of necrosis of the superficial parts of the mucous membrane, the necrosis progressively extending downwards to the mucous coat. No evidence of bacteria was observed in the early stages of infection but they appeared as the areas of necrosis extended. Sections of the ulcers showed large numbers of the Gram-positive bacteria similar to *Bacillus paludis* (later renamed *C. perfringens* (*C. welchii*)) invading the deeper tissues.

Under normal conditions, experimental infection of the adult animal presents considerable problems. Unlike the neonate, which can be infected orally during the first few days of life before the development of stomach acidity, the adult intestine is protected by both stomach acidity and an indigenous microbiological flora. Enteric infections in adult animals can usually only be induced following the neutralization of stomach acidity and interference

FIGURE 5. Fillets of meat prepared awaiting loading into the oven.

with the normal washing action of the intestine by substances such as opium and belladona.[38] Feeding experiments carried out by McEwen[37] did show the possiblity of reproducing the disease experimentally by feeding material rich in *B. paludis* and its toxin. Feeding of the toxin alone was harmless, probably due to the destruction of the toxin in the rumen and reticulum. Some success in reproducing the disease was achieved by feeding large quantities of culture (4 to 6 $\ell$) by stomach tube and symptoms were also reproduced using washed bacteria suspended in nutrient broth. The beta toxin of the organism was demonstrated in the alimentary canal from naturally infected sheep.

The role of *C. perfringens* type C in neonatal disease has been the subject of extensive studies. Acute hemorrhagic enteritis in calves in Colorado and Wyoming attributable to this organism has been described, as has also a similar condition in lambs in Colorado.[39,40] The disease was described in young piglets in Great Britain[41] and Hungary,[42,43] together with a similar disease in neonatal pigs in Minnesota, and subsequently in Iowa when it became evident that the condition had previously been unrecognized in the U.S. More recently, Hogh[44] has carried out extensive studies in Denmark, showing the extent of the disease in that country. The disease has also been reported in chickens.[45] The following is a summary of the etiology and pathology of *C. perfringens* type C enteritis in pigs based on the reports of several workers.[41,44,46,47,48]

The disease in pigs is a profound hemorrhagic necrotic enteritis, less commonly a nectrotic enteritis without appreciable hemorrhage. Piglets are usually affected during the 1st week of life; blood is usually present in the feces and most affected pigs die within 24 to 48 hr

FIGURE 6. Oven base consisting of hot stone.

after the onset of the bloody diarrhea. The essential pathogenesis of the infection is necrosis of the small intestine, particularly severe in the upper jejunum, the lower half being affected with varying degrees of severity. Occasionally, the ileum, cecum, and colon may also be involved. After the initial necrotizing inflammation, necrosis progresses through the mucosal crypts and submucosa and extensive necrosis of the mucosa and muscularis mucosae occurs. The submucosa is described as thickened edematous and occasionally, also emphysematous. In both naturally and experimentally infected animals, the tissue is infiltrated and surrounded by masses of Gram-positive bacteria forming a lining to the intestine. Thrombosis of submucosal vessels is commonly seen in the later stages. Hogh[44] described unique lesions of the central nervous system in pigs affected in Denmark. The lesions were most pronounced in the basal parts of the brain and had the character of an encephalomyelomalacia.

## VIII. PATHOGENICITY OF PIG-BEL

Early work had demonstrated the involvement of *C. perfringens* type C. The organism could be isolated from the jejunal contents from patients with the disease[49] but attempts to detect the organism in the environment using conventional methods of isolation and identification revealed only large numbers of *C. perfringens* type A.[18] With the development of fluorescent labeled antibody techniques to detect the organism, it was possible to apply this more sensitive technique to the problem.[50] The technique used was to culture samples of material, prepare smears, and to stain these with a fluorescent-labeled antiserum prepared

FIGURE 7. Completed oven showing the top mounded with dirt and a hollow tube through which water is poured to give a "pressure cooker" effect.

against one of the New Guinea strains. It had previously been shown that all isolates from New Guinea possessed a common somatic antigen which stained with fluorescent-labeled antibodies. As a result of this study (Table 1), it became evident that the organism was widespread in the community both in the feces of the population, in the soil, and in the ovens in which meat is cooked.[51] The organism thus forms part of the normal flora of the resident population either as an indigenous species or as a transient organism passing through the intestine. The number of organisms in the environment is maintained at a high level through the close association of man and pigs. The coprophagic habits of the pig would tend to keep the number of *C. perfringens* type C in the environment high, making contamination of food more likely.

Unlike Darmbrand, which spontaneously resolved and has not reappeared, Pig-Bel has continued to remain an important disease in New Guinea. In an attempt to understand the reasons for this, attention has been given to the effect of diet on the incidence of the disease and particularly the effects of diet on the levels of pancreatic proteases.

The beta toxin of *C. perfringens* type C is a protein extremely sensitive to proteolysis.[52] This, together with the presence of colostral inhibitors in sow's milk, led to the suggestion that protection of the toxin by the colostral trypsin inhibitors was the basic cause of the disease in young piglets.[53] In the guinea pig it had previously been shown by workers both in Germany[54] and New Guinea[55] that it was possible to produce lesions similar to the human disease by injecting cultures of *C. perfringens* into the small intestine or duodenum at laparotomy but not be feeding contaminated food or by intragastric dosing alone.

Early experiments in New Guinea[56,57] showed that enteritis necroticans could be induced in guinea pigs by introducing a growing culture of *C. perfringens* type C mixed with raw soya bean flour into the stomach through a gastric tube. Feeding of raw soya bean flour alone was harmless and destruction of the trypsin inhibitors present in soya bean by autoclaving prevented the development of the disease. With the recognition of the importance of a heat-stable trypsin inhibitor present in raw kau kau (sweet potato, the staple diet) similar experiments were performed with mixtures of cultures and sweet potato. It was found that

**Table 1**
**PRESENCE AND DISTRIBUTION OF *C. PERFRINGENS* TYPE C IN SAMPLES OF FECES AND SOIL FOR VARIOUS AREAS IN PAPUA NEW GUINEA AS REVEALED BY FLUORESCENT STAINING**

| Place | Area | Sample[a] | No. | Positive[b] | Negative | % Positive[c] | Pig-Bell incidence |
|---|---|---|---|---|---|---|---|
| Kundiawa Hospital | Simbu Highlands | Human feces | 101 | 76 | 25 | 75.2 | High |
| Sina-Sina | Simbu Highlands | Human feces | 77 | 61 | 16 | 79.2 | High |
| Sina-Sina | Simbu Highlands | Pig feces | 42 | 33 | 9 | 78.6 | High |
| Tari | Southern Highlands | Soil samples, pig feces | 12 | 12 | 0 | 100 | High |
| Laiagam Hospital | Enga Highlands | Human feces | 21 | 11 | 10 | 52.4 | Moderate |
| Mt. Wilhelm | Highlands | Soil samples up to 14,780 ft | 10 | 0 | 10 | 0 | No people |
| Okapa and Marawaka | Eastern Highlands | Human feces | 26 | 16 | 10 | 61.5 | Unknown |
| Lae Hospital Morobe | Coastal | Human feces | 31 | 21 | 10 | 67.7 | Low |
| Mandang Hospital | Coastal | Human feces | 120 | 71 | 49 | 59.1 | Low |
| Gogol Valley | Coastal | Human feces | 20 | 13 | 7 | 65.0 | Low |
| Karimui to Lufa | Highlands | Pig feces | 13 | 13 | 0 | 100 | Unknown to high |

[a] Total samples, 473.
[b] Total positive, 327.
[c] Positive, 69%.

From Lawrence, G., Walker, P. D., Garap, J., and Avusi, M., *Papua New Guinea Med. J.*, 22, 69, 1979. With permission.

if the culture was mixed with raw sweet potato flour, lesions similar to those in Pig-Bel were produced in guinea pigs fed a protein-deficient diet but not a normal diet. The effect was abolished by autoclaving the sweet potato.

Following the initial work, a more comprehensive series of experiments designed to explore the complex relationship between organism growth, toxin production, dietary protease inhibitors, and circulating antibodies in the production of Pig-Bel were carried out.[58] Growing cultures of *C. perfringens* type C, and on occasions washed organisms were fed intragastrically to guinea pigs in combination with raw soya bean flour, autoclaved soya bean flour, raw sweet potato, autoclaved sweet potato with or without the addition of proteolytic enzymes, and a defined protease inhibitor. For successful intragastric dosing it was necessary to include substrate in the form of added peptone in the feeds to ensure growth of the organism. The animals were fed either a normal or a low-protein diet and in some cases were previously immunized with a beta toxoid or immunized passively with a *C. perfringens* type C antiserum. Additional experiments were also carried out involving injection of the organisms or toxic filtrates containing antibiotics directly into the jejunum with or without the addition of proteolytic enzymes and in some instances, prior immunization of the guinea pig. Finally, the time course of the disease was followed for up to 7 days, following intragastric dosing with growing cultures mixed with raw soya bean flour.

The disease in guinea pigs was shown to be identical with that in humans regarding its gross pathology and microscopic findings. The time-course experiments clearly showed that the vascular changes which included proliferation of the endothelial layer of the blood vessels occurred subsequent to the necrotic effects of beta toxin, clearly demonstrating the latter as the primary cause of the disease. It should be recalled that vascular changes were implicated as the primary cause of the disease by German workers and similar arguments were used initially in New Guinea.

Intragastric dosing was only successful in causing Pig-Bel if raw soya bean flour or a pure trypsin inhibitor, aprotinin, was fed at the same time. Autoclaving of the soya bean flour abolished this effect as did the addition of excess proteolytic enzymes in the form of pancreatin. The disease was also reproduced using washed organisms and raw soya bean flour. Raw sweet potato was only effective in potentiating the disease in animals which had been fed on a low-protein diet. Previous immunization with beta toxoid totally prevented the disease, although the gut contents were often full of gas bubbles, suggesting that the organism had multiplied but no damage had been done.

Injection of growing cultures directly into the jejunum produced disease as did a toxic filtrate of the organism to which antibiotics had been added to prevent bacterial growth. These effects were abolished if proteolytic enzymes in the form of pancreatin were added to the toxic filtrate prior to injection or the animals had been previously immunized with beta toxoid. Beta antitoxin was effective in preventing the disease providing it was given before or within 7 hr of intragastric dosing. The essential features of these experiments are summarized in Table 2.

These results clearly show the effect of proteolytic enzymes on the course of the disease. It is known that soya bean flour inhibitors inhibit both trypsin and chymotrypsin but that sweet potato contains only trypsin inhibitors.[59] Furthermore, it is known from work with monkeys fed on protein-deficient diets that secretion of chrymotrypsin is inhibited while lower levels of trypsin are still produced.[60] These results explain why soya bean flour is effective in the normal animal and sweet potato only in protein-deficient animals. In the case of protein-deficient animals, no chymotrypsin is produced and sweet potato inhibits the trypsin, thereby protecting the toxin.

The importance of beta toxin in initiating the disease is shown by the fact that the disease can be reproduced by the injection of a toxic filtrate alone and that the symptoms can be prevented by prior active immunization with a beta toxoid or passive immunization with antiserum. The results obtained with passive immunization are similar to those obtained

**Table 2**
**PRODUCTION OR OTHERWISE OF PIG-BEL IN GUINEA PIGS FOLLOWING DOSING OF MIXTURES OF CULTURES OF *C. PERFRINGENS* TYPE C AND VARIOUS AGENTS**

| Dosing (diet) | Mixture | Result |
|---|---|---|
| Intragastric (normal) | Culture or washed organisms plus raw soya bean flour | Pig-Bel |
| | Culture plus raw soya bean flour plus | |
| | Pancreatin | No disease |
| | Immunization | No disease |
| | Culture plus autoclaved raw soya bean flour | No disease |
| | Culture plus Aprotenin plus autoclaved sweet potato | No disease |
| Intragastric (low protein) | Culture plus raw sweet potato | Pig-Bel |
| | Culture plus raw sweet potato plus | |
| | Pancreatin | No disease |
| | Immunization | No disease |
| | Culture plus autoclaved sweet potato | No disease |
| Injection into intestine | Culture (organism plus toxins) | Pig-Bel |
| | Culture plus immunization | No disease |
| | Toxic filtrate plus | |
| | Antibiotics | Pig-Bel |
| | Pancreatin | No disease |
| | Immunization | No disease |

Modified from Lawrence, G. and Cook, R., *Br. J. Exp. Pathol.*, 61, 261, 1980. With permission.

using passive immunization to protect sheep against experimentally induced *C. oedematiens* type A gas gangrene. Antiserum is only effective when given at the same time or within a few hours of challenge.[61] They also serve to explain the apparently conflicting results obtained using *C. perfringens* type C antiserum prophylactically in New Guinea. Although earlier reports suggested some success,[18,19] the use of antitoxin was found to be of no benefit in treatment of the disease in a double-blind trial conducted by Rooney and co-workers.[62]

Lawrence and Cook[58] concluded that a combination of the following factors is necessary for the occurrence of Pig-Bel:

1. Presence of the organism
2. Consumption of nutrients for rapid growth of the organism
3. Low capacity for trypsin secretion
4. Presence of dietary or perhaps parasitic trypsin inhibitors

The following sequence of events would appear to take place in children fed on a staple diet of sweet potato (and thereby compromised as far as their pancreatic protease levels are concerned) who consume a meat meal. In the rich nutrient environment created as a result of the consumption of protein, *C. perfringens* type C organisms consumed either in the meal or present as part of the resident flora multipy rapidly, producing beta toxin which is protected from destruction by the low levels of pancreatic protease activity, trypsin inhibitors present in the sweet potato, and inhibitors secreted by the round worm (*Ascaris lumbricoides*) with which infestation is common. Bacterial growth and toxin production within the intestine is indicated by the need to include further substrate for growth for successful intragastric dosing and the fact that the disease was reproduced in guinea pigs with washed organisms. The toxin causes necrotic changes in the villi and also, in addition, has been shown to interfere with villus motility.[63] Under such conditions of necrosis and intestinal statis, *C. perfringens*

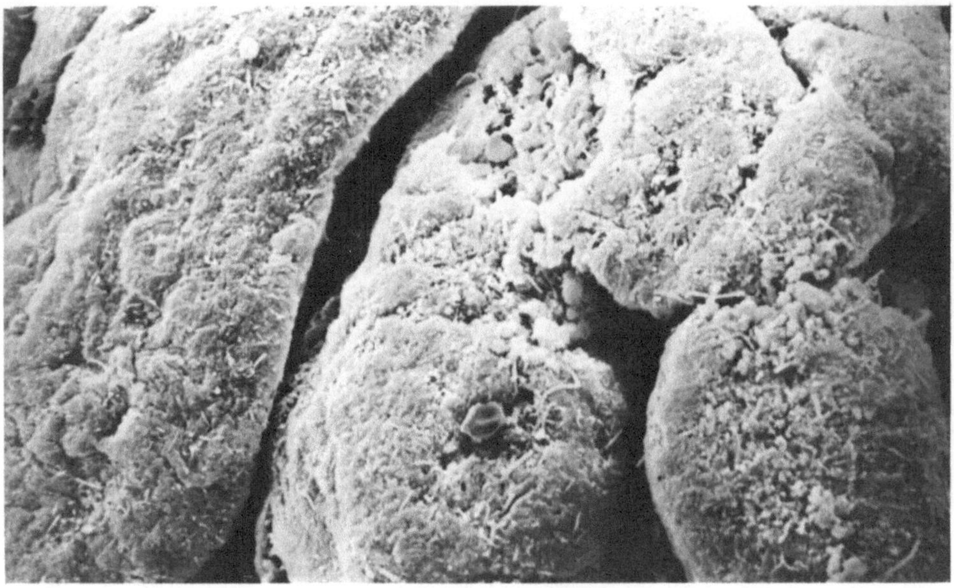

FIGURE 8. Scanning electron micrograph of jejunal villi from acute case of Pig-Bel showing necrosis of the villi. (Magnification × 560.)

type C organisms attach to the villi and proliferate along the villus surface with further production of beta toxin.

The importance of bacterial attachment as a prerequisite in the development of the infectious process in the gut has recently been focused on. This work has largely been stimulated by the elucidation of the role of capsular adhesive antigens of *Escherichia coli* in neonatal diarrhea. The initial attachment of strains of *C. perfringens* type C in artificially infected young piglets has been studied using fluorescent-labeled antibody techniques.[64] It was shown that fluorescent bacilli were present around the outlines of the villi and in the lumen but that the bacilli were more densely aggregated on the villus edge. Degenerative changes began in the distal region of the villi and following attachment, the epithelial layer of the villus was progressively removed from the underlying lamina propria starting in the upper region of the villus. Subsequently, cellular morphology of the lamina propria of the villi was destroyed and this necrosis was associated with the presence of large numbers of Gram-positive bacilli around the villus remnants. Spread of the necrosis from the villi into the crypts occurred and eventually, the complete necrotic mucosa could be sloughed off. He suggested that *C. perfringens*, once adherent, could proliferate along the villus surface and as a result of this attachment the toxins were brought into more intimate contact with the host. These findings are very similar to the observations made recently using the scanning electron microscope (SEM), where similar changes at various stages of infection with Pig-Bel were described and large number of organisms were seen adherent to the villus surface[65] (Figures 8 and 9).

In the young adult, the final outcome will depend on the balance established between the initial growth of the organism, the amounts of toxin produced, the effects of trypsin inhibitors, the levels of proteolytic activity, and the final extent of attachment and invasion of the tissues. Antibodies may be a critical factor in interfering with the final stages of establishment by neutralizing the toxin effect. It is also possible that specific immune mechanisms involving leukocytes may also be involved.

It has recently been shown, using Thiry-Vella loops in piglets, that frequent stimulation of the loops with strains of *E. coli* results in a considerably increased rate of elimination of

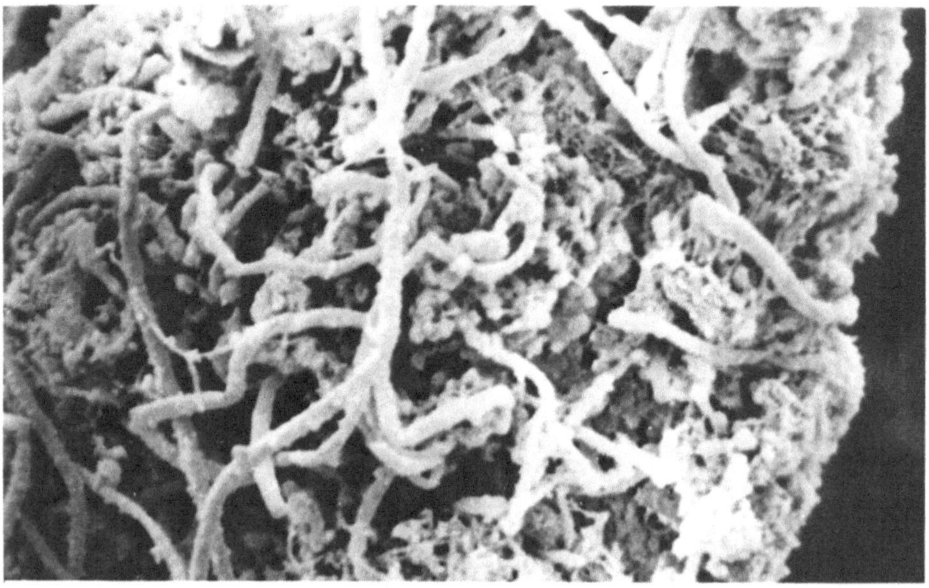

FIGURE 9. Scanning electron micrograph of surface of jejunal villi from acute case of Pig-Bel showing filamentous organisms attached to the distintegrating surface. (Magnification × 4375.)

the organisms from the loops. Microscopic studies have shown that in an efficiently eliminating loop, large numbers of polymorphonuclear leukocytes enter the lumen within 15 min of stimulation and these contain large numbers of bacteria in varying stages of disintegration. This process appears to be immunologically specific in that a loop eliminating one serotype of *E. coli* will not immediately eliminate another serotype introduced for the first time. Several stimulations are required before the second serotype is efficiently eliminated. It could well be that following continual stimulation of the intestine with organisms ingested from the environment, as shown by their presence in the stools of three quarters of the population of the Highlands, such a mechanism is activated and results in an efficiently eliminating system.[66-68]

## IX. PROPHYLAXIS

By analogy with the veterinary situation, the possibility of immunization against this disease was appreciated very early.[18] Unfortunately, for logistic reasons it was not possible to carry out a controlled trial until the mid-1970s. Immunization of sheep, cattle, and pigs against enterotoxemias caused by *C. perfringens* type C is carried out using crude toxoids prepared from toxic filtrates of the organism. The conditions under which beta toxin is elaborated by the organism have been the subject of much research and large quantities of toxoids are produced for incorporation into the multivalent clostridial vaccines.[69] Furthermore, work in animals has determined the level of antigen required to give a satisfactory immune response.

The toxoid developed for use in New Guinea was purified by precipitation of the crude toxoid with ammonium sulfate followed by dialysis against phosphate buffer. Such toxoids are normally standardized in terms of a flocculation (Lf) and total combining power (TCP) in tests. These are carried out according to standard procedures. In selecting a suitable toxoid for immunization, it is preferable to select a toxoid in which the TCP/Lf ratio is as high as possible. Toxoids of various strengths as determined by TCP were adsorbed on aluminium hydroxide adjuvant in borate succinic acid buffer so that each 1-mℓ dose of the blended

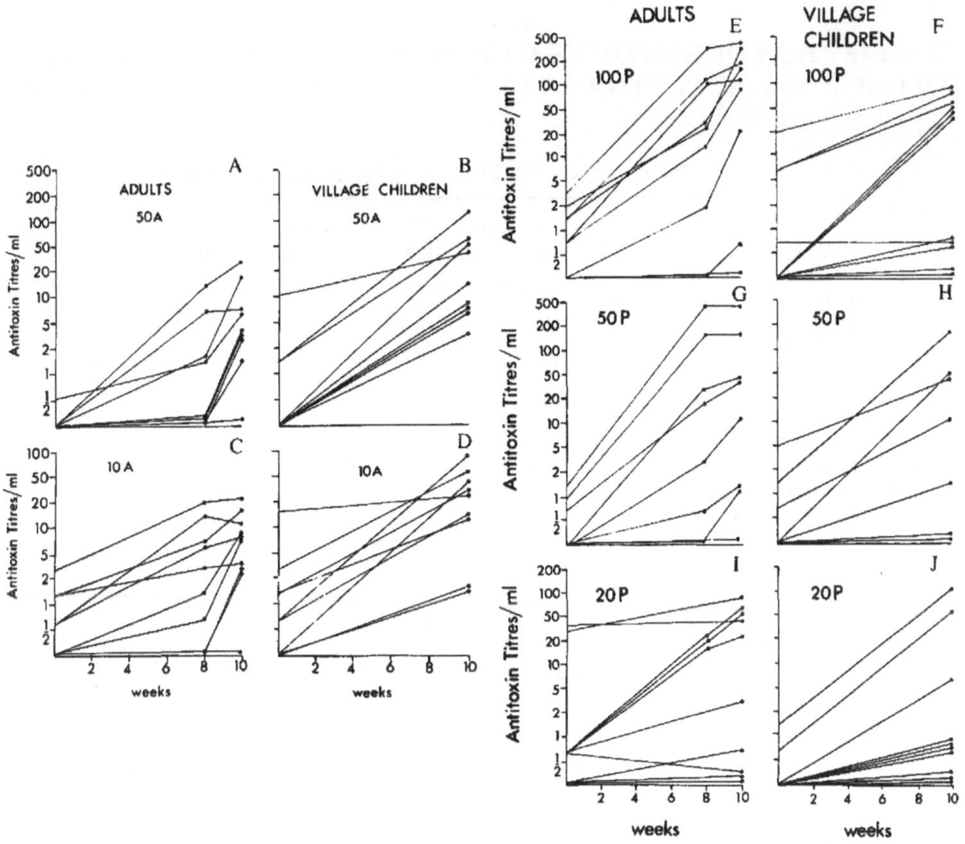


FIGURE 10. The beta antitoxin responses of groups of Melanesian adults and village children following two injections of the appropriate vaccine at 0 and 8 weeks are plotted. (A) Responses of Melanesian adults to two doses of adsorbed vaccine (50 TCP units); (B) responses of village children to two doses of adsorbed vaccine (50 TCP units); (C) responses of Melanesian adults to two doses of adsorbed vaccine (10 TCP units); (D) responses of village children to two doses of adsorbed vaccine (10 TCP units); (E) responses of Melanesian adults to two doses of plain vaccine (100 TCP units); (F) responses of village children to two doses of plain vaccine (100 TCP units); (G) responses of Melanesian adults to two doses of plain vaccine (50 TCP units); (H) responses of village children to two doses of plain vaccine (50 TCP units); (I) responses of Melanesian adults to two doses of plain vaccine (20 TCP units); and (J) responses of village children to two doses of plain vaccine (20 TCP units). (From Walker, B. D., Foster, W. H., Knight, P. A., Freestone, D. S., and Lawrence, G., *J. Biol. Stand.*, 7, 315, 1979. With permission.)

vaccine contained 20 or 100 TCP unit equivalents of *C. perfringens* type C toxoid. Plain vaccines were also prepared in which the aluminium hydroxide was omitted.[70]

Initial tests took place in volunteers in Papua New Guinea consisting of European adults, Melanesian adults, and village children. Volunteers were given two injections of plain or adsorbed toxoid at intervals of 8 weeks in dosages of 20, 50, and 100 TCP units for plain toxoid or 10 and 50 CTP units for adsorbed toxoid. Sera were collected from children and adults for determination of beta antitoxin levels at the time of the first injection and 2 weeks after the second injection. Adults also provided sera at the time the second dose of toxoid was administered. Other criteria examined included hemotological blood picture and various enzyme levels.

The results (Figures 10A to J) clearly show the ability of the toxoid to stimulate high levels of antibody protection. At each level of toxoid dosage and for each type of toxoid, two doses were found to elicit circulating antitoxin responses in most children. In the case of adults who supplied additional blood samples prior to the second injection, it was possible

**Table 3**
**DISTRIBUTION OF RESPONSES TO TWO DOSES OF *C. PERFRINGENS* C TOXOID IN SUBJECTS WITHOUT DETECTABLE BETA ANTITOXIN TITER**

| | | No. of subjects with titer at stated levels after second dose | | | | | | | | | | |
|---|---|---|---|---|---|---|---|---|---|---|---|---|
| Vaccine | Dose | <0.5[a] ≥ | 1.0 0.5 | 2.0 1.0 | 5.0 2.0 | 10 5.0 | 20 10 | 45 20 | >45 | No. | GM titer $m\ell^{-1}$ | %[b] |
| Plain | 20 TCP UE | 8 | 5 | 0 | 0 | 1 | | | | 14 | 0.35 | 43 |
| Plain | 50 TCP UE | 3 | 0 | 3 | 0 | 0 | 1 | 1 | 1 | 9 | 1.61 | 67 |
| Plain | 100 TCP UE | 3 | 3 | 0 | 0 | 0 | 1 | 2 | 2 | 11 | 2.31 | 73 |
| Absorbed | 10 TCP UE | 1 | 0 | 2 | 2 | 2 | 0 | 2 | | 0 | 2.71 | 89 |
| Adsorbed | 50 TCP UE | 1 | 1 | 2 | 5 | 4 | 3 | 2 | | 18 | 3.00 | 95 |

[a] Subjects with titers less than 0.5 units $m\ell^{-1}$ were scored as 0.2 units $m\ell^{-1}$ for calculation of geometric mean.
[b] β units >0.5 units $m\ell^{-1}$. GM, geometric mean.

From Walker, P. D., Foster, W.H., Knight, P. A., Freestone, D. S., and Lawrence, G., *J. Biol. Stand.*, 7, 315, 1979. With permission.

**Table 4**
**DISTRIBUTION OF TITER INCREASES FOLLOWING TWO DOSES OF VACCINE INCLUDING ALL MALANESIAN AND EUROPEAN SUBJECT**

| | | No. of subjects with titers increased | | | | | | | | | | |
|---|---|---|---|---|---|---|---|---|---|---|---|---|
| Vaccine | Dose | 2-fold ≥ | 4-fold 2-fold | 8 8 | 16 16 | 32 32 | 64 64 | 128 64 | 128 | No. | GM[a] increase | % Increase |
| Plain | 20 TCP UE | 8 | 6 | 1 | 0 | 1 | 2 | 4 | | 22 | 4.26 | 64 |
| Plain | 50 TCP UE | 3 | 0 | 4 | 0 | 1 | 3 | 0 | 4 | 15 | 12.7 | 78 |
| Plain | 100 TCP UE | 4 | 4 | 0 | 2 | 0 | 0 | 3 | 8 | 21 | 16 | 81 |
| Adsorbed | 10 TCP UE | 2 | 2 | 2 | 5 | 6 | 0 | 0 | 3 | 20 | 9.8 | 90 |
| Adsorbed | 50 TCP UE | 1 | 2 | 3 | 5 | 5 | 3 | 3 | 1 | 23 | 12.2 | 96 |

[a] GM, geometric mean.

From Walker, P. D., Foster, W. H., Knight, P. A., Freestone, D. S., and Lawrence, G., *J. Biol. Stand.*, 7, 315, 1979. With permission.

to follow primary and secondary responses. In most cases a single dose at each dose level elicited a circulating antitoxin response, although antitoxin response 2 weeks after the second dose was in the majority of cases, superior. Exceptions occurred where there was preexisting circulating antitoxin before vaccination. With plain vaccines a dose response effect was clearly visible. Doses of 100 TCP units of plain vaccine gave greater mean antitoxin responses than those of 50 TCP units which, in turn, were superior to those of 20 TCP units.

Both dose levels of adsorbed toxoid gave greater mean antitoxin responses than the 100 TCP units of plain toxoid, the 50 TCP of adsorbed toxoid being the greatest (Tables 3 and 4). No adverse clinical reactions or hemotological or biological effects were observed.

As the 50 TCP dose per 0.5 mℓ of adsorbed toxoid appeared to give a particularly uniform response, its greater antigenicity and acceptable clinical reactivity resulted in its selection as the dosage to be used in the field study.

The Sina-Sina region of Chimbu, New Guinea was selected for the study because of the known high incidence of Pig-Bel in its population, its high population density, and its ease of access. There are about 10,000 children aged 1 to 15 in Sina-Sina and those children

**Table 5**
**AGE AND SEX OF CHILDREN RECEIVING TWO INTRAMUSCULAR DOSES OF EITHER ADSORBED *C. PERFRINGENS* TYPE C OR TETANUS TOXOID**

| Age (year) | Sex | *C. perfringens* toxoid (%) | Tetanus toxoid (%) | Total |
|---|---|---|---|---|
| 1—5 | M | 427 (16.8) | 492 (19.4) | 1669 |
| | F | 393 (15.5) | 357 (14.1) | |
| 6—10 | M | 593 (23.4) | 593 (23.4) | 2162 |
| | F | 492 (10.4) | 484 (19.1) | |
| 11—15 | M | 343 (13.5) | 309 (12.2) | 1236 |
| | F | 290 (11.4) | 294 (11.6) | |
| Not recorded | Not recorded | | 3 (1) | 3 |
| Totals | | 2538 | 2532 | 5070 |

From Lawrence, G., Shann, F., Freestone, D. S., and Walker, P. D., *Lancet*, 1, 227, 1979. With permission.

**Table 6**
**MORBIDITY AND MORTALITY DUE TO PIG-BEL IN 24 MONTHS AFTER VACCINATION**

| | *C. perfringens* toxoid | | Tetanus toxoid | | |
|---|---|---|---|---|---|
| | 2 doses | 1 dose | 2 doses | 1 dose | Unvaccinated |
| Total cases | 2 | 5 | 17 | 5 | 21 |
| Deaths | 0 | 1 | 1 | 2 | 2 |
| Population | 2,538 | 756 | 2,532 | 733 | 3,500 (approx.) |
| Incidence per 10,000 | 7.9 | 66.1 | 67.1 | 68.2 | 65.7 (approx.) |

From Lawrence, G., Shenn, F., Freestone, D. S., and Walker, P. D., *Lancet*, 1, 227, 1979. With permission.

brought forward by their parents at the time of a government census were invited to enter into the trial. Two vaccines were compared, *C. perfringens* type C toxoid and tetanus toxoid. The population was informed that if either of the vaccines proved effective, then all children would be offered the effective vaccine. The vaccines were of identical appearance and were coded left and right in England. Workers in Papua New Guinea were not aware which vaccine was being demonstrated.[71]

Children were allocated to left and right groups alternately. A tatooing system was used for identification, similar to that previously extensively used as a marker in tuberculosis control. After tatooing, the arm was checked and 0.5 mℓ of the vaccine was injected intramuscularly into the thigh. A second dose was administered 3 to 4 months later.

The instance of Pig-Bel was carefully monitored in the local hospitals both by encouraging people to report all Pig-Bel cases and the employment of a local counselor to collect information about possible cases. Children with a history suggestive of Pig-Bel were examined. All patients had a history of having eaten pig meat or other high-protein food followed by illness in which abdominal pain was the main feature. Patients were diagnosed as having Pig-Bel on the basis of recognized clinical criteria.

The age and sex of the children receiving two doses of either adsorbed *C. perfringens* type C or tetanus toxoids is given in Table 5 and the morbidity and mortality due to Pig-Bel in the 24 months following vaccination is given in Table 6. The results of this control study show that two doses of 50 TCP units of adsorbed *C. perfringens* type C toxoid conferred a high degree of protection against Pig-Bel for at least 24 months. Careful monitoring of the situation over the following 2 years indicated that protection continued but this protection appears to wane completely after 4 years.[72] It is clear that further boosting of immunity may be required after this period, although this may create a number of logistic problems. In the case of veterinary entertoxemias it is customary to boost immunity annually, although in many cases the objective of this is to protect the neonate through passive transfer of antibodies via colostrum and milk, rather than the adult.

## X. DISCUSSION

Apart from diseases such as gas gangrene and *C. perfringens* type A food poisoning, the importance of clostridia as pathogens and particularly, intestinal pathogens has been overlooked. Although one case of the isolation of *C. perfringens* type D from a case of enterotoxemia in man had been recorded,[73] Darmbrand and Pig-Bel provided the first concrete examples of the role of this group as intestinal pathogens. It has always seemed something of an anomaly that these organisms, which play such an important part as causative agents of veterinary enterotoxemias, should have been so rarely implicated in the human situation. Pig-Bel and Darmbrand are examples of how, once the factors are favorable, clostridia can be important intestinal pathogens in a way analogous to their role in veterinary enterotoxemias.

At a conference held in Goroka, New Guinea, in 1965, it was felt that Pig-Bel would disappear with improved nutrition and hygiene much the same as Darmbrand did in Germany. However, if anything the problem has become worse. For example, in the Tari Basin, Pig-Bel has been reported to be responsible for over half the deaths in the 6 to 10 age group. This mortality has continued despite the fact that most of the population eat meat at least once a week, and in this situation there is no indication that the instance of Pig-Bel is decreasing with increasing sophistication and socioeconomic development.[74]

In the short-term, in view of the widespread distribution of the organism, there appears to be no easy solution in terms of public health. Pig-Bel is now a preventable disease by prophylactic immunization with *C. perfringens* beta toxoid and the main problem is one of delivering adequate vaccine to a susceptible population. An improvement in the overall disease situation will, in the short-term, depend on factors such as these.

Since the reports on Pig-Bel, and perhaps even stimulated by the description of this disease, the role of other clostridia in intestinal diseases has become appreciated. *C. butyricum* has been isolated from infection in young children in cases involving congenital heart defects and following artificial feeding.[75] In the case of congenital heart defects it is clear that circulatory problems are experienced, resulting in the creation of an anaerobic environment which favors the growth of any clostridia present. Artificial feeding may well cause interference with the important initial establishment of bifidobacteria in the neonate due to the presence of lactoferrin as well as the lack of antibodies which would be protective. Much more important has been the discovery of the role of *C. difficile* as an intestinal pathogen both in antibiotic-induced enterocolitis and also possibly in chronic bowel disease in general.[76-80] The role of clostridia in these and other diseases of man should not be underestimated and are discussed elsewhere in this volume.

## ACKNOWLEDGMENT

Figures 2 to 7 are black-and-white reproductions of color transparencies originally taken by Dr. G. Lawrence during his stay in New Guinea in the early 1970s. They are reproduced with his kind permission.

# REFERENCES


1. **Shepherd, A.**, Clinical features and operative treatment of Pigbel — enteritis necroticans, *Papua New Guinea Med. J.*, 22, 18, 1979.
2. **Murrell, T. G. C. and Roth, L.**, Necrotising jejunitis; a newly discovered disease in the highlands of New Guinea, *Med. J. Aust.*, 1, 61, 1963.
3. **Shann, F. and Lawrence, G.**, The medical management of Pigbel, *Papua New Guinea Med. J.*, 22, 24, 1979.
4. **Gan, K. H., Sukono, D., Satari, S., Sujudi, R. W., and Njoo, S. N.**, First outbreak of necrotising enteritis caused by *Clostridium welchii* in Indonesia, *Nadj. Kedokt.*, 12, 9, 1962.
5. **Mukherjee, A. P., Foong, W. C., and Ferguson, B. R.**, Necrotising enteritis, *Med. J. Malaya*, 25(4), 285, 1971.
6. **Shann, F., Lawrence, G., and Pan, J-D.**, Enteritis necroticans in China, *Lancet*, i, 1083, 1979.
7. **Smith, F.**, Enteritis necroticans — an international disease of tropical communities, *Papua New Guinea Med. J.*, 22, 60, 1979.
8. **Foster, W. D.**, The bacteriology of necrotising jejunitis in Uganda, *East Afr. Med. J.*, 45, 550, 1966.
9. **van Enk, A. and van de Esch, E. P.**, Enteritis necroticans in Agogo, Southern Ghana, *Trop. Geogr. Med.*, 25, 313, 1973.
10. **Jeckeln, E.**, Über 'Darmbrand'. I. Das Pathologisch-Anatomische Bild, *Dtsch. Med. Wochenschr.*, 72, 105, 1947.
11. **Ruppert, C.**, Das klinische Bild des Darmbrandes, *Dtsch. Med. Wochenschr.*, 72, 108, 1947.
12. **Murrell, T. G. C.**, A history of enteritis necroticans, *Papua New Guinea Med. J.*, 22, 5, 1979.
13. **Barrough, P.**, The method of phisick, *Field (London)*, 1956.
14. **Shepherd, A.**, Clinical features and operative treatment of Pigbel — enteritis necroticans, *Papua New Guinea Med. J.*, 22, 18, 1979.
15. **Cooke, R.**, The pathology of Pigbel, *Papua New Guinea Med. J.*, 22, 35, 1979.
16. **Murrell, T. G. C.**, Pig-bel: epidemic and sporadic necrotising enteritis in the Highlands of New Guinea, *Aust. Ann. Med.*, 16, 4, 1967.
17. **Murrell, T. G. C.**, Some epidemiological features of pig-bel, *Papua New Guinea Med. J.*, 9, 39, 1966.
18. **Murrell, T. G. C., Egerton, J. R., Rampling, A., Samels, J., and Walker, P. D.**, The ecology and epidemiology of the pigbel syndrome in man in New Guinea, *J. Hyg.*, 64, 375, 1966.
19. **Murrell, T. G. C., Egerton, J. R., Roth, L., Samels, J., and Walker, P. D.**, Pig-bel — enteritis necroticans. A study in diagnosis and management, *Lancet*, 1, 217, 1966.
20. **Beckermann, F. and Laas, E.**, Über Nekrotisierende Enteritis, *Arztl. Wochenschr.*, 1, 329, 1949.
21. **Heine, J.**, Zur pathologischen Anatomie der Nekrotisierenden Enteritis, *Arztl. Wochenschr.*, 1—2, 796, 1947.
22. **Ernst, O.**, Untersuchungen zur Nekrotisierenden Enteritis, *Arztl. Wochenschr.*, 3, 24, 1948.
23. **Fink, K. A. and Wolken, A. P.**, Necrotic jejunitis, *Lancet*, 1, 519, 1949.
24. **Husebye, O. W.**, On rontgenological diagnosis of jejunitis acuta phlegmonosa, *Acta Radiol.*, 29, 71, 1948.
25. **Kloos, K. and Brummond, W.**, Epidemiologische Studien über der Enteritis gravis in Nordostholstein, ausser Lubeck, *Z. Hyg.*, 132, 64, 1951.
26. **Zeissler, J.**, Zur Atiologie der Nekrotisierenden Enteritis, *Med. Klin.*, 43, 341, 1948.
27. **Zeissler, J. and Rassfeld-Sternberg, L.**, Enteritis necroticans due to *Clostridium welchii* type F, *Br. Med. J.*, 1, 267, 1949.
28. **Pietzonka, H. and Rassfeld-Sternberg, L.**, Weiteras bakteriologisch gesicherte falle von enteritis necroticans in Hamburg und Dithmarsche (Marz 1949 bis Marz 1949), *Klin. Wochenschr.*, 28, 425, 1950.
29. **Lodenkamper, H.**, Beitrag zur Ursache der Enteritis Nekroticans (Darmbrand), *Klin. Wochenschr.*, 26, 55, 1948.
30. **Brandt, M.**, Über nekrotisierende Enteritis, *Zentralbl. Allg. Pathol. Pathol. Anat.*, 85, 95, 1949.
31. **Griessmann, H.**, Über den Darmbrand, *Arch. Klin. Chir.*, 265, 1, 1950.
32. **Kulpe, W.**, Nekrotisierende Jejuitis und Allergie intoxication, *Dtsch. Ges. Leitwesen*, 4—5, 194, 1949.
33. **Koslowski, L.**, Literarische und Experimentalle Studien uber Allergische Phonomene am Margen-Darm-Kanal. Zugleich ein Buttag zur Atiologie des Darmbrandes, *Frankfurt Z. Pathol.*, 62, 184, 1951.
34. **Hansen, K., Jeckeln, E., Jochims, J., Lezius, A., Meyerburgdorff, H., and Schutz, F.**, in *Darmbrand — Enteritis Necroticans*, George Thieme Verlag, Stuttgart, 1949.
35. **Siegmund, H.**, Demonstrations von Preparaten dreier Falle von Darmbrand, *Med. Klin.*, 43, 213, 1948.
36. **Dormanns, E.**, Über den sogenannten "Darmbrand", *Med. Klin.*, 43, 13, 1948.
37. **McEwan, A. D. and Roberts, R. S.**, "Struck": enteritis and peritonitis of sheep caused by a bacterial toxin derived from the alimentary canal, *J. Comp. Pathol. Ther.*, 44, 26, 1930.
38. **Bennetts, H. W.**, Infectious enterotoxaemia of sheep in Western Australia, *Bull. Counc. Sci. Ind. Res. Aust.*, No. 57, 1, 1932.

39. **Griner, L. A. and Bracken, F. K.,** *Clostridium perfringens* type C in acute haemorrhagic enteritis in calves, *J. Am. Vet. Med. Assoc.*, 122, 99, 1953.
40. **Griner, L. A. and Johnson, H. W.,** *Clostridium perfringens* type C in haemorrhagic enterotoxaemia of lambs, *J. Am. Vet. Med. Assoc.*, 125, 125, 1954.
41. **Field, H. I. and Gibson, E. A.,** Studies on piglet mortality. II. *Clostridium welchii* infection., *Vet. Rec.*, 67, 31, 1955.
42. **Szent-Ivanyi, T. and Szabo, S.,** Infectious necrotic enteritis of suckling pigs. I. Etiology and pathlogy, *Acta Vet. Hung.*, 6, 217, 1956.
43. **Szabo, S. and Szent-Ivanyi, T.,** Infectious necrotic enteritis in suckling pigs. II. Incidence and control of the disease in Hungary, *Acta Vet. Hung.*, 7, 413, 1957.
44. **Hogh, P.,** Porcine Infectious Necrotising Enteritis Caused by *Clostridium perfringens*, Doctoral dissertation, Royal Veterinary and Agricultural University, Copenhagen, 1974.
45. **Parish, W. E.,** Necrotic enteritis in the fowl, *J. Comp. Pathol. Ther.*, 71, 377, 1961.
46. **Moon, H. W. and Bergeland, M. E.,** *Clostridium perfringens* type C enterotoxaemia of the newborn pig, *Can. Vet. J.*, 6, 159, 1965.
47. **Bergeland, M. E.,** Pathogenesis and immunity of *Clostridium perfringens* type C enteritis in swine, *J. Am. Vet. Med. Assoc.*, 160, 568, 1972.
48. **Moon, H. W. and Dillman, R. C.,** Comments on clostridia and enteric disease in swine, *J. Am. Vet. Med. Assoc.*, 160, 572, 1972.
49. **Egerton, J. R. and Walker, P. D.,** The isolation of *Clostridium perfringens* type C from necrotic enteritis of man in Papua New Guinea, *J. Pathol. Bacteriol.*, 88, 275, 1964.
50. **Walker, P. D., Batty, I., and Thomson, R. O.,** The localisation of bacterial antigens by use of the fluorescent and ferritin-labelled antibody techniques, in *Methods of Microbiology*, Vol. 5A, Norris, J. R. and Ribbons, D. W., Eds., Academic Press, London, 1971, 219.
51. **Lawrence, G., Walker, P. D., Garap, J., and Avusi, M.,** The occurrence of *Clostridium welchii* type C in Papua New Guinea, *Papua New Guinea Med. J.*, 22, 69, 1979.
52. **Sakurai, J. and Duncan, C. L.,** Some properties of beta toxin produced by *Clostridium perfringens* type C, *Infect. Immun.*, 21, 678, 1978.
53. **Griner, L. A.,** Some factors influencing the incidence of enterotoxaemia in domestic animals, *Bull. Off. Int. Epizoot.*, 59, 1143, 1963.
54. **Shutz, F.,** Atiologie und Pathogenese, in *Darmbrand-Enteritis Necroticans*, George Thieme, Stuttgart, 1949, 146.
55. **Egerton, J. R.,** Bacteriology of enteritis necroticans in New Guinea Highlanders, *Papua New Guinea Med. J.*, 9, 55, 1966.
56. **Lawrence, G. W.,** Pigbel, *Med. Soc. Papua New Guinea 10th Ann. Symp.*, 1974, 207.
57. **Lawrence, G. W. and Walker, P. D.,** Pathogenesis of enteritis necroticans in Papua New Guinea, *Lancet*, 1, 125, 1976.
58. **Lawrence, G. and Cooke, R.,** Experimental pigbel: the production and pathology of necrotising enteritis due to *Clostridium welchii* type C in the guinea-pig, *Br. J. Exp. Pathol.*, 61, 261, 1980.
59. **Sugiura, M., Ogiso, T., Kazuyuki, T., Semihiro, T., and Akira, I.,** Studies on trypsin inhibitors in sweet potato. I. Purification and some properties, *Biochim. Biophys. Acta*, 328, 407, 1973.
60. **Gyr, K., Wolf, R. H., Imondi, A. R., and Felsenfeld, O.,** Exocrine pancreatic function in protein deficient patas monkeys studied by means of a test meal and in indirect pancreatic function test, *Gastroenterology*, 68, 488, 1975.
61. **Boyd, N. A., Walker, P. D., and Thomson, R. O.,** The prevention of experimental *Clostridium novyi* gas gangrene in high velocity missile wounds by passive immunisation, *J. Med. Microbiol.*, 5, 459, 1972.
62. **Rooney, J., Shepherd, A., and Suebu, A.,** *Clostridium welchii* type C antitoxin in the treatment of "Pig Bel" (Enteritis necroticans): a controlled trial in Papua New Guinea, *Papua Med. J.*, 22, 1, 1979.
63. **Parnas, J.,** The effects of *Clostridium perfringens* beta toxin (type C) on the motility of intestinal segments, *in vitro, Abstr. Zentralbl. Bacteriol. Hyg.*, 234, 243, 1976.
64. **Arbuckle, J. B. R.,** The attachment of *Clostridium welchii (C. perfringens)* type C to intestinal villi of pigs, *J. Pathol.*, 106, 65, 1972.
65. **Walker, P. D., Murrell, T. G. C., and Nagy, L. K.,** Scanning electron microscopy of the jejunum in enteritis necroticans, *J. Med. Microbiol.*, 13, 445, 1980.
66. **Bhogal, G. S.,** A Study of some of the Immunological Mechanisms Involved in the Protection of the Gastrointestinal Tract of the Pig Against *Escherichia coli* Infection, Doctor of Philosophy thesis, Council for National Academic Awards, London, 1978.
67. **Bhogal B. S., Nagy, L. K., and Walker, P. D.,** Mechanisms of elimination of *Escherichia coli* 0149 from the porcine intestinal mucosal surface, in preparation.
68. **Walker, P. D.,** Interaction among micro-organisms and the host, in *Gastroenterology*, Vol. 2, Chadwick, V. S. and Phillips, S. F., Eds., Butterworths, London, 1982, 142.

69. **Walker, P. D. and Foster, W. H.,** Bacterial vaccine production, *Essays in Applied Microbiology*, Norris, J. R. and Richmond, M. H., Eds., John Wiley & Sons, 1981,
70. **Walker, P. D., Foster, W. H., Knight, P. A., Freestone, D. S., and Lawrence, G.,** Development, preparation and safety testing of a *Clostridium welchii* type C toxoid. I. Preliminary observations in man in Papua New Guinea, *J. Biol. Stand.*, 7, 315, 1979.
71. **Lawrence, G., Shann, F., Freestone, D. S., and Walker, P. D.,** Prevention of necrotising enteritis in Papua New Guinea by active immunisation, *Lancet*, 1, 227, 1979.
72. **Davis, M., Lawrence, G., Shann, F., and Walker, P.D.,** Longevity of protection by active immunisation against necrotising enteritis in Papua New Guinea, in preparation.
73. **Warrack, H. G.,** personal communication, 1964.
74. **Smith, D.,** Mortality from Pigbel (enteritis necroticans) in children in Tari 1971—1976, *Papua New Guinea Med. J.*, 22, 74, 1979.
75. **Howard, F. M., Flynn, D. M., Bradley, J. M., Noone, P., and Szawatkowski, M.,** Outbreak of necrotising enterocolitis caused by *Clostridium butyicum*, *Lancet*, 2, 1099, 1977.
76. **Bartlett, J. G., Chang, T. W., Gurwith, M., Gorback, S. L., and Onderdonk, A. B.,** Antibiotic-associated pseudomembranous colitis due to toxin-producing clostridia, *N. Engl. J. Med.*, 298, 531, 1978.
77. **George, R. H., Symonds, J. M., Dimock, F., Brown, J. D., Arabi, Y., Shinagawa, N., Keighley, M. R. B., Alexander-Williams, J., and Burdon, D. W.,** Identification of *Clostridium difficile* as a cause of pseudomembranous colitis, *B. Med. J.*, 1, 695, 1978.
78. **Larson, H. E., Price, A. B., Honour, P., and Borriello, S. P.,** *Clostridium difficile* and the aetiology of pseudomembranous colitis, *Lancet*, 1, 1063, 1978.
79. **LaMont, J. T. and Trnka, Y. M.,** Therapeutic implications of *Clostridium difficile* toxin during relapse of chronic inflammatory bowel disease, *Lancet*, 1, 381, 1980.
80. **Bolton, R. P., Sherriff, R. J., and Read, A. E.,** *Clostridium difficile* associated diarrhoea: a role in inflammatory bowel disease?, *Lancet*, 1, 383, 1980.

Chapter 6

# *CLOSTRIDIUM PERFRINGENS* TYPE A FOOD POISONING

**M. F. Stringer**

## TABLE OF CONTENTS

## I. HISTORICAL OVERVIEW

It is only comparatively recent that *Clostridium perfringens* type A has been established as an important cause of food-borne illness throughout the world. Reasons for this probably include the fact that mortality is rare and the effects of some other agents of food-borne disease are more severe and longer-lasting. The first account of a suspected outbreak of *C. perfringens* food poisoning was recorded by Klein[1] in 1895. This incident occurred in St. Bartholomew's Hospital in London, but the organisms isolated from fecal specimens showed characteristics unlike *C. perfringens* and the presence of a mixed culture was suspected. In 1899, Andrewes[2] reported three more outbreaks in the wards of the same hospital. Although suspected as a cause of outbreaks of enteritis for many years, it was not until 1943 when Knox and MacDonald[3] described an episode of food poisoning in a school associated with gravy prepared the previous day, that *C. perfringens* was again seriously considered as an agent of food-borne disease. Further evidence describing the involvement of *C. perfringens* in food poisoning came in 1945, when McClung[4] reported on three incidents in which patients showed typical symptoms of abdominal pain and diarrhea 8 to 12 hr after consumption of chicken. On all three occasions large numbers of *C. perfringens* were isolated from samples of chicken. Osterling[5] investigated 33 outbreaks of food poisoning in Sweden and concluded that *C. perfringens* was probably responsible for 15 of them. The organism was isolated from incriminated foods and when inoculated into similar foods and fed to volunteers, produced typical symptoms within 10 to 12 hr.

## II. CLINICAL FEATURES

*C. perfringens* food poisoning is characterized by the onset of diarrhea and abdominal pain 8 to 24 hr after the ingestion of food containing large numbers of vegetative cells. Vomiting is uncommon and pyrexia, shivering and headache are rare. The duration of illness is short and symptoms usually disappear within 12 to 24 hr. Fatalities occur occasionally among debilitated persons, particularly the elderly. In the 12-year period between 1969 and 1980, death occurred in only 15 (0.07%) of the 20,513 reported cases of *C. perfringens* food poisoning in England and Wales.[6-10] The results of the post-mortem examination of two of the three fatal cases of food poisoning among geriatric patients in a hospital revealed distension of the large bowel, congestion, and damage to the wall of the jejuneum and ileum.[11]

## III. INCIDENCE OF *C. PERFRINGENS* FOOD POISONING

Food poisoning in England and Wales is assessed annually by the Public Health Laboratory Service Communicable Disease Surveillance Centre and the statistics are published in the medical press. In England and Wales, *C. perfringens* food poisoning is a notifiable disease and the published figures are based on reports made by microbiologists in public health and hospital laboratories to the Public Health Laboratory Service and also on reports made by Environmental Health Departments. Food-borne disease is an international problem, and according to the World Health Organization (WHO), is on the increase. However, there is relatively little information available to assess the world situation. Although food-borne disease surveillance systems exist in several countries, few publish their data on a regular basis. During 1970 to 1980, there were 105,365 reported cases of food poisoning in England and Wales. *Salmonella* accounted for 82,413 (78%) of the cases; however, it must be pointed out that all *Salmonella* infections are included irrespective of whether or not they are known to have been food-borne. *C. perfringens* was responsible for 18,970 (18%) of cases. Almost all of the cases were associated with general and family outbreaks of which there were 567 (average of 51 per year). The vehicle of infection was established in 85% of the *C. perfringens* outbreaks, but in only 20% of the *Salmonella* outbreaks could a specific food be incriminated. During 1973 to 1979, 63 (23%) of the 271 general outbreaks of food poisoning reported in Scotland occurred in hospitals, and 39 (62%) of these were caused by *C. perfringens.*[12]

Annual reports of food poisoning are published by the relevant authorities in Canada and the U.S. *C. perfringens* was involved in 87 (9.5%) of the 916 outbreaks reported in the U.S. from 1970 to 1978,[13-21] and in 40 (14.5%) of the 275 outbreaks which occurred in Canada from 1973 to 1976.[22-23] In addition to the annual figures published by certain countries there are an increasing number of reports of outbreaks from many other countries throughout the world, e.g., New Zealand,[24] Japan,[25] Sweden,[26] Finland,[27] and Italy.[28]

Although it is generally accepted that there is no seasonal prevalence with *C. perfringens* food poisoning, a study of 587 outbreaks by Stringer[29] showed that higher numbers occurred in the warmer months of the year from July to September. Sharp et al.[30] also found that 68% of outbreaks of *C. perfringens* food poisoning in hospitals in Scotland occurred during May to October.

Illness often follows the preparation of meat or poultry dishes to be served to large numbers of people in restaurants, hospitals, factories, and schools. Food is frequently precooked the day before serving and then subjected to inadequate refrigeration or allowed to stand overnight at room temperature. Willarsden et al.[31] reported on a strain of *C. perfringens* with a generation time of 7.8 min — it will be readily appreciated that only a few hours will be required at optimum temperatures for such a strain of *C. perfringens* to germinate and multiply to hazardous levels in a food. During 1970 to 1979, 171 (29%) of the 587 outbreaks investigated by Stringer[29] occurred in hospitals. Schools, canteens, hotels, and old peoples' homes accounted for 246 (59%) of the remaining 416 outbreaks.

## IV. DISTRIBUTION OF *C. PERFRINGENS* IN FOOD, FECES, AND OTHER SOURCES

Smith[32] considered that *C. perfringens* was more widely distributed than any other pathogenic bacterium. The principal habitats are soil, sewage, and the intestinal tract of man and animals, and it can be readily isolated from a wide range of foodstuffs including meat and poultry, fish, vegetables, and spices.

Only type A strains occur as part of the microflora of soil and in a study of fertile soils, Smith and Gardner[33] found samples to contain between $1 \times 10^3$ and $5 \times 10^4$ organisms per gram. The presence of *C. perfringens* in soil, predominantly as vegetative cells, led Smith to conclude that they were actively multiplying and that soil was a true primary habitat. Strains of types B, C, D, and E when added to soil, could not be recovered after a few months. In addition, Miwa[34] examined soil samples taken from the Antarctic and showed that where there was little evidence of contamination by man and animals; 63% of the samples contained *C. perfringens.*

In studies on the ventilation of operating theaters, Blowers and Crew[35] found that although the bacterial content of outdoor air varied a great deal depending on the neighborhood, weather, and height above ground, *C. perfringens* was detected in all of the air samples investigated. In an attempt to identify the sources of bacterial contamination of hospital food, McKillop[36] found *C. perfringens* in 90% of the samples of dust collected from a hospital kitchen.

### A. Foods

#### *1. Meat and Meat Products*

Many reports have been published describing the incidence of *C. perfringens* in both raw and cooked meats. In 1965, Hall and Angelotti[37] isolated *C. perfringens* from 43% of samples of a wide range of meat and meat products. A study by Foster et al.[38] showed that *C. perfringens* was present in 50% of the samples of ground beef. This was an interesting observation in view of the fact that it has been estimated that 50% of the beef consumed in the U.S. is in the form of ground beef or hamburgers, and much of it is consumed in a partially cooked state.[39] Smart et al.[40] isolated *C. perfringens* from 29 of 100 beef, 66 of 100 pork, and 85 of 100 lamb carcasses sampled before chilling at a commercial abattoir at the end of a slaughter line. An area of 100 $cm^2$ adjacent to the anus was examined and the numbers recovered were generally low ($<20/100\ cm^2$).

The degree of bacterial contamination and its significance in the deep tissues of healthy animals at post-mortem has been a subject of debate for many years. Much of the uncertainty lies in the difficulty of obtaining samples of internal tissues without cross-contamination from the carcass surface. Narayan[41] found that adequate rest and preslaughter care and management of animals resulted in a reduction of the internal contamination of carcasses with clostridia from 85.6 to 7.6%. In a later study with emergency-slaughtered cattle, Narayan and Takacs[42] isolated clostridia from nearly all samples of muscle and muscle lymph nodes.

Post-mortem invasion of carcass tissue by intestinal bacteria does not occur until at least several hours after death, when the breakdown of intestinal tissues allows the release of bacteria into the body cavity.[43]

### 2. *Poultry*

In the U.K. between 1971 and 1977, poultry and poultry products were involved in 20% of the outbreaks of *C. perfringens* food poisoning. In recent years the increase in intensive rearing of poultry and the parallel increase in the utilization of processed poultry products has seen chicken and turkey become a major source of protein, replacing more expensive meats such as beef. Lillard[44] showed that on entering a processing plant live birds can carry large numbers of clostridia, but following scalding and plucking relatively few organisms remain on the carcass. Gibbs[45] studied the incidence of clostridia on poultry carcasses and in poultry processing plants and found that between 27.5 and 83.5% of chicken carcasses were contaminated with clostridial spores. *C. perfringens* was isolated from a number of sites within all four processing plants investigated. During the commercial processing of broiler chickens, polyphosphate solutions are frequently added to poultry meat in an effort to increase water retention, improve the color and texture of the meat, and reduce cooking losses. In the past, broilers were immersed in large tanks of polyphosphate solutions for long periods of time. However, this was uneconomical and led to the development of injection procedures whereby polyphosphate could be introduced directly into the deep muscle while the carcasses remained on the processing line. Using a "marker strain" of *C. perfringens*, Mead and Adams[46] showed that during injection, cross-contamination from carcass to carcass was possible and also that organisms could be transferred from the skin to the deep muscle.

### 3. *Fish and Seafoods*

Although fish and fish products are rarely involved in outbreaks of *C. perfringens* food poisoning in the U.K., both raw and cooked fish have been shown to contain *C. perfringens*. Taniguti and Zenitani,[47] in studies on several species of fish, found that 84% of body surface samples and 82% of samples taken from the alimentary canal were positive for *C. perfringens*. Hobbs et al.[48] surveyed commercially vacuum-packed fish products and found 67% of the samples to be positive for clostridia; the predominant species was *C. perfringens*.

### 4. *Other Foods*

Although the number of *C. perfringens* in dried foods, herbs, spices, and stuffings is generally low, the potential hazard exists when they are added to large bulks of meat and poultry, where, with inadequate cooking and slow cooling, the conditions might be ideal for large numbers of *C. perfringens* to develop. In a survey by Sutton and Hobbs,[49] herbs, curry powder, and animal feeding stuffs had noticeably higher counts than other dried foods.

## B. Feces

Most studies on the intestinal flora have been primarily concerned with the bacterial content of feces. Drasar and Hill[50] considered that "such studies were relevant to the discussion on the significance of the intestinal flora, in so much as the findings reflected the bacterial content of the colon". *C. perfringens* is not a common organism of the human intestinal flora when compared to the numbers of strict anaerobes which may be isolated when suitable anaerobic techniques are employed. However, it is an organism which can be isolated with relative ease from nearly all fecal specimens. The complex nature of the flora and the factors which influence it have been reviewed by Drasar and Hill.[50]

Although *C. perfringens* can be isolated from almost 100% of fecal specimens following repeated examination and enrichment techniques, the human excretor does not play a significant role in the spread of *C. perfringens* in the kitchen. At the time of the report by Hobbs et al.[51] on the importance of *C. perfringens* as an agent of food-borne disease, emphasis was placed on the isolation of heat-resistant (HR) strains (i.e., those organisms whose spores could survive heating at 100°C for 60 min) since it was thought more likely that these organisms would survive cooking. Hobbs and colleagues showed that only 2.2% of feces from normal healthy people yielded HR strains, whereas 15.1% of hospitalized old people were positive. A similar finding was made by Dische and Elek,[52] who in attempting to gather more information concerning the role of *C. perfringens* in food poisoning, found that HR strains were carried by 20% of the hospital staff and their families. Leeming et al.[53] examined the feces of over 500 patients on admission to the hospital and found that 22% of males and 14.5% of females carried HR strains. A second fecal specimen taken some days later revealed that this rate had increased in both sexes to 30%. It was suggested that the initial higher carriage rate in males was due to the greater frequency of communal feeding. Similar findings led Collee et al.[54] to conclude that the higher carriage rate of HR strains observed in certain institutions would lessen the significance of isolating "these typical food poisoning strains" from patients and carriers during recognized outbreaks unless individual strains could be identified.

Five groups of the population in New South Wales, Australia were studied by Sutton[55] for the incidence of HR strains and the effects of age, sex, season, and hygiene on the carrier state. The percentage frequency of the HR strains was: primary school children 1.5%, boarding-school students (males — ages 10 to 18 years) 15.1%, general public 6.0%, hospital patients on admission 4.2%, the same patients 8 days later 25%, and aborigines 19%. An age-distribution analysis within the group of aborigines revealed that within ages 0 to 5 years there was a 6.3% incidence of HR strains, but with ages >40 this rose to 33%. One of the hospital patients who entered the hospital carrying a serotype 6 strain was sampled 8 days later and shown to be carrying a type 9 strain — a serotype also found in 3 other patients in the same ward at the same time.

It was soon appreciated that it was important to try and estimate the number of *C. perfringens* in feces, especially with the report of heat-sensitive (HS) strains being equally capable of causing outbreaks of food poisoning.[56,57] Sutton[58] carried out a quantitative investigation on the number of *C. perfringens* in the feces of three of the five groups reported in his earlier study. In primary school children, the general population, and aborigines 69, 70, and 71%, respectively, carried HS strains, and 1, 3, and 10% HR strains; 50% of all fecal specimens were within the range 0 to $5 \times 10^4$ organisms per gram of feces. Sutton[59] carried out viable counts on feces from healthy persons, cases of diarrhea due to causes other than *C. perfringens*, and people associated with outbreaks of food poisoning. The results showed that the feces of people involved in outbreaks of *C. perfringens* food poisoning contained significantly higher numbers of this organism than other groups. Certain nonhemolytic HR strains required heat-activation (HA) at 80°C for 10 min for the spores to germinate and give a true estimate of the numbers present.

An interesting study by Nakagawa and Nishida[60] revealed that 77% of feces from healthy Japanese subjects had counts of $>10^5$ *C. perfringens* per gram and 13% of these were $>10^7$/g — the higher counts tended to be from older people. In 1976, Yamagishi et al.[61] found that the number of *C. perfringens* in the feces of some healthy elderly adults in a Japanese Home for the Aged were constantly of the order of $10^7$ to $10^9$/g. Of the 30 persons studied, 5 had counts of $10^7$ to $10^9$/g all of the time, 5 yielded relatively low counts on each occasion, and the other 20 fluctuated from low to high. It was assumed that the diet was not responsible for the differences as they had all eaten the same food throughout the study period, and samples of food had revealed that the daily intake of *C. perfringens* was $<5 \times 10^2$ organisms. Since none of the persons carrying high numbers of *C. perfringens* were ill, Yamagishi et al.[61] concluded that the clinical symptoms of food poisoning must be related to the site of multiplication rather than the number of organisms present. Similar findings were observed by Stringer[29] in a study of geriatric and young, "long-stay" hospital patients. Some geriatric patients were shown to carry large numbers ($\geq 10^8$/g of feces) of *C. perfringens* for several weeks. It was also observed that some of these patients could also carry the same serotype(s) for a number of weeks.

## V. USE OF SEROLOGICAL TYPING AS AN EPIDEMIOLOGICAL TOOL IN THE INVESTIGATION OF OUTBREAKS

### A. Nature of the Type-Specific Antigens

Attempts to subdivide or classify *C. perfringens* into groups have been based on biochemical reactions, production of toxins, serological methods using whole cells, fractions of cells and spores, and sensitivity to bacteriophages and bacteriocins. Some of these approaches have been reviewed by Mahony.[62] At the present time, only serological division based on differences in the constitution of the capsular polysaccharides has proven sufficiently discriminating to be of use in epidemiological investigations.

Until 1970, there had been several reports in the literature of studies on the soluble antigens of *C. perfringens*, but none had dealt with the homogenous capsular polysaccharides free of contaminating material. Baine and Cherniak[63] therefore attempted to characterize in detail the capsular polysaccharides of Hobbs serotype 5. Isolation, partial purification, and precipitation resulted in a small neutral fraction (peak 1) containing mainly mannose and a large acidic fraction (peak 2) containing glucose, mannose, galactose, glucosamine, and glucuronic acid. Both peaks 1 and 2 formed precipitin bands with antiserum to type 5 but they did not react with specific antisera prepared against other types of *C. perfringens*. Cherniak and Henderson[64] carried out a detailed analysis of the immunochemical relationships of *C. perfringens* serotypes 1, 5, 9, and 10. The isolated polysaccharides formed a major single precipitin line with their homologous antiserum. Hughes et al.[65] showed that when decapsulated cells of *C. perfringens* were used in absorption experiments they did not lower the titers of antisera prepared against the respective parent smooth (capsulated) forms, and also that type-specific antibodies were not produced in rabbits inoculated with decapsulated cells. Paine and Cherniak[66] used gas-liquid chromatography (GLC) as an analytical procedure to quantitatively examine the component monosaccharides of the capsules of four chemically and immunochemically defined strains of *C. perfringens*. The technique involved extraction, hydrolysis, *N*-acetylation, and conversion of the monosaccharides to their trimethylsilyl ether derivates. The composition data obtained by GLC was consistent with previous results found by the study of isolated polysaccharides. The qualitative and quantitative differences were consistent between runs as much as a year apart. Paine and Cherniak[66] suggested that this chemotype method could be used as a means of differentiation and classification.

**B. Development of Serological Typing**

Although outbreaks of *C. perfringens* food poisoning have been reported in many countries throughout the world, only two countries (U.S. and Japan) other than England have developed serological typing schemes for use in the epidemiological investigation of outbreaks.

Serological typing was first used to investigate outbreaks in 1953, when Hobbs et al.[51] prepared antisera against eight HR strains associated with incidents of food poisoning which occurred from 1949 to 1951. The first 13 antisera prepared in the Food Hygiene Laboratory (FHL) were known as the "Hobbs-types". These sera were in use until 1966, when Sutton[59] prepared antisera against a further 11 HR strains and also against 18 HS strains, increasing the total number of antisera to 42. In 1974, a further 6 antisera were prepared and by 1975 the present total of 75 was established. In England a new antiserum was usually prepared only when a large number of serologically nontypable strains were isolated from an episode of food poisoning in which the evidence suggested that *C. perfringens* was the causative agent. In this way the set of 75 antisera has been produced almost entirely against outbreak-related strains. However, this is not so with the strains used for antiserum production in the U.S. and Japan which originate from a wide variety of sources, many of which are unrelated to food poisoning. At the 10th International Congress of Microbiology held in Mexico in 1970, an informal meeting of microbiologists from Australia, Canada, the U.S., England, and Japan decided that an international serotyping scheme would be a worthwhile development and an aid to the investigation of food-borne disease caused by *C. perfringens*. It was agreed that the FHL should receive the type strains from the U.S. and Japan to establish the degree of overlap between the three typing schemes and to assess the possibility of an enlarged set of antiserum. Details of this work have been published by Stringer et al.[67] All the Japanese strains were HR organisms isolated from the feces of healthy people. The American strains had been isolated from a wide range of sources including food poisoning outbreaks, clinical infections, and foods and feces not associated with outbreaks. Table 1 summarizes the use of the 75 English, 34 American, and 34 Japanese antisera to serotype isolates of *C. perfringens* associated with outbreaks of food poisoning which occurred in the U.K. between 1970 and 1980. Using the English set of antisera, a causative serotype was established in 414 (64%) of the 646 outbreaks. The use of the additional 68 Japanese and American sera resulted in the type being identified in a further 32 outbreaks — an increase in the overall percentage typability to 69%. This clearly indicates the value of selecting for antiserum production only those strains involved in well-defined outbreaks. Of the English types, 8 (1, 3, 4, 3/4, 11, 11/13, 29, and 41) were involved in 254 (57%) of the typable incidents. Other English types were responsible for 232 incidents and 40 outbreaks were caused by Japanese or American serotypes. During 1970 to 1980, isolates of *C. perfringens* associated with 41 outbreaks of food poisoning which occurred in countries other than the U.K. were serotyped with the English set of antisera. The 8 common outbreak types listed above were responsible for 57% of the 41 outbreaks. An attempt to estimate the demand for an internationally standardized serotyping scheme indicated that laboratories were keen to participate, with a view to establishing national typing centers holding antisera to all the known types.

Antisera prepared against the original 13 Hobbs-types have been used in several countries including the U.S., Japan, and India. Hayashi et al.[68,69] described two outbreaks in Japan, one due to Hobbs type 2 and the other caused by Hobbs type 6. In a study of 24 outbreaks of *C. perfringens* food poisoning which occurred in Tokyo during 1963 to 1978, 17 involved one of the 13 Hobbs-types.[70] Hatheway et al.[71] reported on 324 outbreaks of *C. perfringens* food poisoning which were notified to the Center for Diseases Control in Atlanta; 163 were laboratory-confirmed and in 57 (34%) of them a causative serotype was established. Many countries have sent isolates of *C. perfringens* to the FHL for serological typing, and if the strains are typable with one of the English antisera, the type strain is often sent to that country for antiserum production. In this way, several countries are establishing sets of antisera for use in the epidemiological investigation of outbreaks of food poisoning.

**Table 1**
**USE OF THE ENGLISH, AMERICAN, AND JAPANESE SETS OF ANTISERA TO SEROTYPE STRAINS OF *CLOSTRIDIUM PERFRINGENS* ISOLATED FROM FOOD POISONING OUTBREAKS IN THE U.K., 1970—1980**

Outbreaks of food poisoning

| Years | No. investigated | No. in which causative serotype established using antisera from | | | Total[a] |
|---|---|---|---|---|---|
| | | England | U.S. | Japan | |
| 1970 | 38 | 22(58)[b] | — | — | 22(58) |
| 1971 | 45 | 18(40) | — | — | 18(40) |
| 1972 | 36 | 18(50) | — | — | 18(50) |
| 1973 | 37 | 24(65) | — | — | 24(65) |
| 1974 | 48 | 26(54) | — | — | 26(54) |
| 1975 | 76 | 50(66) | 1 | 3 | 54(71) |
| 1976 | 96 | 73(76) | 1 | 3 | 77(80) |
| 1977 | 94 | 67(71) | 3 | 2 | 72(77) |
| 1978 | 54 | 34(63) | 1 | 2 | 37(68) |
| 1979 | 63 | 44(70) | 7 | 4 | 55(87) |
| 1980 | 59 | 38(64) | 3 | 2 | 43(73) |
| Totals | 646 | 414(64) | 16 | 16 | 446(69) |

[a] Results using the 75 English, 34 American, and 34 Japanese antisera.
[b] Percentage in brackets.

Updated information from Stringer, M. F., Turnbull, P. C. B., and Gilbert, R. J., *J. Hyg.*, 84, 443, 1980.

**Table 2**
**SEROTYPES ISOLATED FROM FECES DURING THE INVESTIGATION OF A LARGE OUTBREAK OF FOOD POISONING IN A HOSPITAL**

| Serotype | No. of patients | % |
|---|---|---|
| 10 | 1 | 2 |
| 13 | 2 | 4 |
| 17 | 1 | 2 |
| 23 | 5 | 10 |
| 23/24 | 1 | 2 |
| 25 | 3 | 6 |
| 27/31/68 | 39 | 78[a] |
| 28 | 1 | 2 |
| 32 | 14 | 28[a] |
| 33 | 1 | 1 |
| 34 | 22 | 44[a] |
| 35 | 1 | 1 |
| 51 | 1 | 1 |
| 65/69 | 1 | 1 |
| Nontypable | 9 | 18 |

[a] The serotypes involved as causative serotypes occur far more frequently than other types.

From Stringer, M. F., Watson, G. N., and Gilbert, R. J., in *Methods for the Isolation and Identification of Food Poisoning Organisms*, Correy, J. E. L., Roberts, D., and Skinner, F. A., Eds., Academic Press, London, 1982, 111. With permission.

## C. Aims of Serological Typing

In the investigation of outbreaks it is important to establish that:

1. Strains isolated from different patients are the same type.
2. Strains isolated from the incriminated food and fecal specimens are the same type.
3. A particular food is responsible when several foods have been eaten.
4. Occasionally, outbreaks are related.

In general, those persons involved in an outbreak of food poisoning tend to carry large numbers ($\geq 10^6$/g of feces) of the same serotype of *C. perfringens*. Those who did not eat the incriminated food yielded a variety of serotypes which represented the normal flora present at the time of the incident. Table 2 shows the distribution of serotypes isolated from the feces of patients associated with a large outbreak of food poisoning in a hospital. In this outbreak, three serotypes were responsible for the illness and they were isolated from significantly more fecal specimens than other types.

During the investigation of 587 outbreaks in the U.K. (1970 to 1979) a number of serotypes were isolated which were either clearly not the causative type or were isolated from so few fecal specimens that it was not possible to conclusively incriminate them. Table 3 lists these serotypes and the number of times they were isolated from a fecal specimen. Obviously, many strains represent "normal flora" organisms and often they were isolated in addition to or instead of the outbreak serotype. All of the 8 "common" outbreak types (1, 3, 4, 3/4, 11, 11/13, 29, and 41) were isolated on 20 or more occasions, but as they are so often involved in outbreaks it is likely that in a small percentage of incidents they were actually the causative type of an outbreak although not isolated from a significant number of persons to be conclusive. However, more interesting is the frequency of isolation of a number of other serotypes which are isolated relatively infrequently as the cause of outbreaks but as "normal flora" types were much more common. On 20 or more occasions, 13 serotypes (5, 7, 10, 17, 23, 25, 28, 30, 31, 34, 38, 39, and 68) were isolated and 5 of the types more than 30 times. These 13 types accounted for 380 (43%) of the 890 strains isolated from feces and not established as the causative type of an outbreak. It would seem that some types are better able to colonize the intestine.

Serological typing results from a series of outbreaks in three hospitals have indicated the possibility that certain serotypes may predominate in a particular community. Therefore, the significance of isolating common serotypes from a number of ill patients in a suspected outbreak may be doubtful. Table 4 shows the serotypes of *C. perfringens* isolated from cases of diarrhea and suspected outbreaks of food poisoning which occurred in a hospital during a 6-month period. Incident 1 was a confirmed outbreak of *C. perfringens* food poisoning caused by serotype 28. Incident 2 involved a patient who was ill in incident 1 and recovered within 24 hr, then 2 weeks later developed acute diarrhea. Serotype 28 was subsequently isolated in large numbers ($>10^8$/g) from a fecal specimen. Another outbreak of illness occurred 2 months later and serotype 28 was isolated from 4 of the 6 patients who were ill; no food was incriminated. A month later, a further outbreak of diarrhea was investigated and serotype 28 was isolated from 6 of 7 patients, including the patient involved in the first 2 incidents. One of the patients involved in the last outbreak also developed diarrhea a few weeks later when serotype 28 was found yet again in a fecal specimen with a count of $10^8$/g.

As a result of such findings and the inconclusive typing results obtained from a number of other outbreaks often involving institutionalized persons, investigations have been carried out on the carriage of numbers and serotypes of *C. perfringens* in the feces of young adult and geriatric "long-stay" hospital patients, mental hospital patients, and healthy laboratory workers.[29] Laboratory workers carried between $10^3$ and $10^4$ organisms per gram of feces and a wide range of serotypes. An individual did not carry a particular type for more than 2 consecutive weeks and it was extremely rare to find 2 persons with the same type. Similar findings were observed with the young adult hospital patients. However, with the geriatric patients, large numbers of *C. perfringens* were not unusual and the feces of 6 patients often contained $>10^8$ organisms per gram. Some patients carried certain serotypes for up to 10 weeks of a 12-week study period. In addition, it was common for a significant proportion of the geriatric patients to carry the same type(s) in any 1 week. It is clear from these studies that the "normal flora" of institutionalized old people may be significantly different to that of the community at large. It is important, therefore, when investigating suspected outbreaks of food poisoning affecting geriatric hospital patients or aged people in homes to interpret with care the significance of higher numbers of *C. perfringens* in feces and the carriage of common types.

## VI. ENTEROTOXIN

It is now well established that an enterotoxin is responsible for the typical symptoms of *C. perfringens* food poisoning. Ingestion of large numbers of vegetative cells in a food

**Table 3**
**SEROTYPES OF *C. PERFRINGENS* OTHER THAN THE 'CAUSATIVE TYPES' ISOLATED FROM FECES DURING THE INVESTIGATION OF 587 OUTBREAKS OF FOOD POISONING**

| Serotype | Incidence[a] | Serotypes | Incidence[a] |
|---|---|---|---|
| 1 | 21 | 33 | 17 |
| 2 | 9 | 33/61 | 1 |
| 3 | 7 } 70 | 34 | 38 |
| 4 | 8 } 70 | 35 | 7 |
| 3/4 | 55 } 70 | 36 | 7 |
| 4/22 | 1 | 37 | 9 |
| 5 | 34 | 38 | 27 |
| 6 | 7 | 38/39 | 1 |
| 7 | 27 | 38/67 | 4 |
| 7/13 | 1 | 39 | 20 |
| 7/13/20 | 1 | 40 | 3 |
| 8 | 10 | 41 | 29 |
| 9 | 7 | 42 | 2 |
| 10 | 29 | 43 | 7 |
| 11 | 18 } 20 | 44 | 7 |
| 11/13 | 2 } 20 | 45 | 1 |
| 12 | 6 | 46 | 3 |
| 12/23 | 3 | 47 | 2 |
| 12/69 | 1 | 48 | 3 |
| 13 | 6 | 50 | 10 |
| 14 | 2 | 50/74 | 2 |
| 15 | 6 | 51 | 14 |
| 15/30 | 1 | 52 | 8 |
| 16 | 2 | 53 | 4 |
| 17 | 24 | 54 | 3 |
| 18 | 1 | 55 | 7 |
| 19 | 16 | 56 | 1 |
| 20 | 9 | 57 | 3 |
| 21 | 3 | 58 | 2 |
| 21/36 | 1 | 58/64 | 1 |
| 22 | 1 | 59 | 1 |
| 23 | 41 | 61 | 1 |
| 24 | 9 | 62 | 12 |
| 24/3/4/69 | 6 | 63 | 4 |
| 25 | 39 | 64 | 2 |
| 25/31 | 1 | 65 | 2 |
| 25/31/68 | 2 | 65/69 | 4 |
| 26 | 4 | 66 | 3 |
| 27 | 19 | 67 | 2 |
| 28 | 34 | 68 | 24 |
| 28/37 | 1 | 68/70 | 1 |
| 29 | 22 | 69 | 3 |
| 30 | 23 | 69/72 | 1 |
| 30/31 | 1 | 70 | 2 |
| 31 | 20 | 71 | 13 |
| 31/37 | 1 | 73 | 3 |
| 31/68 | 1 | 74 | 6 |
| 32 | 12 | 75 | 8 |

[a] Non-typable strains were isolated from 261 outbreaks. Incidence — the number of times each serotype was isolated from feces during the investigation of food poisoning outbreaks.

**Table 4**
**SEROTYPES OF *C. PERFRINGENS* ISOLATED FROM CASES OF DIARRHEA AND AN OUTBREAK OF FOOD POISONING WHICH OCCURRED IN A HOSPITAL DURING 1977**

| Incident no. | Nature of illness | No. of patients ill | Serotypes islated from feces (no.) |
|---|---|---|---|
| 1 | Outbreak of food poisoning | 28 | 28[a] (13), including patient J. E. |
| 2 | Diarrhea (patient J. E.) 2 weeks after above outbreak | 1 | 28 |
| 3 | Outbreak of diarrhea | 6 | 28 (4)<br>25/31/68 (1)<br>NT[b] (1) |
| 4 | Outbreak of diarrhea | 7 | 28 (6), including patients J. E. and M. G.; 25 (1) |
| 5 | Diarrhea (patient M. G.) | 1 | 28 |

[a] Serotype considered to be the cause of illness.
[b] Not typable.

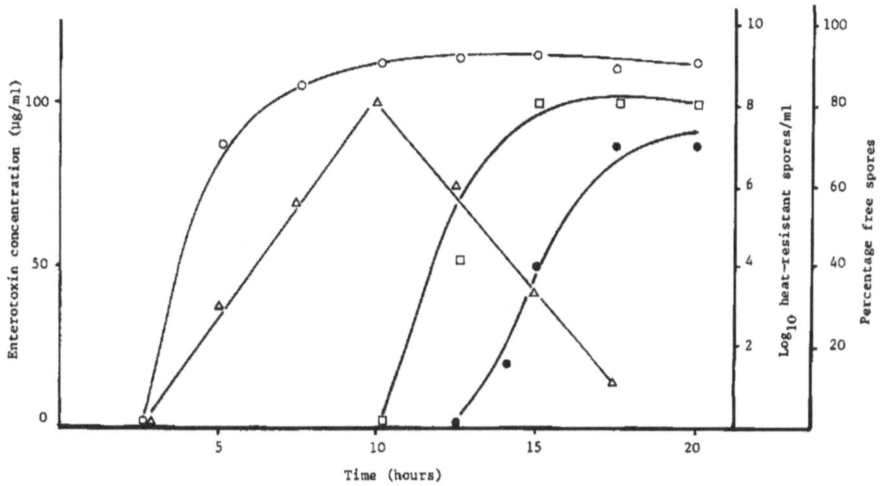


FIGURE 1. Sporulation and enterotoxin production following the inoculation of vegetative cells of *C. perfringens* into the sporulation medium of Tsai, Torres-Anjel, and Riemann, and incubation at 37°C for 20 hr ○, Heat-resistant spores; △, intracellular enterotoxin; □, free spores; and ●, extracellular enterotoxin.

results in multiplication and sporulation in the intestine. Following the release of mature spores from the sporangia, there is an accompanying release of enterotoxin.

## A. Relationship Between Sporulation and Enterotoxin Production

In the laboratory, enterotoxin production is achieved by inoculating vegetative cells of *C. perfringens* into a suitable sporulation medium. Many different types of media have been developed for this purpose but it seems that no single medium is suitable for all strains. Indeed, it is frequently found that different cultures of the same organism may vary markedly in their ability to sporulate even when inoculated into the same sporulation medium.

Figure 1 details the sequence of events which occur when a sporulation medium is inoculated with vegetative cells of *C. perfringens*. After about 3 hr, HR spores develop, followed closely by the accumulation of intracellular enterotoxin. Maximum numbers of HR spores are obtained after 10 hr, and at this time free spores can be detected and reach a maximum at 14 to 15 hr. With the liberation of mature spores from the sporangium, enterotoxin is released and hence, the concentration of enterotoxin in cell extracts decreases and the concentration of extracellular enterotoxin increases in parallel with the increase in numbers of free spores.

Duncan et al.[72] isolated mutants of *C. perfringens* with an altered ability to sporulate from a sporulation positive ($sp^+$) — enterotoxin positive ($ent^+$) strain, both as spontaneous mutants and following mutation with acridine orange or nitrosoguanidine; all $sp^-$ mutants were also $ent^-$. Some mutants blocked at an intermediate stage in sporulation were capable of producing enterotoxin, but not HR spores. Revertants were capable, on regaining the ability to sporulate, of also producing enterotoxin. Duncan et al.[72] concluded from these studies that enterotoxin was a sporulation-specific gene product and that the formation of mature, HR spores was not a prerequisite for enterotoxin production. Although sporulation is essential for enterotoxin production, there are many nonenterotoxin-producing strains which sporulate equally well if not better than some $ent^+$ strains.

Frieben and Duncan[73] examined spores of three $ent^-$ and three $ent^+$ strains and extracted a protein from each which contained a fraction that could identify both biologically and serologically with *C. perfringens* enterotoxin. Also, highly purified enterotoxin was precipitated by an antiserum prepared against the spore protein. The extractable toxin-like protein was considered to be a structural component of the spore coat. Duncan et al.[74] noticed that a paracrystalline inclusion was formed only by $ent^+$ strains and they suggested that the enterotoxin of *C. perfringens* may accumulate intracellularly in certain strains during sporulation as a result of loosely regulated spore coat synthesis. They concluded that failure to detect enterotoxin by some strains may be due to tight regulation of spore coat synthesis. Frieben and Duncan[75] were able to show that the different types of enterotoxin-like protein could be characterized biochemically and that their differences were due primarily to size.

The choice of sporulation medium is very important when attempting to produce enterotoxin. Studies by Tsai et al.[76] compared a modified sporulation medium with that of Ellner[77] and Duncan and Strong.[78] The optimum inoculum size was 1 to 2% and all of the strains tested produced a 100- to 1000-fold increase in number of spores when a heat-activated (75°C for 20 min) culture was used instead of a nonheated activated culture. The mechanism of heat-activation of spores is unknown although it has been suggested that heat increases the germination rate by changing the permeability of spores, or it selects those cells with a high inherent ability to produce spores. It is not known whether the increase in sporulation induced by heat-activation is related to these factors. Several workers, including Uemura et al.[79] and Tsai et al.[76] have observed increased enterotoxin production with cultures subjected to heat-activation. Uemura[80] used a triple heat procedure to test strains of *C. perfringens* isolated from the feces of healthy humans for enterotoxigenicity. In this way, 11 (31%) of 35 individuals were shown to carry enterotoxigenic strains.

Uemura[81] described a solidified sporulation medium for sporulation and enterotoxin production by *C. perfringens*. Using *C. perfringens* strain NCTC 8798 (serotype 9) he found that two distinct colony types existed and that the larger, opaque colonies of this strain produced more enterotoxin than the small, flat, translucent colony types.

### B. Biological Effects of Enterotoxin

**Ligated ileal loop test** — Duncan et al.[82] used the rabbit ligated ileal loop test to study the effects of *C. perfringens* food poisoning. They found that the method of preparation of the challenge inoculum was important in determining whether a given strain provoked a response. When cultures were grown at 37°C for 4 hr in skim milk, 14 of 29 strains associated with outbreaks of food poisoning produced fluid accumulation in the ligated loops. However, 15 of 18 strains from other sources failed to give a positive response. By varying the procedure for inoculum preparation, nearly all of the strains, irrespective of source, were capable of producing a positive loop. The ligated ileal loop test has also been carried out in calves,[83] sheep,[84] and chickens.[85] A rapid ligated intestinal loop technique in the rabbit has been described by Hauschild et al.[86] for purified enterotoxin preparations or extracts from sporulated cells in which results are obtained after 90 min. In control loops the original volume of fluid in the loop of about 2.5 mℓ decreased to 0.7 mℓ (± 0.7 mℓ) within 90 min, whereas the loops injected with enterotoxin preparations showed little change in volume. More recently, Yamamoto et al.[87] described a technique for the detection of enterotoxin based on fluid accumulation in the mouse ligated intestine. With concentrations of enterotoxin between 1 and 16 μg, the increase in loop weight was proportional to the amount of toxin injected.

**Oral challenge** — Volunteer experiments with humans fed strains of *C. perfringens* isolated from fecal specimens associated with outbreaks led to diarrhea and other symptoms typical of *C. perfringens* food poisoning.[51] Dische and Elek[52] confirmed the findings of Hobbs et al.[51] when cultures, supernatants, and cooked meat culture supernatant fluids of *C. perfringens* serotypes 1, 2, and 3 were fed to volunteers. Bacteria-free filtrates had no effect. Dische and Elek concluded that *C. perfringens* food poisoning was a mild, transient infection in which a "living-agent" was required. Strong et al.[88] fed humans strains of *C. perfringens* which had been grown in beef stew for 3 hr at 46°C. The strains which had previously caused fluid accumulation in the rabbit ligated ileal loop test caused a 61% incidence of diarrhea, whereas rabbit-negative strains administered in a similar way failed to cause diarrhea. Work by Skjelkvåle and Uemura[89] has shown that an oral dose of between 8 and 10 mg of purified enterotoxin is required to produce typical symptoms of diarrhea and abdominal pain in humans. Uemura et al.[90] found that a dose of 5 mg of enterotoxin was required to cause both diarrhea and vomiting in monkeys. Both Uemura et al.[90] and Skjelkvåle and Uemura[89] found it necessary to neutralize the gastric juice with sodium bicarbonate before administration of the enterotoxin. Tsai and Riemann[91] showed that an oral dose of enterotoxin >12 μg was lethal to young mice weighing between 9 to 13 g and enterotoxin could be detected in the blood of these mice.

**Intravenous challenge** — A dose of >3 μg administered intravenously in mice resulted in death and a hyperemic small intestine with congestion in the liver, lungs, spleen, and kidneys. Lethal doses of enterotoxin injected intravenously into young fowls by Niilo[92] caused immediate lassitude with partial recovery followed by death 7 to 35 hr later.

**Erythemal activity** — Hauschild[93] showed that extracts from sporulated cells that were capable of producing diarrhea and fluid accumulation in ligated intestinal loops of rabbits and lambs also caused erythema in guinea pigs and rabbits when injected intradermally. Areas with diameters of erythema between 0.5 and 1.2 cm were proportional to the concentration of enterotoxin in the test sample.

**Increased capillary permeability** — Stark and Duncan[94] found that a transient increase in capillary permeability was induced in the skin of guinea pigs by *C. perfringens* enterotoxin. Sample volumes of 0.05 mℓ were injected intradermally, followed 10 to 20 min later by an intravenous injection of 1.0 mℓ of Evans blue dye in 0.85% sodium chloride. The intensity and diameter of the blue lesions were measured after 1 hr.

### C. Purification of Enterotoxin

Hauschild and Hilsheimer[95] extracted enterotoxin from sporulated cells and purified it by gel- and ion-exchange chromatography. Electrophoresis on cellulose polyacetate revealed a single protein band, and a single precipitin line was obtained in immunodiffusion tests with a crude antiserum. Disc gel electrophoresis resulted in two minor bands, both of which were biologically active and appeared to derive from a major band. The molecular weight was shown to be 36,000 ± 4,000 and the isoelectric point occurred at pI 4.3. Frieben and Duncan[75] extracted enterotoxin-like protein from both ent$^+$ and ent$^-$ strains of *C. perfringens*. Only in ent$^-$ strains could they show a relationship between the formation of paracrystalline inclusions and the synthesis of enterotoxin. Stark and Duncan[96] treated a crude cell extract with ribonuclease for 30 min, followed by sequential chromatography on Sephadex® G-100, Cellex T cellulose, and hydroxylapatite. Properties of the toxin were similar to those observed by Hauschild and Hilsheimer.[95] Sakaguchi et al.[97] described a simplified method for purification in which the toxin was extracted from sporulated cells by sonication and then precipitated by the addition of 40% ammonium sulfate. Differential solubilization followed by repeated gel filtration on Sephadex® G-200 resulted in a toxin of 98% purity.

Niilo[98] examined the enterotoxins produced by 15 different serotypes of *C. perfringens* and showed that their immunochemical and biological properties were identical. Skjelkvåle and Duncan[99] examined 69 strains of *C. perfringens* belonging to the various toxin types and showed that only type A and C strains produced enterotoxin. The enterotoxigenic type C strains were associated with cases of Pig-Bel in New Guinea. The enterotoxin from the type C strains was immunochemically and biologically identical to that produced by the type A strains. However, when subjected to examination by polyacrylamide electrophoresis, the enterotoxin produced by some type C strains gave two precipitation bands — one with a relative mobility of 0.53 and a faster migrating component with a mobility of 0.82. Both of these bands were serologically active and Skjelkvåle and Duncan[99] concluded that the enterotoxins of type A and C strains were identical in most respects and that the "double-band" effect probably existed in type A strains as well. Indeed, Bartholomew and Stringer[100] have shown this to be the case.

Following the preparation of a crude sporulated cell extract antibody, affinity chromatography was used by Barnhart et al.[101] to purify the enterotoxin. Scott and Duncan's method[102] for affinity chromatography utilized purified antienterotoxin immunoglobulins prepared against highly purified enterotoxin. Affinity chromatography has also been used to purify an enterotoxin produced by a *C. perfringens* type D strain. The toxin was immunologically identical to that produced by type A and C strains. The organism was subjected to heat treatment (75°C for 20 min) on 5 occasions before it finally produced enterotoxin in vitro at a concentration of 1.3 $\mu$g/m$\ell$.[103]

Granum and Whitaker[104] described an improved method of purification involving two ammonium sulfate precipitation stages at 40 and 15% saturation, respectively. They found that the procedure of Sakaguchi et al.[97] worked only if the enterotoxin concentration in the cell sonicate was >2 mg/m$\ell$. The advantage of Granum and Whittaker's method is that it works well with all concentrations of enterotoxin. Enders and Duncan's[105] work on the polyacrylamide gel electrophoresis of highly purified enterotoxin revealed electrophoretic microheterogenicity of the enterotoxin apparently due to shift charge differences in the peptides. Recently, Skjelkvåle[106] described a method for radiolabeling *C. perfringens* enterotoxin utilizing chloramine T. This will facilitate studies on the uptake and binding of the toxin to cellular receptor sites where a toxin of high specific activity is required. Important observations by Granum et al.[107] showed that the enterotoxin increased threefold in biological activity when trypsinized, but immunochemical (serological) activity remained unchanged. The activated enterotoxin had the same molecular weight as the untreated toxin but the number of available amino groups had increased. These findings are significant since activation of enterotoxin by limited proteolysis is likely to occur in the small intestine during food poisoning and hence, biological and serological in vitro tests for toxin may not always indicate true enterotoxigenic potential. Naik and Duncan[108] showed that biological activity was destroyed within 5 min at a temperature of 60°C whereas about 10% or less residual serological activity could be detected even after 80 min of exposure at 60°C. Loss of serological activity also occurred at 60°C in foods. About 12% of heat-inactivated serological activity could be recovered by treating the toxin in food with urea, but biological activity could not be restored.

### D. Methods for the Detection and Assay of Enterotoxin

Methods for the detection and assay of *C. perfringens* enterotoxin can be divided into one of three groups: (1) biological — tests or responses involving various animal species, (2) serological — tests involving the use of a specific antiserum prepared against purified enterotoxin, and (3) tissue culture assays — cytotoxic effects on various cell lines. The biological effects have been discussed earlier and nowadays the methods of choice tend to be serological with the advantages of increased sensitivity, specificity, and reproduceability.

**Preparation of antienterotoxin** — Stark and Duncan[94] prepared an antiserum by injecting rabbits i.m. with a 1:1 mixture of enterotoxin and Freund's complete adjuvant at 7-day intervals for a period of 6 weeks. Sakaguchi et al.[97] injected 20 μg of toxin in 2 doses at a 5-week interval. The first dose was given with an equal volume of Freund's complete adjuvant. Genigeorgis et al.[109] administered an initial series of doses in rabbits (half i.m. and half s.c.) followed by a 5th s.c. dose 28 days later. A total dose of 217 μg of enterotoxin was administered. More recently, Stringer[29] used a short schedule of i.v. inoculations based on the method described for staphylococcal enterotoxins by Bradstreet et al.[110] A comparison of these methods[100] indicated that there is often greater variation between individual rabbit responses than the different methods, but i.m. procedures generally produce higher serum titers.

**Double gel diffusion** — Full details of this method have been described by Stringer et al.[111] The test is capable of detecting 0.5 μg/mℓ of enterotoxin.

**Electroimmunodiffusion** — Duncan and Somers[112] described a technique in which an electric charge induced rapid migration of enterotoxin from a well into a gel containing antibodies against enterotoxin. Cones of precipitation are produced along the path of antigen migration wherever antigen-antibody equivalence is reached. The cone length is proportional to the concentration of enterotoxin in the sample.

**Counterimmunoelectrophoresis (CIEP)** — Naik and Duncan[113] described a method in which doubling dilutions of enterotoxin are placed in wells at the cathode end of a plate coated in agar. Antiserum is placed in wells at the anode end (5 mm from toxin wells). Electrophoresis using a barbital acetate buffer induces a precipitation line between toxin and serum within 30 min. A linear relationship exists between the concentration of enterotoxin and the CIEP titer of toxin as determined by the highest dilution of toxin exhibiting a visible precipitation band.

**Reversed passive hemagglutination** — The use of this technique for the detection of *C. perfringens* enterotoxin was first described by Uemura et al.[79] Formalin-treated sheep erythrocytes (FSRBC) were sensitized with specific immunoglobulin G antienterotoxin. Test samples were then mixed with a suspension of the sensitized FSRBC. Although reported to have a sensitivity of 0.001 μg of toxin, Naik and Duncan[113] encountered a number of variable results due to nonspecific agglutination and variations with different batches of sheep erythrocytes.

**Fluorescent antibody (FA)** — Niilo[114] used fluorescein isothiocyanate-conjugated antibody prepared against purified enterotoxin to study the intracellular formation of toxin in single cells. Using this method, enterotoxin can only be detected intracellularly. When the toxin is released from sporulated cells, the antigen can no longer be demonstrated by FA as it is soluble and washes off during the staining process.

**Tissue culture effects** — The effects of *C. perfringens* enterotoxin on morphology, viability, and macromolecular synthesis in Vero cells have been described by McClane and McDonel.[115] After 30 min of exposure to enterotoxin, considerable morphological damage was done to the cells, and within 40 min approximately 75% of the cells had detached from the wells of tissue culture plates. More recently, McDonel and McClane[116] have described a sensitive assay technique for enterotoxin based on the ability of the toxin to inhibit the plating efficiency of Vero cells grown in culture. As little as 0.001 μg of toxin could be detected in this way. These workers defined a new unit of biological activity called the "plating efficiency unit", i.e., the amount of enterotoxin that caused a 25% inhibition of the plating of 200 cells inoculated into 100 μℓ of medium in a microwell culture system; 1 mg of purified enterotoxin contained 400,000 plating efficiency units. "The biological activity" of enterotoxin preparations stored at 37°C disappeared within 7 days as did the activity when stored frozen at −20°C for 28 days. However, serological activity of the enterotoxin preparations subjected to similar storage conditions remained unaffected. McDonel and McClane concluded that biological and serological activities were separable entities. Bradshaw et al.[117] used Vero cells to study the thermal inactivation of *C. perfringens* enterotoxin in buffer and chicken gravy. The assay method was based on the ability of enterotoxin to induce cytopathic morphological effects in Vero cell monolayers seeded in wells of microtiter trays. In commenting on Naik and Duncan's[108] work on the thermal stability of enterotoxin detected by biological and serological tests, they discuss their own work in which the biological activity (assayed by Vero cell cytopathy) was more stable at a low acid pH range — results in sharp contrast to those of Naik and Duncan who found that enterotoxin heated at 60°C was more resistant above neutral pH.

Giugliano et al.[118] compared Vero cells with double gel diffusion as methods for the detection of *C. perfringens* enterotoxin — Vero cells were considerably more sensitive.

## VII. SENSITIVITY OF METHODS FOR THE DETECTION OF ENTEROTOXIN

The sensitivities of various methods for the detection of enterotoxin are shown in Table 5. The gel diffusion and electrophoretic techniques can detect between 0.2 and 1.0 μg/mℓ of toxin, whereas RPHA, radioimmunoassay, and Vero cells detect as little as 0.001 μg/mℓ. It is interesting to note that some strains of *C. perfringens* associated with outbreaks produce very low levels of enterotoxin in vitro and this has raised the question as to what is a significant level of toxin production in relation to in vivo disease potential.

Niilo[119] examined 19 rabbit loop and erythemal activity positive strains and 22 rabbit loop and erythemal activity negative strains for the ability to produce enterotoxin detectable by fluorescent antibody (FA), reverse passive hemagglutination (RPHA), and double gel diffusion tests (ID). The 19 biologically positive strains were positive by all three methods. In addition, the FA test detected five positive strains which were not positive by the other two tests. The suggestion by several workers that serological and biological activities are separate entities raises doubt as to the significance of positive findings with the various enterotoxin detection methods. An attempt was made by Niilo[120] to study this problem when he examined 114 strains of *C. perfringens* isolated from the intestinal contents of cattle, sheep, and chickens with enteritis or other diseases. Strains were grown in Duncan and Strong medium and examined for toxin production by RPHA, FA, and gel diffusion; 18 strains were negative (i.e., RPHA titers of <10) and 15 strains had a titer of 10. All of these 33 strains were FA-negative. A further 54 strains had RPHA titers of 20 to 40 of which 14 were minimally FA-positive. RPHA titers of 80 to 5120 (i.e., 0.08 to 5.12 μg/mℓ of enterotoxin) was seen in 27 strains, and this level of toxin production was considered significant. Of these 27 strains, 25 (92%) were FA-positive, thus 13 strains were FA- and RPHA-positive (>0.8 μg/mℓ of toxin) but ID-negative. Niilo considered either (1) impurities present in the enterotoxin preparation used for antiserum production or (2) enterotoxin-like antigenic protein in spore or cellular components of nontoxigenic strains may contribute to the nonspecific reactions observed with the RPHA test. Although Niilo correctly states that there is no evidence to

**Table 5**
**SENSITIVITY OF BIOLOGICAL AND SEROLOGICAL METHODS FOR THE DETECTION AND ASSAY OF *C. PERFRINGENS* ENTEROTOXIN**

| Method | Enterotoxin (μg) | Dose volume (mℓ) | Extrapolated minimal concentration of enterotoxin (μg/mℓ) |
|---|---|---|---|
| Biological | | | |
| Erythemal activity | 0.06—0.125 | 0.05 | 1.25—2.5 |
| Mouse lethality ($LD_{50}$) | 1.8 | 0.25 | 6.8 |
| Rabbit ileal loop (standard) | 29.0 | 3.0 | |
| Rabbit ileal loop (90 mm) | 6.25 | 2.5 | |
| Chicken ileal loop | 20.0—30.0 | | |
| Calf ileal loop | 42.0—80.0 | | |
| Sheep ileal loop | 26.0—80.0 | | |
| Serological | | | |
| Single gel diffusion | 0.3 | 0.3 | 0.9 |
| Double gel diffusion | 0.013 | 0.025 | 0.5 |
| Electroimmunodiffusion | 0.01 | 0.01 | 1.0 |
| Counterimmunoelectrophoresis | 0.002 | 0.02 | 0.2 |
| Reversed passive hemagglutination | 0.00005 | 0.05 | 0.001 |
| Radioimmunoassay | 0.001 | | 0.001 |
| Fluorescent antibody | For spore-associated toxin | | |

Based on Genigeorgis, C., Sakaguchi, G., and Riemann, H. P., *Appl. Microbiol.*, 26, 111, 1973.

suggest that negative or weakly toxigenic strains of *C. perfringens* can become strong toxin producers, it is clear that since enterotoxin is a spore-related toxin, the composition of the sporulation medium will have a considerable effect on the amount of toxin produced. It is also evident that both the culture conditions and the heat treatment prior to inoculation into the sporulation medium are important factors. For example, Uemruama[81] showed that enterotoxin production was greatly enhanced in two of five food poisoning strains of *C. perfringens* when the cultures were subjected to three successive heat treatments. The optimum temperature for heat treatment varied with different strains; however, using the triple heat procedure, 11 (31%) of 35 individuals were found to be carriers of enterotoxigenic strains — the isolates producing between 0.2 and 6.4 μg/mℓ of enterotoxin.

Experiments with ligated loops in sheep, cattle, and chickens have shown that 20 to 40 μg of toxin per 5- to 20-cm loop are required to produce a positive reaction. In Niilo's work,[119] the biologically positive strains produced RPHA titers equivalent to between 0.64 and 10.24 μg/mℓ of toxin. It is reasonable to assume, therefore, that only four of Niilo's 114 strains would have produced biological or pathological effects. Niilo also states that it is not possible to relate the degree of enterotoxin production in a culture to the enterotoxigenic potential in the intestinal tract. Niilo attempted to group his enterotoxin-producing strains into (1) high toxin producers and (2) potentially enterotoxigenic strains.

## VIII. DETECTION OF ENTEROTOXIN IN FECES AND FOODS AND ANTIENTEROTOXIN IN SERUM

In the monkey feeding studies carried out by Uemura et al.,[90] enterotoxin could be detected in most of the fecal specimens, but neither enterotoxin nor antienterotoxin could be detected in serum samples taken up to 21 days after challenge. It was concluded that a significant amount of enterotoxin was not absorbed from the intestine into the bloodstream. Skjelkvåle and Uemura[89] investigated two incidents of food poisoning in which one person was involved in both. In the first outbreak, 13 to 16 μg of enterotoxin per gram of feces could be detected and in the second episode (8 months later), 3 to 4 μg of toxin per gram of feces was found. Enterotoxin was not detected in serum during the acute stage of the illness, but it was detected as a rise in antibody titer during the following 2 months. In a later study, Skjelkvåle and Uemura[121] fed purified enterotoxin to 6 human volunteers which resulted in diarrhea and abdominal pain in 1 to 2.5 hr. A dose of 8 mg of enterotoxin caused diarrhea in 1 of 2 volunteers, and all 5 subjects who received 10 to 12 mg of toxin developed symptoms. Doses of 10 to 12 mg resulted in a detectable enterotoxin concentration in feces of between 2 and 16 μg/g. Before challenge, antienterotoxin titers in serum varied from <160 to 640; 2 weeks after the administration, toxin titers ranged from <160 to 5000. It was concluded that serum titers of antienterotoxin were of little value in the diagnosis of *C. perfringens* food poisoning.

In 1975, Dowell et al.[122] suggested that the detection of enterotoxin direct in feces may serve as a diagnostic test for implicating *C. perfringens* as the cause of food-borne illness. Dowell and co-workers compared the use of serological typing with the detection of toxin in feces while investigating an outbreak of *C. perfringens* food poisoning in a federal prison. All of those (13 persons) who had eaten the suspect food had serotype 1 in their feces, whereas only 10 (77%) had a detectable *C. perfringens* enterotoxin (CPE) titer of 2000 (and 3 of these were after concentration of the specimen). The severe cases with profuse, watery diarrhea had a diluted CPE level in their stools. A similar study was undertaken by Naik and Duncan[123] who used counterimmunoelectrophoresis to detect enterotoxin in the feces of individuals involved in two outbreaks of food poisoning. In the first outbreak, seven of eight fecal specimens were positive for toxin; in the second outbreak however, enterotoxin was not detected in the feces of the six subjects examined although *C. perfringens* strains isolated from five of the six individuals were subsequently shown to be enterotoxigenic in vitro. It was thought that the late collection of fecal specimens in outbreak 2 (4 to 5 days) was responsible for the absence of CPE. Studies on antienterotoxin levels in the serum of high-risk persons (i.e., laboratory staff working with *C. perfringens*) and low-risk individuals showed no correlation between the titers of CPE and risk category. Itoh et al.[124] examined 130 fecal isolates associated with 13 outbreaks of food poisoning. Of the 130 fecal specimens, 109 (84%) contained toxin (levels ranged from 0.1 to 25.6 μg/mℓ). Toxin was found 1 to 2 days after illness but not after 5 days and none of the fecal specimens from healthy subjects contained toxin. More recently, Uemura et al.[125] investigated a large outbreak of *C. perfringens* food poisoning in which 427 employees of 96 different factories were affected. Serotype 65/69 was isolated from the incriminated food and a number of fecal specimens. However, by in vitro tests, 42 (44%) of the 95 isolates were enterotoxigenic (>0.1 μg/mℓ) and the remaining 53 were nonenterotoxigenic (<0.1 μg/mℓ). Of 16 isolates from the food, only 9 (56%) were toxigenic. They showed that in some cases heat treatment stimulated the ability to produce enterotoxin.

Work by Uemura et al.[126] indicated the presence of antienterotoxin antibody in the sera of 82% of Brazilian and 65% of American subjects; titers ranged from <160 to 5120. Antibody levels observed in 5 individuals over a period of 17 weeks remained stable. In 1980, Niilo and Bainborough[127] examined sera from humans, cattle, sheep, swine, and horses for the presence of *C. perfringens* enterotoxin antibody using the passive hemagglutination test, ID, and CIEP. The 224 human sera ranged in titer from 0 to 128; 3 of the sera were positive in all 3 tests. Niilo and Bainborough discussed the differences between their results and those of Uemura et al. and suggested that the sensitivity of the PHA test may vary markedly between different laboratories.

**Table 6**
**RESULTS FROM SOME STUDIES ON THE ISOLATION OF ENTEROTOXIGENIC STRAINS OF *C. PERFRINGENS* FROM DIFFERENT SOURCES**

| Report | Source of strain and no. examined | | No. enterotoxin positive (%) | Method for toxin detection | Ref. |
|---|---|---|---|---|---|
| Skjelkvale et al. | Feces (food poisoning) | 65 | 56 (86) | Electroimmunodiffusion | 130 |
| | Meat carcasses | 68 | 0 | Radioimmunoassay | |
| | Minced beef | 26 | 0 | | |
| | Pig feces | 60 | 2 (3) | | |
| | Foods and feces not associated with food poisoning | 14 | 11 | | |
| | Clinical infections | 6 | 0 | | |
| Yasukawa et al. | Feces (food poisoning) | 66 | 51 (77) | Gel diffusion | 131 |
| | Human feces | 117 | 1 | | |
| | Food | 120 | 1 | | |
| | Soil | 85 | 0 | | |
| | River water | 12 | 0 | | |
| | Swabs from cooking establishments | 13 | 1 | | |
| Gyobu | Feces (food poisoning) | 33 | 20 | Gel diffusion | 132 |
| | Human feces and soil | 53 | 0 | RPHA | |
| | | | | Vero cells | |
| Tsai et al. | Chicken feces | 25 | 15 (60) | Gel diffusion | 133 |
| | Beef cattle | 40 | 27 (67) | RPHA | |
| | | | | Rabbit loop | |
| | | | | Mouse lethality | |
| Chakrabarty and Boro | Blood meal | 4 | 1 | Rabbit loop | 134 |
| | Fish meal | 5 | 3 | Erythemal activity | |

Naik and Duncan[128] studied the ability of *C. perfringens* to grow, sporulate, and produce enterotoxin in foods. A count of $10^7$ to $10^8$ vegetative cells per gram of food was usually obtained within 4 hr and the spore count gradually increased to a level of $10^4$ to $10^5$ per gram. Enterotoxin was detected in cooked chuck roast, ground beef, turkey, and also in moist cooked and dry roasted chicken up to 0.125 μg/g. Enterotoxin was not detected until 10 hr after inoculation in moist, cooked turkey incubated at 37°C. However, in moist, cooked turkey that had been held at 37°C for 6 hr, stored cold for 15 hr, and reheated to 37°C, toxin was detected after 5 hr. This gives some, albeit inconclusive, support to the suggestion that some outbreaks of *C. perfringens* food poisoning in which there is an early onset of symptoms may be caused by a toxin preformed in the food.

Craven et al[129] examined the ability of 17 strains of *C. perfringens* to spore and produce enterotoxin in autoclaved chicken meat and in Duncan and Strongs' sporulation medium (DS). Of 17 strains, 7 produced toxin in the chicken meat (0.21 to 24 μg/mℓ). It was considered that the chicken would become organoleptically unacceptable before spores and enterotoxin were formed and therefore would not be eaten.

## IX. DISTRIBUTION OF ENTEROTOXIGENIC STRAINS

It is clear from Table 6 that a high proportion of strains associated with outbreaks of food poisoning are enterotoxigenic in vitro. It was surprising that, in contrast to the finding of other workers, such a high number of strains from chicken feces and beef cattle were toxigenic.[133] In addition to the isolates described in Table 6, Skjelkvåle et al.[130] also examined five isolates of lecithinase-negative *C. perfringens* which were known to have caused outbreaks of food-borne illness. All 5 strains produced enterotoxin in vitro (0.13 to 105 μg/mℓ).

## X. MODE OF ACTION OF ENTEROTOXIN

Niilo[135] administered cell extracts of an enteropathogenic strain of *C. perfringens* intravenously to lambs, rabbits, and guinea pigs and the results suggested that enterotoxin caused (1) increased capillary permeability, (2) increased vasodilation, and (3) an increase in intestinal mobility. McDonel and Duncan[136] examined the effects of enterotoxin on the metabolic indexes of everted rat ileal sacs. Enterotoxin-treated sacs consumed less oxygen; however, the conversion of glucose to lactate (via glycolysis) was unaffected. The rabbit loop studies of McDonel and Duncan[137] confirmed their earlier findings in rats that excess enterotoxin does not increase fluid loss or tissue damage beyond a certain point. Ileal loops challenged with a range of 50 to 1000 erythermal units of activity produced essentially the same amount of fluid accummulation. Examination of the histopathological effect of enterotoxin on the rabbit ileum indicated that at concentrations of enterotoxin necessary to induce fluid accumulation, epithelium was denuded from the tips of the villi. McDonel et al.[138] used scanning electron microscopy (SEM) to show that morphological damage to the intestinal epithelial caused by enterotoxin was confined to the tips of the villi which were often collapsed and covered with rounded blebs. Some epithelial cells were completely destroyed and damage to the apical membranes led to sloughing, death, and lysis.

Rabbit ileal, jejunal, and duodenal loops were exposed to purified enterotoxin and then perfused for comparative analysis of the effects of enterotoxin on each region of the intestine. Ileal loops responded with enhanced net secretion of fluid and sodium, inhibition of chloride and glucose uptake, and substantial sloughing of epithelial cells. Severity of response was ileum> jejunum> duodenum.[139]

The secretion of fluid and electrolytes was not mediated through elevated levels of cyclic AMP as is the case with cholera and *Escherichia coli* heat-labile enterotoxins. McDonel and Duncan suggested that *C. perfringens* enterotoxin acted either by transport inhibition or by membrane damage, causing subsequent disruption of cellular function. Studies by McDonel and Duncan[140] also found that there was a reduction in the rate of $O_2$ consumption by rat liver mitochondria when certain TCA cycle intermediates were used as subtrates, although P/O rates (phosphate esterification coupled to subtrate oxidation, i.e., ATP production) were unaltered.

In 1979, McDonel and McClane[141] used tissue culture techniques and Vero (African green monkey kidney) cells to study the relationship between binding and biological activity with cells both resistant and sensitive to enterotoxin. Binding studies using $^{125}$I-labeled enterotoxin revealed the existence of high- and low-affinity binding sites for the enterotoxin on both cell types. The binding site density on resistant cells was one tenth that of sensitive cells. They concluded that without specific binding there was little chance of enterotoxin having any biological activity within cells.

McDonel[142] investigated the ability of labeled enterotoxin to bind specifically to rabbit ileal epithelial intestinal cells. The rate and amount of binding was temperature-dependent in that heat treatment which destroyed biological activity also destroyed binding activity. Caution is required in drawing comparisons between the epithelial cell studies and the work with Vero cells because the latter were incubated at 37°C and the former at 24°C. McDonel proposed that enterotoxin binds to a specific receptor, followed by insertion into cell membrane, resulting in a breakdown in membrane structure and function. McClane and McDonel[116] demonstrated that Vero cells were extremely sensitive to *C. perfringens* enterotoxin; within 30 min of exposure, cells were affected. In summary, enterotoxin affects morphology, detachment, viability, plating efficiency, and macromolecular synthesis of Vero cells.

Further studies by McClane and McDonel[143] examined the alterations in plasma membrane permeability induced by *C. perfringens* enterotoxin using Vero cells radioactively labeled with four markers of different molecular size. Results suggested that enterotoxin produced holes of a defined size through which essential cellular fractions leaked out, leading to eventual cell death. McClane and McDonel[144] showed that culture media made hypertonic by the addition of osmotic stabilizers protected Vero cells against changes in morphology and plasma membrane permeability induced by enterotoxin. This work led to a proposed model for the mode of action of *C. perfringens* enterotoxin with the following sequence of events:

1. Enterotoxin binds to susceptible cells.
2. Plasma membrane is made leaky (mechanism unknown) creating ion-imbalance.
3. Net influx of ions and water into cells.
4. Plasma membrane stretches forming membrane blebs which affect permeability barrier.
5. Eventual cell death.

Skjelkvåle et al.[145] injected radiolabeled enterotoxin intravenously into mice and rats and within 10 to 15 min the highest amount of activity was found in the liver. In vitro studies with purified parachymeal liver cells showed a rapid release of lactate dehydrogenase during treatment with enterotoxin, thus supporting the findings of McDonel and colleagues that enterotoxin causes severe membrane damage.

Miwatani et al.[146] investigated the mode of lethal activity when enterotoxin was injected intravenously into mice. Deaths were not due to cardiotoxicity of the enterotoxin since cultured fetal mouse heart cells were still beating in the presence of 4 μg/mℓ of toxin. Studies of EEG, respiration, and heart rates suggested that the enterotoxin acted directly on the brain of the animals. McDonel and Denvers[147] carried out studies on the involvement of the colon in food poisoning. Rabbit colonic loops were exposed to purified toxin and perfused. Results suggested a very limited involvement of the colon in the expression of food poisoning symptoms. Further binding studies showed that although binding was necessary for biological activity it is not always the case in that if binding occurs, biological activity is expressed. McDonel and Denvers therefore suggested that biological activity involves a multistep process in which binding is the first essential step. Matsuda and Sugimato[148] proposed a two-stage mechanism for enterotoxin activity. Using Vero and Hela cells the first step was Ca-independent followed by a Ca-dependent step and the formation of blebs.

## XI. CONTROL AND PREVENTION OF *C. PERFRINGENS* FOOD POISONING

No review of *C. perfringens* food poisoning would be complete without reference to the factors involved in the control and prevention of food-borne illness.[149,150] In a comprehensive study of the factors which contributed to food poisoning in England and Wales during 1970 to 1979, Roberts[150] analyzed the reported data on 387 outbreaks caused by *C. perfringens*. In 338 (87%) of the outbreaks, the food was prepared too far in advance; in 208 (54%) there was storage at ambient temperature after cooking. Inadequate cooling was a factor in 236 (61%) of the outbreaks and ineffectual reheating of food contributed to 215 (55%) incidents. Bryan's[149] earlier study in 1978 of a much smaller number of outbreaks showed the same factors to be the cause of *C. perfringens* food poisoning in the U.S. It is clear that while conditions for cooking, serving, and storage of cooked food are allowed which enable the rapid multiplication of *C. perfringens* to high numbers (i.e., $\geq 10^6$/g), outbreaks will continue to occur.

Following a large outbreak of *C. perfringens* food poisoning in a hospital, Thomas et al.[151] made a number of recommendations in an attempt to prevent similar episodes in other hospitals. These included:

1. Rapid consumption or rapid cooling and refrigeration of cooked food.
2. Reheated cooked foods should be subjected to heating at 100°C throughout their bulk, thereby killing all vegetative cells.
3. Provision of suitable staff training in hygiene and catering practice.

If, as so often occurs, meals are prepared in advance of needs, then as Hobbs[152] states, there is a need for, "education of architects and engineers in the basic causes so that they may design special areas and equipment for faster rates of cooling and spacious cold storage."

## ACKNOWLEDGMENT

The author is grateful to Dr. R. J. Gilbert for helpful advice and discussion.

## REFERENCES

1. **Klein, E.,** Üeber einen pathogenen anaeroben Darmnacillus, *Bacillus enteritidis sporogenes, Zentralbl. Bakteriol Parasitenkd. Infektionskr. Hyg. Abt. I: Orig.* 18, 737, 1895.
2. **Andrewes, F. W.,** On an outbreak of diarrhoea in the wards of St. Bartholomew's hospital, probably caused by infection of rice pudding with *Bacillus enteritidis* sporogenes, *Lancet,* 1, 8, 1899.
3. **Knox, R. and MacDonald, E. K.,** Outbreaks of food poisoning in certain Leicester institutions, *Med. Off.*, 69, 21, 1943.
4. **McClung, L. S.,** Human food poisoning due to growth of *Cl. perfringens Cl. welchii)* in freshly cooked chicken: preliminary note, *J. Bacteriol.*, 50, 229, 1945.
5. **Osterling, S.,** Food poisoning caused by *Clostridium perfringens (welchii)* preliminary note, *Nord. Hyg. Tidskr.*, 33, 173, 1952.
6. **Vernon, E. and Tillett, H. E.,** Food poisoning and *Salmonella* infections in England and Wales, 1969—1972, *Publ. Health London,* 88, 225, 1974.
7. **Vernon, E.,** Food poisoning and *Salmonella* infections in England and Wales, 1973—1975, *Public Health London,* 91, 225, 1977.
8. **Hepner, E.,** Food poisoning and *Salmonella* infections in England and Wales, 1976—1978, *Public Health London,* 94, 337, 1980.
9. **Anon.,** Food poisoning and salmonellosis surveillance in England and Wales: 1979, *Br. Med. J.*, 281, 1360, 1980.
10. **Anon.,** Food poisoning and salmonellosis surveillance in England and Wales: 1980, *Br. Med. J.*, 283, 924, 1981.
11. **Parry, W. H.,** Outbreak of *Clostridium welchii* food poisoning, *Br. Med. J.*, 2, 1616, 1963.
12. Communical Diseases Scotland, Food poisoning in hospitals in Scotland, 1973—79, *Communicable Diseases Scotland,* eoli, (vii), 1980.
13. **CDC,** Foodborne Outbreaks, Annual Summary 1970—1971, Center for Disease Control, Public Health Service, U.S. Department of Health, Education and Welfare, Atlanta.
14. **CDC,** Foodborne Outbreaks, Annual Summary 1971—1972, Center for Disease Control, Public Health Service, U.S. Department of Health, Education and Welfare, Atlanta.
15. **CDC,** Foodborne, Outbreaks, Annual Summary 1972—1973, Center for Disease Control, Public Health Service, U.S. Department of Health, Education and Welfare, Atlanta.
16. **CDC,** Foodborne and Waterborne Disease Outbreaks, Annual Summary 1973—1974, Center for Disease Control, Public Health Service, U.S. Department of Health, Education and Welfare, Atlanta.

17. **CDC,** Foodborne and Waterborne Disease Outbreaks, Annual Summary 1974—1976, Center for Disease Control, Public Health Service, U.S. Department of Health, Education and Welfare, Atlanta.
18. **CDC,** Foodborne and Waterborne Disease Outbreaks, Annual Summary 1975—1976, Center for Disease Control, Public Health Service, U.S. Department of Health, Education and Welfare, Atlanta.
19. **CDC,** Foodborne and Waterborne Disease Outbreaks, Annual Summary 1976—1977, Center for Disease Control, Public Health Service, U.S. Department of Health, Education and Welfare, Atlanta.
20. **CDC,** Foodborne and Waterborne Disease Outbreaks, Annual Summary 1977—1979, Center for Disease Control, Public Health Service, U.S. Department of Health, Atlanta.
21. **CDC,** Foodborne Disease, Annual Summary 1978—1981, Center for Disease Control, Public Health Service, U.S. Department of Health, Education and Welfare, Atlanta.
22. **Todd, E. C. D.,** Foodborne disease in six countries — a comparison, *J. Food Prot.*, 41, 559, 1978.
23. **Health Protection Branch,** Foodborne and Water-Borne Disease in Canada, Annual Summary 1976, 1980, Health and Welfare, Canada.
24. **Alexander, R. and Johnstone, M. C.,** An outbreak of *Clostridium perfringens* Hobbs type 21 food poisoning, *Zentralbl. Bakteriol. Parasitenkd. Infektionskr. Hyg. Abt. 1: Orig. B*, 173, 488, 1981.
25. **Itoh, T.,** Incidence of heat-resistant *Clostridium perfringens* in faeces of healthy subjects serotyping of isolates, and food poisoning caused by new serotype organisms, *Annu. Rep. Tokyo Metropolitan Res. Lab. Public Health*, No. 34, 7, 1972.
26. **Fabiansson, S. and Normark, D.,** *Clostridium perfringens* son matforgiftningsframkallare, *Sven. Vet.*, 28, 687, 1976.
27. **Raevuori, M.,** The occurrence of food poisoning outbreaks in Finland, *Suom. Elanilaakarilehti*, 82, 501, 1976.
28. **Caroli, G., Armani, C., Scinocu, A., Bargagna, M., and Levre, E.,** Considerazion epidemiologiche su due episodi di tossinfezione alimentare du *Clostridium perfringens, Ann. Sclavo.*, 19, 494, 1977.
29. **Stringer, M. F.,** Studies on *Clostridium perfringens* Food Poisoning, Ph.D. thesis, University of London, 1981.
30. **Sharp, J. C. M., Collier, P. W., and Gilbert, R. J.,** Food poisoning in hospitals in Scotland, *J. Hyg.*, 83, 231, 1979.
31. **Willarsden, R. R., Busta, F. F., Allen, C. E., and Smith, L. B.,** Growth and survival of *Clostridium perfringens* during constantly rising temperatures, *J. Food Sci.*, 43, 470, 1978.
32. **Smith, L. D. S.,** *The Pathogenic Anaerobic Bacteria*, Charles C Thomas, Springfield, Ill., 1975, chap. 6.
33. **Smith, L. D. S. and Gardner, M. V.,** Vegetative cells of *Clostridium perfringens* in soil, *J. Bacteriol.*, 58, 407, 1949.
34. **Miwa, T.,** Clostridia in soil of the antartica, *Jpn. J. Med. Sci. Biol.*, 28, 201, 1975.
35. **Blowers, R. and Crew, B.,** Ventilation of operating theatres, *J. Hyg.*, 58, 427, 1960.
36. **McKillop, E. J.,** Bacterial contamination of hospital food with special reference to *Clostridium welchii* food poisoning, *J. Hyg.*, 57, 31, 1959.
37. **Hall, H. E. and Angelotti, R.,** *Clostridium perfringens* in meat and meat products, *Appl. Microbiol.*, 13, 352, 1965.
38. **Foster, J. F., Fowler, J. L., and Ladiges, W. C.,** A bacteriological survey of raw ground beef, *J. Food Prot.*, 40, 790, 1977.
39. A close look at hamburger, *Consumer Reports*, 36, 478, 1981.
40. **Smart, J. L., Roberts, T. A., Stringer, M. F., and Shah, N.,** The incidence and serotypes of *Clostridium perfringens* on beef, pork and lamb carcasses, *J. Appl. Bacteriol.*, 46, 377, 1979.
41. **Narayan, K. G.,** Studies on clostridia — incidence in the beef cattle, *Acta Vet. Hung.*, 16, 65, 1966.
42. **Narayan, K. G. and Takacs, J.,** Incidence of clostridia in emergency — slaughtered cattle, *Acta Vet. Hung.*, 16, 345, 1966.
43. **Gill, C. O., Penney, N., and Nottingham, P. M.,** Tissue sterility in uneviscerated carcasses, *Appl. Environ. Microbiol.*, 36, 356, 1978.
44. **Lillard, H. S.,** Occurrence of *Clostridium perfringens* in broiler processing and further processing operations, *J. Food. Sci.*, 36, 1008, 1971.
45. **Gibbs, P. A.,** The incidence of Clostridia in poultry carcasses and poultry processing plants, *Br. Poult. Sci.*, 12, 101, 1971.
46. **Mead, G. C. and Adams, B. W.,** Microbiological aspects of polyphosphate injection in the processing and chill storage of poultry, *J. Hyg.*, 82, 133, 1979.
47. **Taniguti, T. and Zenitani, B.,** Incidence of *Clostridium perfringens* in fishes. I. On the application of the LAS medium to detection of *Clostridium perfringens, J. Food Hyg. Soc. Jpn.*, 10, 199, 1969.
48. **Hobbs, G., Cann, D. C., Wilson, B. B., and Shewan, J. M.,** The incidence of organisms of the genus *Clostridium* in vacuum packed fish in the United Kingdom, *J. Appl. Bacteriol.*, 28, 265, 1965.

49. **Sutton, R. G. A. and Hobbs, B. C.,** Enumeration of *Clostridium perfringens* in dried foods and feeding stuffs, in *The Microbiology of Dried Foods, Proc. 6th Int. Symp. Food Microbiol.*, Kampelmacher, E. H., Ingram, M., and Mossel, D. A. A., Eds., Grafische Industrie, Haarlem, 1969, 243.
50. **Drasar, B. S. and Hill, M. J.,** *Human Intestinal Flora,* Academic Press, London, 1974, chap. 1.
51. **Hobbs, B. C., Smith, M. E., Oakley, C. L., Warrack, G. H., and Cruickshank, J. C.,** *Clostridium welchii* food poisoning, *J. Hyg.*, 51, 75, 1953.
52. **Dische, F. E. and Elek, S. D.,** Experimental food poisoning by *Clostridium welchii, Lancet,* 2, 71, 1957.
53. **Leeming, R. L., Pryce, J. D., and Meynell, M. J.,** Clostridium *welchii* and food poisoning, *Br. Med. J.*, 1, 501, 1961.
54. **Collee, J. G., Knowlden, J. A., and Hobbs, B. C.,** Studies on the growth, sporulation and carriage of *Clostridium welchii* with special reference to food poisoning strains, *J. Appl. Bacteriol.*, 24, 326, 1961.
55. **Sutton, R. G. A.,** Distribution of heat-resistant *Clostridium welchii* in a rural area of Australia, *J. Hyg.*, 64, 65, 1966.
56. **Hall, H. E. Angelotti, R., Lewis, K. H., and Foter, M. J.,** Characteristics of *Clostridium perfringens* associated with food and food-borne disease, *J. Bacteriol.*, 85, 1094, 1963.
57. **Taylor, C. E. D. and Coetzee, E. F. C.,** Range of heat resistance of *Clostridium welchii* associated with suspected food poisoning, *Mon. Bull. Minist. Health,* 25, 142, 1966.
58. **Sutton, R. G. A.,** Enumeration of *Clostridium welchii* in the faeces of varying sections of the human population, *J. Hyg.*, 64, 367, 1966.
59. **Sutton, R. G. A.,** The Pathogenesis and Epidemiology of *Clostridium welchii* Food Poisoning, Ph.D. thesis, Unversity of London, 1969.
60. **Nakagawa, M. and Nishida, S.,** Heat resistance and toxigenicity of *Clostridium perfringens* strains in normal intestines of Japanese, *Jpn. J. Microbiol.*, 13, 133, 1969.
61. **Yamagishi, T., Serikawa, T., Morita, R., Nakamura, S., and Nishida, S.,** Persistent high numbers of *Clostridium perfringens* in the intestines of Japanese aged adults, *Jpn. J. Microbiol.*, 20, 397, 1976.
62. **Mahony, D. E.,** Bacteriocin, bacteriophage and other epidemiological typing methods for the genus *Clostridium,* in *Methods in Microbiology,* Vol. 13, Bergan, T. and Norris, J. R., Eds., Academic Press, London, 1980, 1.
63. **Baine, H. and Cherniak, R.,** Capsular polysaccharides of *Clostridium perfringens* Hobbs 5, *Biochemistry,* 10, 2948, 1971.
64. **Cherniak, R. and Henderson, B. G.,** Immunochemistry of the capsular polysaccharides from *Clostridium perfringens*: selected Hobbs Strains 1, 5, 9 and 10, *Infect. Immun.*, 6, 32, 1972.
65. **Hughes, J. A., Turnbull, P. C. B., and Stringer, M. F.,** A serotyping scheme for *Clostridium welchii (C. perfringens)* type A, and studies on the type-specific antigens, *J. Med. Microbiol.*, 9, 475, 1976.
66. **Paine, C. M. and Cherniak, R.,** Composition of the capsular polysaccharides of *Clostridium perfringens* as a basis for their classification by chemotypes, *Can. J. Microbiol.*, 21, 181, 1975.
67. **Stringer, M. F., Turnbull, P. C. B., and Gilbert, R. J.,** Application of serological typing to the investigation of outbreaks of *Clostridium perfringens* food poisoning, 1970—1978, *J. Hyg.*, 84, 443, 1980.
68. **Hayashi, K., Kugita, Y., Tawara, M., and Yamagata, H.,** A mass food poisoning affair caused by *Clostridium welchii* Hobbs type 2, *Endem. Dis. Bull. Nagasaki Univ.*, 3, 1, 1961.
69. **Hayashi, K., Kugita, Y., Tawara, M., and Yamagata, H.,** A food poisoning incident of a family caused by *Clostridium welchii* Hobbs' type 6, *Edem. Dis. Bull. Nagasaki Univ.*, 3, 87, 1961.
70. **Itoh, T.** personal communiation.
71. **Hatheway, C. L., Whaley, D. N., and Dowell, V. R.,** Epidemiological aspects of *Clostridium perfringens* foodborne illness, *Food Technol.*, 34(4), 77, 1980.
72. **Duncan, C. L., Strong, D. H., and Sebald, M.,** Sporulation and enterotoxin production by mutants of *Clostridium perfringens, J. Bacteriol.*, 110, 378, 1972.
73. **Frieben, W. R. and Duncan, C. L.,** Homology between enterotoxin protein and spore structural protein in *Clostridium perfringens* type A, *Eur. J. Biochem.*, 39, 393, 1973.
74. **Duncan, C. L., King, J. G., and Frieben, W. R.,** A paracrystalline inclusion formed during sporulation of enterotoxin-producing strains of *Clostridium perfringens* type A, *J. Bacteriol.*, 114, 845, 1973.
75. **Frieben, W. R. and Duncan, C. L.,** Heterogeneity of enterotoxin-like protein extracted from spores of *Clostridium perfringens* type A, *Europ. J. Biochem.*, 55, 455, 1975.
76. **Tsai, C. C., Torres-Anjel, M. J., and Riemann, H. P.,** Improved culture techniques and sporulation medium for enterotoxin production by *Clostridium perfringens* type A, *J. Formosan Med. Assoc.*, 73, 404, 1974.
77. **Ellner, P. D.,** A medium promoting rapid quantitative sporulation in *Clostridium perfringens, J. Bacteriol.*, 71, 495, 1956.
78. **Duncan, C. L. and Strong, D. H.,** Improved medium for sporulation of *Clostridium perfringens, Appl. Microbiol.*, 16, 82, 1968.
79. **Uemura, T., Sakaguchi, G., and Riemann, H. P.,** In vitro production of *Clostridium perfringens* enterotoxin and its detection by reversed passive haemagglutination, *Appl. Microbiol.*, 26, 381, 1973.

80. **Uemura, T.**, Incidence of enterotoxigenic *Clostridium perfringens* in healthy humans in relation to the enhancement of enterotoxin production by heat treatment, *J. Appl. Bacteriol.*, 44, 41, 1978.
81. **Uemura, T.**, Sporulation and enterotoxin production by *Clostridium perfringens* on solidified sporulation media, *J. Food Hyg. Soc. Jpn.*, 19, 462, 1978.
82. **Duncan, C. L.,Sugiyama, H., and Strong, D. H.**, Rabbit ileal loop response to strains of *Clostridium perfringens*, *J. Bacteriol.*, 95, 1560, 1968.
83. **Niilo, L. and Dorward, W. J.**, The effect of enterotoxigenic *Clostridium welchii (perfringens)* type A on the bovine intestine, *Res. Vet. Sci.*, 12, 376, 1971.
84. **Niilo, L., Hauschild, A. H. W., and Dorward, W. J.**, Immunization of sheep against experimental *Clostridium perfringens* type A enteritis, *Can. J. Microbiol.*, 17, 391, 1971.
85. **Niilo, L.**, Response of ligated intestinal loops in chickens to the enterotoxin of *Clostridium perfringens*, *Appl. Microbiol.*, 28, 889, 1974.
86. **Hauschild, A. H. W., Hilsheimer, R., and Rogers, C. G.**, Rapid detection of *Clostridium perfringens* enterotoxin by a modified ligated intestinal loop technique in rabbits, *Can. J. Microbiol.*, 17, 1475, 1971.
87. **Yamamoto, K., Ohishi, I., and Sakaguchi, G.**, Fluid accumulation in mouse ligated intestine inoculated wtih *Clostridium perfringens* enterotoxin, *Appl. Environ. Microbiol.*, 37, 181, 1979.
88. **Strong, D. H., Duncan, C. L., and Perna, G.**, *Clostridium perfringens* type A food poisoning. II. Response of the rabbit ileum as an indication of enteropathogenicity of strains of *Clostridium perfringens* in human beings, *Infect. Immun.*, 3, 171, 1971.
89. **Skjelkvåle, R. and Uemura, T.**, Experimental diarrhoea in human volunteers following oral administration of *Clostridium perfringens* enterotoxin, *J. Appl. Bacteriol.*, 43, 281, 1977.
90. **Uemura, T., Sakaguchi, G., Itoh, T., Okazawa, K., and Sakai, S.**, Experimental diarrhoea in cynomolgus monkeys by oral administration with *Clostridium perfringens* type A viable cells or enterotoxin, *Jpn. J. Med. Sci. Biol.*, 28, 165, 1975.
91. **Tsai, C. C. and Riemann, H. P.**, Food poisoning signs in mice induced orally by *Clostridium perfringens* type A enterotoxin, *J. Formosan Med. Assoc.*, 74, 310, 1975.
92. **Niilo, L.**, The effect of enterotoxin of *Clostridium perfringens* type A on fowls, *Res. Vet. Sci.*, 20, 225, 1976.
93. **Hauschild, A. H. W.**, Erythermal activity of the cellular enteropathogenic factor of *Clostridium perfringens* type A, *Can. J. Microbiol.*, 16, 651, 1970.
94. **Stark, R. L. and Duncan, C. L.**, Transient increase in capillary permeability induced by *Clostridium perfringens* type A enterotoxin, *Infect. Immun.*, 5, 147, 1972.
95. **Hauschild, A. H. W. and Hilsheimer, R.**, Purification and characteristics of the enterotoxin of *Clostridium perfringens* type A, *Can. J. Microbiol.*, 17, 1425, 1971.
96. **Stark, R. L. and Duncan, C. L.**, Purification and biochemical properties of *Clostridium perfringens* type A enterotoxin, *Infect. Immun.*, 6, 662, 1972.
97. **Sakaguchi, G., Uemura, T., and Riemann, H. P.**, Simplified method for purification of *Clostridium perfringens* type A enterotoxin, *Appl. Microbiol.*, 26, 762, 1973.
98. **Niilo, L.**, Antigenic homogenicity of enterotoxin from different agglutinating serotypes of *Clostridium perfringens*, *Can. J. Microbiol.*, 19, 521, 1973.
99. **Skjelkvåle, R. and Duncan, C. L.**, Enterotoxin formation by different toxigenic types of *Clostridium perfringens*, *Infect. Immun.*, 11, 563, 1975.
100. **Bartholomew, B. A. and Stringer, M. F.**, Observations on the purification of *Clostridium perfringens* type A enterotoxin and the production of a specific antiserum, *FEMS Microbiol. Lett.*, 18, 43, 1983.
101. **Barnhart, H. M., Bullerman, L. B., Ball, E. M., and Wagner, F. W.**, Isolation and purification of *Clostridium perfringens* enterotoxin by affinity chromatography, *J. Food Sci.*, 41, 903, 1976.
102. **Scott, V. N. and Duncan, C. L.**, Affinity chromatography purification of *Clostridium perfringens* enterotoxin, *Infect. Immun.*, 12, 536, 1975.
103. **Uemura, T. and Skjelkvåle, R.**, An enterotoxin produced by *Clostridium perfringens* type D. Purification by affinity chromatography, *Acta Pathol. Microbiol. Scand. Sect. B*, 84, 414, 1976.
104. **Granum, P. E. and Whitaker, J. R.**, Improved method for purification of enterotoxin from *Clostridium perfringens* type A, *Appl. Environ. Microbiol.*, 39, 1120, 1980.
105. **Enders, G. L. and Duncan, C. L.**, Anomolous aggregration of *Clostridium perfringens* enterotoxin under dissociating conditions, *Can. J. Microbiol.*, 22, 1410, 1976.
106. **Skjelkvåle, R.**, Radioiodination of enterotoxin from *Clostridium perfringens* type A using Chloramine T, *J. Appl. Bacteriol.*, 48, 283, 1980.
107. **Granum, P. E., Whitaker, J. R., and Skjelkvåle, R.**, Trypsin activation of enterotoxin from *Clostridium perfringens* type A. Fragmentation and some physicochemical properties, *Biochem. Biophys. Acta*, 668, 325, 1981.
108. **Naik, H. S. and Duncan, C. L.**, Thermal inactivation of *Clostridium perfringens* enterotoxin, *J. Food Protect.*, 41, 100, 1978.

109. **Genigeorgis, C., Sakaguchi, G., and Riemann, H. P.,** Assay methods for *Clostridium perfringens* type A enterotoxin, *Appl. Microbiol.*, 26, 111, 1973.
110. **Bradstreet, C. M. P., Dighero, M. W., Gilbert, R. J., and Wieneke, A. A.,** Raising antibodies to staphylococcal enterotoxins A and B, *J. Appl. Bacteriol.*, 42, 117, 1977.
111. **Stringer, M. F., Watson, G. N., and Gilbert, R. J.,** *Clostridium perfringens* type A: serological typing and methods for the detection of enterotoxin, in *Methods for the Isolation and Identification of Food Poisoning Organisms*, Corry, J. E. L., Roberts, D., and Skinner, F. A., Eds., Academic Press, London, 1982, 111.
112. **Duncan, C. L. and Somers, E. B.,** Quantitation of *Clostridium perfringens* type A enterotoxin by electroimmunodiffusion, *Appl. Microbiol.*, 24, 801, 1972.
113. **Naik, H. S. and Duncan, C. L.,** Rapid detection and quantitation of *Clostridium perfringens* enterotoxin by counterimmunoelectrophoresis, *Appl. Environ. Microbiol.*, 34, 125, 1977.
114. **Niilo, L.,** Enterotoxin formation by *Clostridium perfringens* type A studied by the use of fluorescent antibody, *Can. J. Microbiol.*, 23, 908, 1977.
115. **McClane, B. A. and McDonel, J. L.,** The effects of *Clostridium perfringens* enterotoxin on morphology, viability and macromolecular synthesis in Vero cells, *J. Cell Physiol.*, 99, 191, 1979.
116. **McDonel, J. L. and McClane, B. A.,** Highly sensitive assay for *Clostridium perfringens* enterotoxin that uses inhibition of plating efficiency of Vero cells grown in culture, *J. Clin. Microbiol.*, 13, 940, 1981.
117. **Bradshaw, J. G., Stelman, G. N., Jones, V. I., Peeler, J. T., Wimsatt, J. C., Corwin, J. J., and Twedt, R. M.,** Thermal inactivation of *Clostridium perfringens* enterotoxin in buffer and in chicken gravy, *J. Food Sci.*, 47, 914, 1982.
118. **Giugliano, L. G., Stringer, M. F., and Drasar, B. S.,** Detection of *Clostridium perfringens* enterotoxin by tissue culture and double gel diffusion methods, *J. Med. Microbiol.*, 16, 233, 1983.
119. **Niilo, L.,** Efficacy of laboratory tests for the detection of enterotoxigenic *Clostridium perfringens*, *Can. J. Microbiol.*, 24, 633, 1978.
120. **Niilo, L.,** Enterotoxigenic *Clostridium perfringens* type A isolated from intestinal contents of cattle, sheep and chickens, *Can. J. Comp. Med.*, 42, 357, 1978.
121. **Skjelkvåle, R. and Uemura, T.,** Detection of enterotoxin in faeces and anti-enterotoxin in serum after *Clostridium perfringens* food poisoning, *J. Appl. Bacteriol.*, 42, 355, 1977.
122. **Dowell, V. R., Torres-Anjel, M. J., Riemann, H. P., Merson, M., Whaley, D., and Darland, G.,** A new criterion for implicating *Clostridium perfringens* as the cause of food poisoning, *Rev. Lat. Am. Microbiol.*, 17, 137, 1975.
123. **Naik, H. S. and Duncan, C. L.,** Detection of *Clostridium perfringens* enterotoxin in human feed samples and anti-enterotoxin in sera, *J. Clin. Microbiol.*, 7, 337, 1978.
124. **Itoh, T., Saito, K., Inaba, M., and Sakai, S.,** Detection of *Clostridium perfringens* enterotoxin from feces of food poisoning cases, in *Proc. 14th Joint Conf. US-Jpn. Coop. Med. Sci. Prog. Cholera Panel Symp. Cholera*, Takeay, K. and Zinnaka, Y., Eds., Fiji Printing, Tokyo, 1979, 310.
125. **Uemura, T., Kunitoh, D., and Sakaguchi, G.,** Variability of enterotoxigenicity of *Clostridium perfringens*, in *Proc. 8th Int. Symp. World Assoc. Vet. Food Hyg.*, Collins, J. D. and Hannan, J., Eds., E. T. A. Publications, Ireland, 1981, 155.
126. **Uemura, T., Genigeorgis, C., Riemann, H. P., and Franti, C. E.,** Antibody against *Clostridium perfringens* type A enterotoxin in human sera, *Infect. Immun.*, 9, 470, 1974.
127. **Niilo, L. and Bainborough, A. R.,** A survey of *Clostridium perfringens* enterotoxin antibody in human and animal sera in western Canada, *Can. J. Microbiol.*, 26, 1162, 1980.
128. **Naik, H. S. and Duncan, C. L.,** Enterotoxin formation in foods by *Clostridium perfringens* type A, *J. Food Safety*, 1, 7, 1977.
129. **Craven, S. E., Blankenship, L. C., and McDonel, J. L.,** Relationship of sporulation, enterotoxin formation, and spoilage during growth of *Clostridium perfringens* type A in cooked chicken, *Appl. Environ. Microbiol.*, 41, 1184, 1981.
130. **Skjelkvåle, R., Stringer, M. F., and Smart, J.,** Enterotoxin production by lecithinase-positive and lecithinase-negative *Clostridium perfringens* isolated from food poisoning outbreaks and other sources, *J. Appl. Bacteriol.*, 47, 329, 1979.
131. **Yasukawa, A., Okada, Y., Kitase, T. M., and Miyamoto, S.,** Distribution of enterotoxin-producing strains of *Clostridium perfringens* type A in humans, foods and soils, *J. Food Hyg. Soc. Jpn.*, 16, 313, 1975.
132. **Gyobu, Y.,** Enterotoxin production of *Clostridium perfringens* originated from food-poisoning and characteristics of enterotoxin-producing strains, *J. Food Hyg. Soc. Jpn.*, 18, 375, 1977.
133. **Tsai, C. C., Torres-Anjel, M. J., and Riemann, H. P.,** Characteristics of enterotoxigenic *Clostridium perfringens* type A isolated from cattle and chickens, *J. Formosan Med. Assoc.*, 73, 501, 1974.
134. **Chakrabarty, A. K. and Boro, B. R.,** Prevalence of food poisoning (enterotoxigenic) *Clostridium perfringens* type A in blood and fish meal, *Zbl. Bakt. Hyg., I. Abt. Orig. B.*, 172, 427, 1981.

135. **Niilo, L.,** Mechanism of action of the enteropathogenic factor of *Clostridium perfringens* type A, *Infect. Immun.*, 3, 100, 1971.
136. **McDonel, J. L. and Duncan, C. L.,** Effects of *Clostridium perfringens* enterotoxin on metabolic indexes of everted rat ileal sacs, *Infect. Immun.*, 12, 274, 1975.
137. **McDonel, J. L. and Duncan, C. L.,** Histopathological effect of *Clostridium perfringens* enterotoxin in the rabbit ileum, *Infect. Immun.*, 12, 1214, 1975.
138. **McDonel, J. L., Chang, L. W., Pounds, J. G., and Duncan, C. L.,** The effects of *Clostridium perfringens* enterotoxin on rat and rabbit ileum, *Lab. Invest.*, 39, 210, 1978.
139. **McDonel, J. L. and Duncan, C. L.,** Regional localization of activity of *Clostridium perfringens* type A enterotoxin in the rabbit ileum, jejunum and duodenum, *J. Infect. Dis.*, 136, 661, 1977.
140. **McDonel, J. L. and Duncan, C. L.,** Effect of *Clostridium perfringens* enterotoxin on mitochondrial respiration, *Infect. Immun.*, 15, 999, 1977.
141. **McDonel, J. L. and McClane, B. A.,** Binding versus biological activity of *Clostridium perfringens* enterotoxin in Vero cells, *Biochem. Biophys. Res. Commun.*, 87, 497, 1979.
142. **McDonel, J. L.,** Binding of *Clostridium perfringens* $^{125}$I enterotoxin to rabbit intestinal cells, *Biochemistry*, 21, 4801, 1980.
143. **McClane, B. A. and McDonel, J. L.,** Characterisation of membrane permeability alterations induced in Vero cells by *Clostridium perfringens* enterotoxin, *Biochim. Biophys. Acta*, 600, 974, 1980.
144. **McClane, B. A. and McDonel, J. L.,** Protective effects of osmotic stabilisers on morphological and permeability alterations induced in Vero cells by *Clostridium perfringens* enterotoxin, *Biochim. Biophys. Acta*, 641, 401, 1981.
145. **Skjelkvåle, R., Tolleshang, H., and Jarmund, T.,** Binding of enterotoxin from *Clostridium perfrigens* type A to liver cells *in vivo* and *in vitro*, *Acta Pathol. Microbiol. Scand Sect. B*, 88, 95, 1980.
146. **Miwatani, T., Honda, T., Arai, S., and Nakayama, A.,** Mode of lethal activity of *Clostridium perfringens* type A enterotoxin, *Jpn. J. Med. Sci. Biol.*, 31, 221, 1978.
147. **McDonel, J. L. and Denvers, G. W.,** In vivo effects of enterotoxin from *Clostridium perfringens* type A in the rabbit colon: binding vs. biologic activity, *J. Infect. Dis.*, 145, 490, 1982.
148. **Matsuda, M. and Sugimoto, N.,** Calcium-independent and dependent steps in action of *Clostridium perfringens* enterotoxin on Hela and Vero cells, *Biochem. Biophys. Res. Commun.*, 91, 629, 1979.
149. **Bryan, F. L.,** Factors that contribute to outbreaks of foodborne disease, *J. Food Protect.*, 41, 816, 1978.
150. **Roberts, D.,** Factors contributing to outbreaks of food poisoning in England and Wales, 1970—1979, *J. Hyg.*, 89, 491, 1982.
151. **Thomas, M., Noah, N. D., Male, G. E., Stringer, M. F., Kendall, M., Gilbert, R. J., Jones, P. H., and Phillips, K. D.,** Hospital outbreak of *Clostridium perfringens* food-poisoning, *Lancet*, 1, 1046, 1977.
152. **Hobbs, B. C.,** *Clostridium perfringens* gastroenteritis, in *Food-Borne Infections and Intoxications*, Riemann, H. and Bryan, F. L., Eds., Academic Press, London, 1979, 131.

Chapter 7

# PSEUDOMEMBRANOUS AND ANTIBIOTIC-ASSOCIATED COLITIS

**S. P. Borriello and H. E. Larson**

## TABLE OF CONTENTS

# I. INTRODUCTION

The most extreme form of antibiotic-associated gut disease is pseudomembranous colitis (PMC). This is not a new disease. In fact, early reports of it antedate the antibiotic era by nearly two generations.[1] Initially it was recognized following major GI tract surgery; subsequent observations established a list of medical and toxic conditions which could also accompany it. A prevalent view supported by some pathological and experimental evidence was that vascular damage complicated hypotensive episodes, causing the intestinal mucosal lesions.

From the 1950s onwards, case reports mentioned the disease as a side effect of antibiotic treatment. Antecedent shock was no longer a consistent clinical event and attention turned to a possible infective cause. Staphylococci were incriminated because they often colonized patients receiving antibiotics and because clusters of them appeared in Gram stains of pus and stools from some colitis patients. However, staphylococci also failed to be consistently associated with cases. The introduction of lincomycin and clindamycin into wide clinical use focused interest on an entirely different etiological mechanism.

A number of reports linked the occurrence of PMC uniquely to the administration of clindamycin and lincomycin.[2,3] Several centers described what could fairly be termed epidemics of the disease coincident with a rise in prescription of these antibiotics.[4] However, this association also proved to be inconsistent, for it became evident that large groups of patients had received these drugs without incident. At this point the argument was overtaken by the novel microbiological observations which are described below. These findings have clearly established that PMC is an infection and provide grounds for explaining much of the perplexing epidemiology of the disease. At the same time it can be said that PMC very rarely occurs in persons whose health and medical history is completely normal. It is now possible to find some common threads among all the disparate associated medical and surgical conditions and to define susceptibility to infection in terms of GI microbial ecology.

# II. ETIOLOGY AND PATHOGENESIS

Infection with *Clostridium difficile* is implicated as the specific cause of PMC for the following reasons:

1. Almost every patient with bona fide PMC is found to have *C. difficile* and free toxin in his feces. The data from three studies are given in Table 1. The amount of free toxin in the stool can be very large, sometimes as much as $1 \times 10^6$ tissue culture doses per gram of feces.
2. Completely healthy adults almost never carry *C. difficile* in their stools in so far as the most sensitive culture methods available can determine. Patients in a hospital appear to be colonized asymptomatically at a higher rate than the general population and some individuals may even excrete free toxin for a short period after antibiotic therapy without having diarrhea. These exceptions aside, *C. difficile* is not part of the autochthonous colonic flora and free toxin is not a normal finding in healthy persons.
3. The synergy between antibiotic treatment and *C. difficile* infection can be fully reproduced in an animal model. Clindamycin-treated hamsters develop a fatal enterocecitis if they are given viable *C. difficile* by mouth. Uncontaminated animals remain well. Every type of antibiotic associated with the human disease has been shown to have an action similar to clindamycin in the hamsters.
4. Many patients who have PMC develop antibodies to *C. difficile* toxin, detected using ELISA methods. A few of these patients also develop neutralizing antibodies. In contrast, ELISA antibodies to *C. difficile* toxin occur in only 7% of the general population and neutralizing antibodies are not detected.[5]

**Table 1**
**PREVALENCE OF *C. DIFFICILE* TOXIN IN FECES OF PATIENTS WITH PMC**

| Toxin | | Isolation | | Either | | |
|---|---|---|---|---|---|---|
| Tested | Positive | Tested | Positive | Tested | Positive | Ref. |
| 53 | 38 | 58 | 44 | 58 | 50 | 7 |
| 35 | 29 | 35 | 31 | | | 154 |
| 43 | 42 | | | | | 155 |

**Table 2**
**ANTIBIOTICS ADMINISTERED TO PATIENTS DEVELOPING *C. DIFFICILE* COLITIS**

| | PMC | | All toxin positive cases | |
|---|---|---|---|---|
| Antibiotic[a] | Single | Multiple | Single | Multiple |
| Ampicillin/Amoxycillin | 6 | 13 | 14 | 57 |
| Clindamycin/Lincomycin | 5 | 3 | 9 | 5 |
| Cephalexin | 2 | 1 | 9 | 15 |
| Cephradine | 1 | 4 | 4 | 14 |
| Cefuroxine | | 1 | 2 | 12 |
| Cloxacillin/Flucloxacillin | 1 | 12 | 3 | 43 |
| Penicillin | | 6 | 2 | 21 |
| Cephazolin | | 1 | 1 | 5 |
| Metronidazole | | 6 | 1 | 29 |
| Tetracycline | 1 | 1 | 1 | 3 |
| Cefoxitin | | | 1 | 1 |
| Neomycin | | | 1 | 1 |

[a] Only those antibiotics which were associated with at least one case in single use are listed.

5. Oral treatment of patients with PMC with a nonabsorbed antibiotic specifically active against *C. difficile* results in dramatic clinical improvement as well as the disappearance of both toxin and organism from the stool. Untreated cases show disappearance of the toxin and the organism with clinical recovery.

The above logic has never been satisfied for the other etiologies suggested for PMC such as *Staphylococcus aureus*, hypotension, shock, or vascular occlusion. The role of antibiotics in the development of the colitis is thus adjunctive rather than direct. Many different antibiotics have been associated[6-9] and data on the antibiotics associated with *C. difficile* or its toxin carriage from our own studies over the last 4 years are presented in Table 2. Clindamycin, ampicillin, and the cephalosporins are the antibiotics most frequently associated with disease. However, almost any antibiotic, regardless of its activity against *C. difficile*, can be involved. There are a number of explanations for how an antibiotic to which *C. difficile* is uniformly sensitive might allow colonization. Firstly, the levels reaching the gut may be very low or the antibiotic may be inactivated by the other components of the gut flora. In animals, the administration of ampicillin produces an increase in fecal β-lactamase activity. This results in a substantial decrease in the fecal levels of ampicillin, allowing *C. difficile* to colonize the gut.[10] Thirdly, exposure to and infection with *C. difficile* may take place after cessation of antibiotic treatment but before the normal flora is restored.

There is no correlation between the incidence of PMC and the dose or duration of antibiotic treatment. However, there is evidence to suggest that the disease occurs more frequently with the oral than the parenteral route of administration.[9]

PMC also occurs in patients who have not been given antibiotics; recently, *C. difficile* and its toxin have been detected in these cases as well.[6,11] The mean age of the nonantibiotic-associated cases was 72.[6] Age-associated changes in the bowel flora of older people, similar to those noted for *C. perfringens* carriage[12] may facilitate colonization by *C. difficile*. This may parallel age-associated changes in the gut flora of infants. *C. difficile* can be isolated frequently from infants and both cytotoxin and *C. difficile* enterototoxin detected in the absence of disease.[13-16] A number of infant strains have been examined and most are pathogenic in the hamster model of disease.[17] This implies that the lack of disease in infants or in the occasional asymptomatic adult carrier has to do with the host rather than the organism. It is possible that other, as yet unknown, components of the gut flora must be present for the full expression of pathogenicity or, conversely, that asymptomatic carriers possess sufficient components of their normal flora to suppress pathogenicity.

It has been shown that *C. difficile* can become intimately associated with the gut mucosa in man[14] and in the hamster model.[18] It is possible, therefore, that transient and asymptomatic carriage may be due to exclusion of *C. difficile* from the gut wall by normal components of the mucosal flora. Other factors could facilitate susceptibility to colonization by *C. difficile* such as stress-associated changes in the gut flora,[19] colonization with other pathogens,[20] or diet as shown in the hamster model.[21] Antineoplastic agents have also been shown to be associated with this disease.[22]

An individual's stable microflora is a barrier between the host and potential pathogens. The inhibition expressed against these organisms by the normal flora is known as colonization resistance. It is when the gut flora is disrupted and colonization resistance is lowered that a patient risks infection by *C. difficile*. The importance of colonization resistance can be inferred by the low carriage rate of *C. difficile* in healthy adults, and by the lack of infection in patients who have not had any treatment that is known to interfere with the gut flora. Unless first treated with an antibiotic, the hamster is resistant to colonization with *C. difficile* after an oral challenge of the organism.[23] The growth of *C. difficile* is inhibited in fecal emulsions prepared from the stools of healthy subjects, but no such inhibition is noted in some of the emulsions prepared from subjects with antibiotic-associated diarrhea.[24-26] The fecal emulsions from healthy volunteers did not inhibit if the emulsion was first sterilized by heat or by filtration; this shows that viable bacteria must be present. It would be of obvious interest to identify the minimum components that confer this resistance to colonization. Some work in this respect has already been done. Representative strains of six genera, in particular, lactobacilli and enterococci, inhibit the growth of *C. difficile* in vitro.[27,28] It has also been shown that *C. beijerinckii* will inhibit *C. difficile*, that the inhibition is bactericidal, and that it is almost totally specific for *C. difficile*.[29] There are reports that colonization resistance can be reconstituted by feeding microoganisms.[30]

### A. Other Causes of Antibiotic-Associated Diarrhea (AAD)

Although *C. difficile* can be implicated as the causative organism of PMC and many cases of AAD, there remain a fairly high proportion of patients with diarrhea for which the etiology remains unknown. Some of these cases could be due to an as yet to be discovered pathogen. There is some clinical evidence that these patients have an infection.[31] It is also probable that some are due to enterotoxigenic *C. perfringens*, as we have recently found high fecal levels of both organism and enterotoxin in 12 such patients.

Not all cases of AAD need necessarily be due to the presence of toxigenic or invasive bacteria. It is becoming increasingly evident that what is accepted as normal gut function is the end product of a series of complex interrelationships between the host and his normal gut flora. Any disruption of this flora may lead to alterations in GI secretion, absorption, and motility.[32] The most extreme cases of this are the frequently noted chronic mild diarrhea in germ-free animals and the increased size and hydration of the cecum, which is also a feature of antibiotic-treated animals. It is likely that antibiotic treatment in man may affect the gut flora in such a way that results in an increase in osmotically active compounds (e.g., carbohydrates) and an increase in known secretagogues such as chenodeoxycholic acid that would otherwise be metabolized to inactive or less active compounds.

## III. PATHOLOGY

The chief pathological features of the disease are defined in its name. The pseudomembranes consist of focal, cream to yellow-green colored plaques adhering to intestinal mucosa. These may be only a few millimeters in diameter and just barely visible, or they may be confluent, resulting in a substantial mucosal slough. When PMC was chiefly a postoperative complication, the lesions were more often discovered in the small intestine at post-mortem examination. When PMC is diagnosed as a complication of antibiotic treatment the lesions more often involve the colon and may be recognized at endoscopy. Their distribution can be patchy and failure to observe them in the rectum and sigmoid colon does not guarantee their absence more proximally.[33]

From the histological perspective, the disease is also defined by the presence of the focal plaque.[34] This consists of a spray of fibrin interlacing polymorphonuclear leukocytes which emerge from a site where the mucosal epithelium has become necrotic. Several of the morphological features of the disease suggest a locally acting toxic cause.[35] The small, early lesions indicate that the mucosal surface is the first tissue site affected. Subsequent evolution of the lesion involves progressively deeper layers of the mucosa. Gram stain of the pseudomembranes fails to demonstrate an appreciable accumulation of microorganisms. Thus, the principal and characteristic histological features are entirely consistent with a toxin-mediated etiology.

Recently, Price[36] has described histological changes in cases of antibiotic-associated colitis that lack pseudomembranes. These patients have generally had clinically milder disease and *C. difficile* and toxin have been shown to be present. Although the mucosal cells are not necrotic, their morphology is abnormal. The findings resemble those in colitis caused by *Campylobacter jejuni/coli*.[37] As in other types of colitis, the feces of patients infected with *C. difficile* contain large numbers of polymorphonuclear leukocytes and red blood cells.

## IV. CLINICAL FEATURES AND MANAGEMENT

PMC can be a serious and life-threatening illness. A fulminating complication of surgery, it may mimic intestinal anastomotic dehiscence.[38] Even cases without previous surgery may present as an acute surgical abdomen. The diagnosis is often suggested in a patient with diarrhea, severe abdominal tenderness, and fever, who has recently received antibiotic treatment. The history can be one of troublesome diarrhea that has persisted in spite of symptomatic measures. There may be evidence of weight loss and moderate dehydration. The abdomen may be distended with signs of peritonism. Tenderness is almost invariably present. Some patients experience a more acute onset and the first symptoms noted can be a chill with fever followed by signs suggesting an abdominal catastrophe. Diarrhea may not begin until 24 to 48 hr later. In fact, the severity of diarrhea is very variable, ranging from profuse to none. Studies reporting only patients with moderate or severe diarrhea have probably excluded part of the spectrum of disease.

In the elderly, the presenting findings may be of abdominal distension and ileus; a plain abdominal film shows a gas-distended colon that approaches megacolon proportions. Diarrhea may have been slight or intermittent or patients may even be obstipated. PMC is, in fact, one of the causes of the toxic megacolon syndrome.[39] These represent the most serious cases; others resemble *Campylobacter* and *Salmonella* infections so closely that they cannot be differentiated on purely clinical grounds. Grossly bloody stools are not characteristic of PMC although occult testing is invariably positive. In fact, the acute onset of grossly bloody diarrhea following ampicillin or penicillin treatment is a syndrome usually not associated with the presence of pseudomembranes or the finding of *C. difficile* toxin.[40]

The history of previous antibiotic treatment has to be evaluated carefully. The duration of such therapy does not have to have been lengthy; well-documented cases have followed only a few doses. The onset of symptoms may follow completion of the treatment by up to 3 or 4 weeks and an antibiotic history may thus not be offered readily. Patients who have treated themselves may be reluctant to recall the details. A positive history must be viewed in the light of the clinical findings since only one fifth to one third of patients with simple AAD have evidence of *C. difficile* infection. However, AAD without *C. difficile* is almost invariably short-lived. Of course, PMC continues to occur in the absence of any previous antibiotic treatment; a recent paper noted that half of the patients who present from the community have no such history.[41]

Wider recognition of PMC and employment of toxin testing and *C. difficile* isolation methods has recently highlighted the variety of presentations and clinical settings of this disease. Colitis can occur after only three doses of perioperative prophylactic antibiotics and as a complication of renal and bone marrow transplantation. *C. difficile* bowel infection in neutropenic patients may present with such atypical features as ascites and jaundice.[42] Isolated pseudomembranes have been seen on rectal polyps and on a "blind loop" segment of small bowel.[43] *C. difficile* infection can complicate inflammatory bowel disease; improvement follows treatment directed specifically against the infection.[44,45]

The first investigation to be performed in a patient with severe colitis is often a plain X-ray of the abdomen. Other than finding a hugely dilated colon as noted above, this is most helpful for ascertaining absolute surgical indications such as obstruction or free air. The next investigation is often sigmoidoscopy and this can provide a rapid definitive diagnosis if pseudomembranes are seen; $^3/_4$ of patients have pseudomembranes visible within the terminal 25 cm of the colon.[33]

Other investigations may be helpful in confirming a diagnosis of PMC. The peripheral blood may show a neutrophilic leukocytosis, occasionally to 30 to 40,000 cells per cubic milliliter. The serum urea is often raised, usually out of proportion to the degree of acute dehydration. Sterile pyuria has been reported.[44] If the patients' clinical course has been protracted, the serum albumin level may be reduced. In addition to endoscopy and rectal biopsy, analyses of feces for *C. difficile* and free toxin are the most helpful diagnostic maneuvers. These results, however, have to be interpreted in the light of other clinical findings. Isolation of *C. difficile* in the absence of free toxin or any symptoms can probably be ignored. It appears that hospitalized patients have a higher carrier rate than healthy adults.[46] Young children and infants especially may be asymptomatic excretors of both organism and toxin, although many isolates from infants fail to produce toxin in vitro.[17] It should be borne in mind that PMC, though rare, does occur in this age group. The isolation of *C. difficile* from an adult patient with diarrhea is likely to be of etiological significance, especially if supported by finding free toxin and by confirming that the isolate produced toxin in vitro.

Vancomycin was suggested as effective treatment for PMC in 1966,[47] but the recent findings regarding *C. difficile* explain why this is so and have greatly encouraged its use. Previously, physicians may have been reluctant to use antibiotics as treatment for a condition which antibiotics presumably caused. Vancomycin is administered orally and 125 mg 4 times a day gives good levels in the feces. This treatment significantly shortens the clinical course of the illness by reducing and usually eliminating *C. difficile* and its toxin from the feces.[48] Recently, oral metronidazole has been compared to vancomycin using random patient selection and was also found to give good results.[49] Oral Metronidazole is of course, absorbed, whereas vancomycin is not, but this appears to be only a theoretical disadvantage against the fact that metronidazole is much cheaper. It is likely that a cheaper oral formulation of vancomycin will become available in the near future. Bacitracin has also been used to treat individual patients successfully,[50] but this drug is not available in the U.K. and it has not been studied in a controlled fashion. Occasional strains of *C. difficile* are resistant to bacitracin while resistance to vancomycin and metronidazole has not been reported.

The principal stumbling block in the treatment of PMC is relapse infection. The incidence of relapse after successful treatment is variable but some large series mention it as occurring in 10 to 15% of the cases. Particularly troublesome are patients who experience multiple relapses. Up to five and more episodes have been described. Relapse infections remain responsive to antibiotic therapy and there has never been any evidence that a change in treatment regime is helpful. Vancomycin treatment fails to eliminate *C. difficile* from the stools of some patients, although persistent carriage does not seem to predispose to a clinical relapse. Both Vancomycin and Metronidazole can predispose experimental animals to *C. difficile* enterocecitis and may possibly serve as inciting antibiotics for relapse in the same manner as clindamycin and ampicillin do for original episodes.

Supportive measures such as fluid electrolyte and blood replacement may of course be indicated. There is no evidence that steroids or toxin-binding resins such as cholestyramine offer any benefit. Many patients recover with no specific treatment if the inciting antibiotic is discontinued. If continued antibiotic therapy is necessary, clinical experience now suggests that vancomycin treatment is still likely to be successful. There is always a temptation to change to a different antibiotic in these circumstances, but there is no reason why this should be advantageous. For example, antibiotics active against Gram-positive rods are, as a class, all capable of inciting colitis.

The treatment of patients who cannot take vancomycin or metronidazole by mouth is also difficult. Parenteral use of both vancomycin and metronidazole has resulted in treatment failure. Installation of antibiotic through a nasogastric tube or a cecostomy tube is probably the best one can do. Theoretically, administration of the specific antitoxin ought to be useful under these circumstances, but such a product is not likely to be widely available. Active immunization against *C. difficile* disease also ought to be possible, but proof that antibodies could prevent enteric disease would be needed.

## V. LABORATORY METHODS

### A. Isolation and Identification of *C. difficile*

Isolation remains the most sensitive and reliable means of detecting *C. difficile*. Its culture has been greatly facilitated by the medium developed by George et al.[51] using 500 μg/mℓ of cycloserine and 16 μg/mℓ of cefoxitin as selective agents. Wiley and Bartlett[52] modified this to 250 and 10 μg/mℓ, respectively, and showed that the antibiotics could be successfully used with a routine blood agar base media. These are the concentrations now used in a commercially produced media. The further addition of sodium taurocholate in a final concentration of 0.1% improves growth.[53,54] The selective agar is seeded with a low dilution of the fecal material under investigation and the plates are incubated anaerobically at 37°C for 24 to 48 hr. An alternative to the use of a selective medium is the alcohol treatment method[55] where an aliquot of the fecal dilution is mixed with an equal volume of absolute ethanol and left at room temperature for 1 hr on the bench. The sample can then be used to seed an ordinary blood plate which is incubated under the same conditions as the selective medium. After incubation, the plates are screened for presumptive *C. difficile* colonies, a procedure which is simplified by the use long wave UV light (360 nm). Colonies of *C.*

*difficile* exhibit a characteristic green or chartreuse fluorescence when exposed to this wavelength.[51,55] A number of workers have used the fact that *C. difficile* can produce *p*-cresol from L-tyrosine[56] as a selective and presumptive identification procedure.[57] However, *p*-cresol in agar plates proved to inhibit the growth of *C. difficile* and there is now little to recommend this technique.[51,52]

Identification of presumptive *C. difficile* isolates needs to be confirmed. In practice ability to grow on the selective agar, the characteristic fluorescence and volatile fatty acid profile, and typical cellular and colonial morphology are adequate criteria for identification. Testing isolates for specific cytotoxin production can also be used to assist identification since this is unique in our experience for organisms resembling *C. difficile*. However, not all *C. difficile* isolates produce cytotoxin. If toxin production is the major criterion used for identification, more than one isolate from each patient should be tested; occasionally, both toxigenic and nontoxigenic variants may be carried simultaneously.[58] A number of workers have developed an immunofluorescent technique designed for the rapid screening of stool specimens for the presence of *C. difficile*.[59,60] However, the technique yields a high proportion of false-positive results. A full, detailed review of all the methods available for the detection, isolation, and identification of *C. difficile* can be found elsewhere.[61]

**B. Toxin Detection**

Two physically and biologically distinct toxins are produced by *C. difficile*. Only one is tested for routinely; however, as no strain of *C. difficile* has yet been found which produces only a single type. The available methods for detection of each type have recently been reviewed.[62] Toxin present in stools or produced in vitro by isolates of *C. difficile* can be detected by the use of tissue cultures. In this test, an aliquot of filter-sterilized stool suspension is added to a tissue culture tube containing a confluent monolayer of cells. The inoculated tubes are incubated at 37°C on a tissue culture roller drum and examined at 24 and 48 hr for the presence of a cytopathic effect. Any positive sample can be repeated after neutralization by the cross-reacting *C. sordellii* antitoxin, or by *C. difficile* antitoxin. If large numbers of stool specimens or culture filtrates are to be screened, a micromethod can be used.[63] Not all laboratories have tissue culture facilities and this has stimulated interest in the development of alternative methods for the detection of this toxin.

Various workers have investigated the use of counterimmunoelectrophoresis (CIE) for the detection of cytotoxin.[64-70] The method described by Welch et al.[70] was for the detection of toxin in culture filtrates only. The use of CIE for the detection of toxin in feces was described by Ryan et al.[64] An evaluation of the results from 6 different studies comparing CIE with tissue culture for the detection of *C. difficile* cytotoxin showed that of a total of 842 samples analyzed, 96 were positive by tissue culture, of which only 41 were detected by CIE.[62] In addition, CIE gave positive results for 105 samples which were negative by tissue culture. The remaining 641 samples were negative by both methods. A true evaluation of the usefulness of this technique is difficult as in most cases, cross-reacting *C. sordellii* antitoxin and not *C. difficile* antitoxin was used.

The most recent method described for the detection of toxin is an enzyme immunoassay (EIA) system.[71] The immunoassay was shown to be a rapid and sensitive technique in a comparison of EIA with tissue culture cytotoxicity assay for the detection of *C. difficile* toxin in stool specimens. Of 277 specimens analyzed, 84 samples were positive and 190 negative by both techniques, with 3 specimens positive only by EIA. Of these three specimens, *C. difficile* was isolated from two and the third specimen was from a patient with PMC whose previous stool sample was cytotoxin positive.

Despite the development of these alternatives for the detection of *C. difficile* cytotoxin, they have not been adopted to any large extent in microbiology laboratories. This is largely due to concern about the accuracy of CIE and the large throughput required to make the

**Table 3**
**EXPERIMENTAL ANIMAL MODELS**

| Animal | Ref. |
|---|---|
| Hamster | 13, 23, 73, 75, 76 |
| Guinea pig | 93—96 |
| Mouse | 98 |
| Rat | 100, 101 |
| Rabbit | 102—105 |
| Chinchilla | 92 |

setting up of a specific ELISA method worthwhile. We are currently evaluating a simplified tissue culture system for the detection of cytotoxin that could be easily maintained and used in any routine microbiology laboratory that lacks tissue culture facilities. We feel that the introduction of this type of system would greatly increase the number of centers that could perform the service.

### C. Detection of *C. perfringens* and Its Enterotoxin

Occasionally, cytopathic effects other than that typical for *C. difficile* toxin are produced in tissue cultures. Recently we have identified one type of effect as being due to *C. perfringens* enterotoxin. This toxin is known to produce cytopathic effects in Vero cells.[72] The effects can be neutralized by *C. perfringens* antienterotoxin. The presence of the enterotoxin correlates with high counts of *C. perfringens* in the stools, detected using the selective alcohol method described above. Detection of *C. perfringens* enterotoxin illustrated the specificity some bacterial toxins have for cell cultures. The cytopathic effect of *C. perfringens* enterotoxin in fibroblast cell cultures is very much less distinct than in Vero cells. On the other hand, *C. difficile* produces distinct effects in both cell lines. We currently employ Vero cells to screen all fecal suspensions for toxins.

## VI. EXPERIMENTAL ANIMAL MODELS

Small[73] observed that a single subcutaneous injection of lincomycin often resulted in enterocolitis and death in hamsters. Following the demonstration of a fecal cytotoxin in a patient with antibiotic-associated PMC,[74] there was a reemergence of interest in this model. It was soon shown that clindamycin and other antibiotics induced a similar disease and that the enterocecitis was caused by *C. difficile*.[13,23,75-77] Although there are now other animal models of antibiotic-associated colitis (Table 3), the hamster model remains the most extensively studied.

In order to induce disease in the hamster, at least two events are required: firstly, disruption of the normal gut flora, e.g., by antibiotic administration, and secondly, exposure to *C. difficile*. It has been well documented that neither administration of organism alone nor antibiotic alone, if the animals are housed in isolators, is sufficient to cause disease.[23] If animals are not housed in isolators, then administration of antibiotic alone may induce disease — the number of animals affected depending on the degree of exposure to *C. difficile* in their immediate environment. The more frequently a room is used for experiments, the higher the number of deaths obtained until the situation may be reached where all animals receiving antibiotic die with enterocolitis. The only way to prevent uncontrolled exposures is to house the animals in individual isolators. These measures are essential if nonpathogenic or less pathogenic strains of *C. difficile* are being compared, especially as colonization with nonpathogenic strains has been shown to protect animals against disease.[18,24,78,79] Only barrier-maintained animals are defined, controlled, and reproducible.

This model has been used to obtain information about the epidemiology and treatment of the disease. Vancomycin[80,81] and passive and active immunization[82] have all been shown to be protective. One of the most recent uses of the hamster model has been as a tool for research into effective bacterioprophylaxis. Administration of feces from normal hamsters via the orograstric route or by enema to antibiotic-treated hamsters protected them from development of enterocolitis.[83] This protection has also been achieved by administration of a lactobacillus preparation[84] or by cultures of nonpathogenic clostridia.[18,24,78,79] It is hoped that experiments of this sort will enable workers to define the minimum number of components needed to reestablish colonization resistance to *C. difficile* and yield clues for effective bacterioprophylaxis in man.

The model has been less useful for the study of the pathogenesis of the lesions as there are a number of major differences between those seen in man and those in the hamster.[85-87] Unlike human PMC, hamsters do not regularly develop the classical yellow plaques. Also, the disease is restricted almost entirely to the cecum. Morphologically, the comparative microscopy has only limited similarities.[87] However, the changes present in early disease in the hamster are similar to the minor mucosal abnormalities present in antibiotic-associated colitis in man,[88] as are those present in hamsters challenged with a pathogenic strain of *C. difficile* after protection with a nonpathogenic strain.[89] Despite the limitations of this model, it greatly aided the isolation and identification of the organism responsible for toxin production in the human disease.[18,75,90,91]

A variety of animal models other than the hamster have been used (Table 3). Chinchillas developed a pseudomembranous enterocolitis associated with chlortetracycline in food.[92] The disease was credited to overgrowth of *Staphylococcus aureus*. However, disease in the colony stopped abruptly following administration of neomycin and bacitracin, despite the fact that *S. aureus* was not eliminated from the stools of all animals.

A variety of antibiotics have been shown to induce colitis in guinea pigs and *C. difficile* and its toxin are implicated.[93-97] This model has not been studied extensively. Mice are variably colonized by *C. difficile*, with Dabard and colleagues[98] reporting no effect in young mice despite the presence of the organism and its toxin in the intestines. Onderdonk et al.[99] showed development of diarrhea and intestinal inflammation in gnotobiotic mice, though deaths were rare. A similar pattern of a low incidence of death (17%) was noted in germ-free rats,[100] which may prove to be a useful model to the pathologist as there is widespread colonic involvement. However, it must be noted that the ulceration seen in the rats is not quite typical of human PMC. Monoassociation of gnotobiotic rats with *C. difficile* by Gustafsson et al.[101] resulted in death very rarely. These results may have been due to the use of a poorly pathogenic strain of *C. difficile*.

Antibiotic administration to rabbits can lead to diarrhea, colitis, and death. Although *C. difficile* has been shown to cause colitis in a rabbit receiving penicillin and ampicillin,[102] in most cases the disease is associated with the presence of a toxin that can be neutralized by *C. perfringens* anti-iota toxin.[103,104] It is now known that *C. spiroforme* produces this toxin and is the cause of the disease.[105,106] Although there is no known natural disease due to *C. difficile* in rabbits, this organism has been associated with diarrhea in young conventional hares and has been used to induce disease in axenic hares.[98]

Many animals will develop GI disease after exposure to antibiotics and in many cases the cause is due to clostridia, with a variety of different species being implicated (see Chapter 10). It is also evident that *C. difficile* can cause disease in a variety of animals and that although much has been learned from the use of these models, there is still a need to find an animal model in which a more exact replica of the colonic disease seen in man may be produced.

## VII. *C. DIFFICILE* TOXINS

At the time of its earliest description, *C. difficile* was considered to be pathogenic because

### Table 4
### COMPARISON OF THE TWO *C. DIFFICILE* TOXINS

| | Enterotoxin | | Cytotoxin |
|---|---|---|---|
| Molecular weight | 440,000—600,000 | | 107,000—550,000 |
| Heat labile | + | | + |
| Estimated isoelectric point | 5.4—5.6 | | 5.0—6.2 |
| Inactivation by acid | − | | + |
| Inactivation by alkali | − | | + |
| Inactivation by proteolytic enzymes | + | | + |
| Hydrophobic | + | | + |
| Cytotoxic | + | << | + |
| Positive fluid response in ligated intestinal loops | + | | − |
| Positive fluid response in infant mice | $+^-$ | | $+^-$ |
| Lethality on intracecal injection (hamster) | + | | + |
| Lethality on i.p. injection (mice) | + | | $+^-$ |
| Increases vascular permeability | + | | + |
| Neutralization by *C. sordellii* antitoxin | + | | + |
| Stimulation of adenylate cyclase | − | | − |
| Stimulation of guanylate cyclase | + | | + |

*Note:* $+^-$: some workers found a negative result for this test.

it produced a potent exotoxin.[107] Injections of cell-free filtrates into guinea pigs and rabbits resulted in edema, respiratory arrest, and death. The production of a toxin by *C. difficile* was confirmed in 1937 by Snyder[108] who showed that broth filtrates of spent culture medium were also lethal to the cat, rat, dog, and pigeon. However, it was not until the demonstration of the role of *C. difficile* in antibiotic-associated GI disease that any detailed work on the toxins produced was performed. Much of the early work on crude culture filtrates or partially purified toxin is now difficult to interpret as it is known that at least two distinct toxins are produced. The presence of a second toxin was first demonstrated by Taylor and Bartlett[109] and Banno et al.[110] After ion-exchange chromatography of the toxin preparations, cytotoxic fractions failed to cause fluid accumulation in the ligated ileal loop assay. Another fraction eluted earlier did have this property. A comprehensive review of purification procedures and properties of the two toxins is to be found elsewhere.[111] A comparison of the different properties of these two toxins is presented in Table 4.

## A. *C. difficile* Enterotoxin

This heat-labile toxin is referred to as toxin A because it elutes prior to the cytotoxin (toxin B) on ion-exchange chromatography. It is a large, extracellular molecule with an estimated molecular weight of between 440,000 and 600,000. An isoelectric point of 5.4 to 5.6 has been reported; however, it is difficult to achieve a sharp zone in isoelectric focusing gels as the toxin precipitates in the pH range of 5 to 5.5.[111] This toxin retains activity after exposure to acid (pH 4) or alkaline (pH 9) conditions[112-114] but loses activity on reaction with proteolytic enzymes.[110,113,114] The toxin appears to be hydrophobic as it can only be eluted from hydrophobic gels with high concentrations of ethylene glycol.[114]

Purified toxin A is much less cytotoxic than toxin B.[114] It is unlikely that this low level of activity is due to contamination because the enterotoxin cytotoxicity is not neutralized by antiserum to toxin B.[111,114] The fluid accumulation in ligated rabbit ileal loops is accompanied by hemorrhage and necrosis in the intestinal mucosa.[110] These changes are rather different from the classical use of this system to demonstrate the action of enterotoxins from

*Vibrio cholera* and *Escherichia coli*. *C. difficile* filtrates and purified enterotoxin also produce a positive fluid response in infant mice after intragastric challenge,[116,117] although this has not been confirmed by other workers.[115] Intracecal injection of enterotoxin in the hamster causes necrosis of the cecal mucosa with acute inflammation, congestion, moderate hemorrhage, fluid accumulation, and death.[118] Injection of nanogram amounts into the skin of rabbits produces erythematous and hemorrhagic lesions and increases vascular permeability.[110,115,117] Intraperitoneal injection into mice causes death[110,114,115] though the lethal dose in mice ranged from 10 ng to 5 μg.

### B. *C. Difficile* Cytotoxin

This heat-labile toxin (toxin B) is a large extracellular molecule with an estimated molecular weight of between 107,000 to 550,000. It has recently been shown that a homogeneous cytotoxin preparation contained 5 immunologically related cytotoxic proteins with estimated molecular weights of 70,000, 160,000, 250,000, 280,000, and 340,000.[111] The isoelectric point has been reported as 6.2[119] and 5.0,[113] although other workers think that the true value is much lower than this.[111] The cytotoxin is inactivated after exposure to acid (pH 4) or alkaline (pH 9) conditions,[112-114] or after reaction with proteolytic enzymes.[13,110,113,114] It has been reported that the toxin is inactivated by large amounts of amylase,[113] leading to the suggestion that cytotoxin may be a glycoprotein. However, others have not found this,[114] showing that the cytotoxin does not bind to gels which bind most glycoproteins like concanavalin A-sepharose and glycoboronate. The toxin is not inactivated by RNase, DNase, or lipase, but may be inactivated by several heavy-metal compounds.[112,113] The cytotoxin, like the enterotoxin, appears to be relatively hydrophobic.[114]

The cytotoxin is so called as it is considerably more active with respect to cytotoxicity than the enterotoxin in tissue culture. Although the toxin will not cause a positive fluid response in the ligated ileal loop assay, it does elicit a positive response in infant mice.[112] However, Taylor et al.[115] found the cytotoxin to be negative in this assay also. Intracecal injection of partially purified cytotoxin in hamsters is lethal and causes severe superficial inflammation with some submucosal congestion and hemorrhage.[118] In addition, it has been shown that intraperitoneal injection into mice causes death,[114,115] though the amounts of toxin required by the two groups to achieve this differed. It is interesting to note that Banno et al.[110] found their cytotoxin preparation to be nonlethal in this assay. The cytotoxin will also cause an erythematous, hemorrhagic lesion, and increase vascular permeability when injected into the skin of rabbits.[110,115,117]

### C. Other Toxins

A high molecular weight protein produced by *C. difficile* has recently been reported which causes changes in the electrical potential in isolated segments of rabbit intestine.[120] This is distinct from cytotoxin and enterotoxin and may act as a "motility — altering factor".

### D. Mechanisms of Action

Much of the early work on mechanisms of action was performed with crude toxin preparations. Most of the subsequent work with purified toxins has been done with the cytotoxin and very little is known about the enterotoxin. To date, no report on the mechanism for the cytopathic effect of the enterotoxin has been published. It is known that neither toxin stimulates adenylate cyclase,[110,121] but that both will stimulate colonic guanylate cyclase.[111,121] The relevance of these observations in the disease process is unknown.

The cytotoxin causes a cytopathic effect in all of the cell lines so far tested. In addition, it has been shown to inhibit growth of suspensions of human cervic carcinoma (HeLa), mouse adrenocortical tumor (YI), ascites sarcoma, and human lymphoblastoid cells.[122,123] The toxin binds to receptors on the cell surface which have yet to be identified, though it seems likely to be a lipid receptor. The toxin does not appear to be membrane-damaging, and once bound has to be internalized. Evidence to date indicates that internalization takes place by endocytosis.[123] It is not known what the primary intracellular target is, although guanylate cyclase would be a good candidate since purified cytotoxin has been shown to activate it. Microfilaments are also a likely primary target. Both crude toxin and purified toxin disrupt the microfilament bundles in fibroblasts.[123,124] Cytochalasins, known to interfere with the microfilaments, produce similar effects. In addition, Wedel et al.[125] have shown that purified toxin induces disorganization of the actin-containing myofilament bundles in rabbit smooth muscle cells. The interested reader is directed to a recently published full and comprehensive review of the information available on the mechanism of action of the cytotoxin in cultured cells.[123]

## VIII. EPIDEMIOLOGY

In early work on PMC, toxin production was thought to result from the overgrowth of endogenous clindamycin-resistant strains of *C. difficile*. This hypothesis was consistent with the understanding that *C. difficile* was a common fecal organism in infants and the supposition that this was also true of adults. However, it did not explain some of the unusual epidemiological features of the clinical disease. Subsequent observations now allow a more complete picture of how this organism finds its way into the GI tract and causes disease.

### A. Endogenous Carriage

*C. difficile* is commonly carried in healthy infants and cytotoxin and enterotoxin may be present as well.[13,14,17,107,110,126,127] The prevalence is very variable and high rates are due to cross-infection within a ward rather than acquisition of the organism from any maternal source.[17] Although in the majority of cases carriage is asymptomatic, this bacterium may not always be a harmless commensal in the pediatric population.[128-130] In contrast, fecal carriage of *C. difficile* in adults is uncommon. This conclusion is based on studies of healthy European[13,14,131] and North American adults[51,132] and Indians.[133] However, Nakamura and colleagues[134] have shown a much higher carriage rate in healthy Japanese adults with isolation rates of 15.4% in healthy young adults in their 20s, 7% in healthy elderly adults, and 15.9% in elderly adults with cerebrovascular disease. A carriage rate of 12% has recently been shown in a hospitalized geriatric population where no antibiotics had been received for at least 8 weeks prior to sampling.[135] These findings imply age-associated differences in carriage and the possibility that there may be differences in carriage among different ethnic populations. Overall, the findings argue that *C. difficile* is not typically a permanent resident of the human GI tract, but that carriage occurs under special conditions such as the extremes of age and during hospitalization.

### B. Exogenous Infection

A number of clinical reports prior to 1977 described temporal and geographical case clustering that suggested an infectious agent.[4,136] Since the discovery of *C. difficile* as an etiological agent, there has been increasing circumstantial evidence of cross-infection based on the occurrence of outbreaks, case clustering, and asymptomatic infant carriage.[17,132,137-147] Although a common effective typing scheme for *C. difficile* has yet to be developed, a number of studies have shown evidence for nosocomial transmission based on distinctive antimicrobial resistance patterns[137,139] and the use of agarose gel, polyacrylamide gel electrophoresis, and crossed immunoelectrophoresis techniques.[140,148,149] In addition, the recent isolation of phage from *C. difficile* has allowed an attempt at phage typing to be made. Use of this system showed that isolates from 16 patients taken 1 to 14 days apart were similar in their phage sensitivity patterns, whereas isolates from separate geographic

**Table 5**
**ISOLATION OF *C. DIFFICILE* FROM HOSPITAL ENVIRONMENTS**

| Group | Positive case environments | Control environments | Ref. |
|---|---|---|---|
| Children | +[a] | –[b] | 137 |
| Geriatrics | + | – | 158 |
| Infants | 79/415 | 66/1050 | 140 |
| Infants | 5/201 | 0/241 | 17 |
| Infants | + | – | 127 |
| Adults and infants | 13/319 | 1/78 | 139 |
| Adults and infants | 111/1159 | 13/620 | 152 |
| Adults | 9/120 | – | 157 |
| Adults | 37/114 | 6/445 | 151 |
| Total | 254/2328 | 86/2534 | |
| % Positive | 10.9 | 3.4 | |

[a] +, *C. difficile* isolated but number of sites not given.
[b] –, No information available.

**Table 6**
**ISOLATION OF *C. DIFFICILE* FROM SITES OTHER THAN HOSPITALS**

| Site | Ref. |
|---|---|
| Soil, sand, and mud | 57 |
| Animal dung (camel, cow, donkey, horse, and pig) | 158—160 |
| Household pets | 153 |
| Neonatal hare | 98 |
| Laboratory animals (rat, mouse, and hamster) | 161, see Chapter 10 |
| Weddel seal | 162 |
| Surfaces in a veterinary clinic | 153 |
| Microbiology laboratory | 163 |
| Surfaces in animal holding room | 163 |
| Patients' homes | 152 |

locations showed a great variety of patterns.[150] Furthermore, experiments with the hamster model have shown that exposure to *C. difficile* is an important prerequisite for induction of disease after antibiotic administration.[23]

There is as yet no good evidence on the methods of transmission of this organism, although recent work in a study of carriage in infants yielded evidence for both person-to-person transmission as well as common source spread.[17] There is evidence for common source spread in adults, with good evidence of transmission of disease from one patient to another via a shared commode chair.[147] Despite the inability to specifically identify incriminating fomites, there is a great deal of evidence to show that *C. difficile* can contaminate and survive on surfaces in hospitals (Table 5) as well as in the homes of patients (Table 6). *C. difficile* can be detected on room surfaces up to 40 days after a patient with *C. difficile* diarrhea has vacated it.[151] In addition, the organism has been isolated from the hands of nursing personnel.[127,139,152] As expected, the organism can be isolated much more frequently from the environment of patients who are excreting *C. difficile* than from other areas (Table 5).

Not all infections occur in the hospital setting, and attempts have been made to identify possible sources of infection in the community (Table 6). In two of the more interesting studies it was shown that household pets can carry the organism[153] and that *C. difficile* could be readily isolated from the homes of two patients known to be excreting the organism.[152]

Overall, the evidence is in favor of the disease being a true infection in the majority of cases. However, this does not exclude the possibility that some cases are due to overgrowth of small numbers of *C. difficile* resident transiently or permanently in the gut, and that these types of patients can contribute to infection of susceptible patients. Because of this evidence, it would seem prudent to adopt enteric isolation precautions for patients with the disease.

## REFERENCES


1. **Finney, J. M. T.,** Gastroenterostomy for cicatrizing ulcer of the pylorus, *Bull. John Hopkins Hosp.*, 45, 53, 1893.
2. **Scott, A. J., Nicholson, G. I., and Kerr, A. R.,** Lincomycin as a cause of pseudomembranous colitis, *Lancet*, 2, 1232, 1973.
3. **Tedesco, F. J., Barton, R. W., and Alpers, D. H.,** Clindamycin-associated colitis: a prospective study, *Ann. Intern. Med.*, 81, 429, 1974.
4. **Kabins, S. A. and Spira, T. J.,** Outbreak of clindamycin-associated colitis, *Ann. Intern. Med.*, 83, 830, 1975.
5. **Aronsson, A., Granstrom, M., Mollby, R., and Nord, C. E.,** Enzyme-linked immunosorbent assay (ELISA) for antibodies to *Clostridium difficile* toxins in patients with pseudomembranous colitis and antibiotic-associated diarrhoea, *J. Immun. Methods*, 60, 341, 1983.
6. **Borriello, S. P. and Larson, H. E.,** Antibiotics and pseudomembranous colitis, *J. Antimicrob. Chemother.*, 7(Suppl. A), 53, 1981.
7. **Bartlett, J. G.,** Antimicrobial agents implicated in *Clostridium difficile* toxin-associated diarrhoea or colitis, *Johns Hopkins Med. J.*, 149, 6, 1981.
8. **Larson, H. E., Borriello, S. P., Honour, P., and Welch, A.,** *Clostridium difficile* and its toxin in human disease, in *Bacterial Protein Toxins*, Jeljaszewicz, J. and Alouf, J., Eds., Academic Press, London, 1984, 339.
9. **Tedesco, F. J.,** Antibiotics associated with *Clostridium difficile* mediated diarrhoea and/or colitis, in *Antibiotic Associated Diarrhoea and Colitis: Clostridium difficile in Gastrointestinal Disorders*, Borriello, S. P., Ed., Martinus Nijhoff, The Hague, 1984, 3.
10. **Rolfe, R. D. and Finegold, S. M.,** Intestinal β-lactamase activity in ampicillin-induced, *Clostridium difficile*-associated ileocecitis, *J. Infect. Dis.*, 147, 227, 1983.
11. **Wald, A., Mendelow, H., and Bartlett, J. G.,** Non-antibiotic associated pseudomembranous colitis due to toxin producing clostridia, *Ann. Intern. Med.*, 92, 798, 1980.
12. **Yamagishi, T., Serikawa, T., Morita, R., Nakamura, S., and Nishida, S.,** Persistent high numbers of *Clostridium perfringens* in the intestines of Japanese aged adults, *Jpn. J. Microbiol.*, 20, 397, 1976.
13. **Larson, H. E., Price, A. B., Honour, P., and Borriello, S. P.,** *Clostridium difficile* and the aetiology of pseudomembranous colitis, *Lancet*, 1, 1063, 1978.
14. **Borriello, S. P.,** *Clostridium difficile* and its toxin in the gastrointestinal tract in health and disease, *Res. Clin. Forums*, 1, 33, 1979.
15. **Smith, M. F., Borriello, S. P., Claydon, G. S., and Casewell, M. W.,** Clinical and bacteriological findings in necrotising enterocolitis: a controlled study, *J. Infect.*, 2, 23, 1980.
16. **Libby, J. M., Donta, S. T., and Wilkins, T. D.,** *Clostridium difficile* toxin A in infants, *J. Infect. Dis.*, 148, 606, 1983.
17. **Larson, H. E., Barclay, F. E., Honour, P., and Hill, I. D.,** Epidemiology of *Clostridium difficile* in infants, *J. Infect. Dis.*, 146, 727, 1982.
18. **Borriello, S. P.,** *Clostridium difficile* and gut disease, in *Microbes and Infections of the Gut*, Goodwin, C. S., Ed., Blackwell, Oxford, 1984, 327.
19. **Holdeman, L. V., Good, I. J., and Moore, W. E. C.,** Human faecal flora: variation in bacterial composition within individuals and a possible effect of emotional stress, *App. Environ. Microbiol.*, 31, 359, 1976.
20. **Falsen, E., Kaijser, B., Nehls, L., Nygren, B., and Svedhem, A.,** *Clostridium difficile* in relation to enteric bacterial pathogens, *J. Clin. Microbiol.*, 12, 297, 1980.

21. **Michelich, V. J., Nunez-Montiel, O., Schuster, G. S., Thompson, F., and Dowell, V. R., Jr.,** Diet as a coadjuvant for development of antibiotic-associated diarrhoea in hamsters *(Mesocricetus auratus)*, *Lab. Anim. Sci.*, 31, 259, 1981.
22. **Fainstein, V., Bodey, G. P., and Fekety, R.,** Relapsing pseudomembranous colitis associated with cancer chemotherapy, *J. Infect. Dis.*, 143, 865, 1981.
23. **Larson, H. E., Price, A. B., and Borriello, S. P.,** Epidemiology of experimental enterocecitis due to *Clostridium difficile*, *J. Infect. Dis.*, 143, 408, 1980.
24. **Borriello, S. P., Larson, H. E., Honour, P., and Barclay, F. E.,** Antibiotic-associated diarrhoea and colitis, in *Clinical Research Centre Report*, Medical Research Council, London, 1982, 96.
25. **Borriello, S. P., Barclay, F. E., and Welch, A. R.,** Colonization resistance of the gastrointestinal tract — a potential *in vitro* method for the determination of susceptibility to infection with *C. difficile*, in *Models of Anaerobic Infection*, Hill, M. J., Ed., Martinus Nijhoff, The Hague, 1984, 241.
26. **Barclay, F. E. and Borriello, S. P.,** *Clostridium difficile* and colonization resistance, in *Antibiotic Associated Diarrhoea and Colitis: Clostridium difficile in Gastrointestinal Disorders*, Borriello, S. P., Ed., Martinus Nijhoff, The Hague, 1984, 79.
27. **Rolfe, R. D., Halebian, S., and Finegold, S. M.,** Bacterial interference between *Clostridium difficile* and normal fecal flora, *J. Infect. Dis.*, 143, 470, 1981.
28. **Malamu-Ladas, H. and Tabaqchall, S.,** Inhibition by *Clostridium difficile* of faecal Streptococci, *J. Med. Microbiol.*, 15, 569, 1982.
29. **Barclay, F. E. and Borriello, S. P.,** *In vitro* inhibition of *C. difficile*, *Eur. J. Chemother. Antibiot.*, 2, 155, 1982.
30. **Bowden, T. A., Mansberger, A. R., and Lykins, L. E.,** Pseudomembranous enterocolitis: a mechanism of restoring floral homeostasis, *Am. Surg.*, 47, 178, 1981.
31. **Jewkes, J., Larson, H. E., Price, A. B., Sanderson, P. J., and Davies, H. A.,** The aetiology of acute diarrhoea in adults, *Gut*, 22, 388, 1981.
32. **Borriello, S. P.,** Bacteria and gastrointestinal secretion and motility, *Scand. J. Gastroenterol.*, 19 (Suppl. 93), 115, 1984.
33. **Tedesco, F. J., Corless, J. K., and Brownstein, R. E.,** Rectal sparing in antibiotic-associated pseudomembranous colitis: a prospective study, *Gastroenterology*, 83, 1259, 1982.
34. **Price, A. B. and Davies, D. R.,** Pseudomembranous colitis, *J. Clin. Pathol.*, 30, 1, 1977.
35. **Goulston, S. J. M. and McGovern, V. J.,** Pseudomembranous colitis, *Gut*, 6, 207, 1965.
36. **Price, A. B.,** Pseudomembranous colitis, in *Recent Advances in G.I. Pathology*, Wright, R., Ed., W. B. Saunders, London, 1980, 160.
37. **Price, A. B., Jewkes, J., and Sanderson, P. J.,** Acute diarrhoea: campylobacter colitis and the role of rectal biopsy, *J. Clin. Pathol.*, 32, 990, 1979.
38. **Kappas, A., Shinagawa, N., Arabi, Y., Thompson, H., Burdon, D. W., Dimock, F., George, R. H., Alexander-Williams, J., and Keighley, M. R. B.,** Diagnosis of pseudomembranous colitis, *Br. Med. J.*, 1, 675, 1978.
39. **Cone, J. B. and Wetzel, W.,** Toxic megacolon secondary to pseudomembranous colitis, *Dis. Colon Rectum*, 25, 478, 1982.
40. **Larson, H. E.,** Adverse effects of antibiotics on the bowel, in *Present Advances in Infection*, Vol. 2, Reeves, D. S. and Geddes, A. M., Eds., Churchill Livingstone, London, 1982, 51.
41. **Brettle, R. P., Poxton, I. R., Murdoch, J. M., Brown, R., Byrne, M. D., and Collee, J. G.,** *Clostridium difficile* in association with sporadic diarrhoea, *Br. Med. J.*, 284, 230, 1982.
42. **Rampling, A., Warren, R. E., Berry, P. J., Swirsy, D., Hoggarth, C. E., and Bevan, P. C.,** Atypical *Clostridium difficile* colitis in neutropenic patients, *Lancet*, 2, 162, 1982.
43. **Goulston, S. J. M. and McGovern, V. J.,** Clinical settings in pseudomembranous colitis, *Aust. N.Z. J. Med.*, 10, 139, 1980.
44. **McKinley, M. J., Troncale, F., Sangree, M. H., Scholhamer, C., and Brand, M.,** Antibiotic-associated colitis: clinical and epidemiological features, *Am. J. Gastroenterol.*, 77, 77, 1982.
45. **La Mont, J. T. and Trnka, Y. M.,** Therapeutic implications of *Clostridium difficile* toxin during relapese of chronic inflammatory bowel disease, *Lancet*, 1, 381, 1980.
46. **Varki, N. M. and Aquino, T. I.,** Isolation of *Clostridium difficile* from hospitalized patients without antibiotic-associated diarrhea or colitis, *J. Clin. Microbiol.*, 16, 659, 1982.
47. **Khan, M. Y. and Hall, W. M.,** Staphylococcal enterocolitis: treatment with oral vancomycin, *Ann. Intern. Med.*, 65, 1, 1966.
48. **Keighley, M. R. B., Burdon, D. W., Arabi, Y., Alexander-Williams, J., Thompson, H., Youngs, D., Johnson, M., Bentley, S., George, R. H., and Mogg, G. A. G.,** Randomised controlled trial of vancomycin for pseudomembranous colitis and post-operative diarrhoea, *Br. Med. J.*, 2, 1667, 1978.
49. **Teasley, D. G., Gerding, D. N., Olson, M. M., Peterson, L. R., Gebhard, R. L., Schwartz, M. J., and Lee, J. T., Jr.,** Prospective randomised trial of metronidazole versus vancomycin for *Clostridium-difficile*-associated diarrhoea and colitis, *Lancet*, 2, 1043, 1983.

50. **Chang, T-W., Gorbach, S. L., Bartlett, J. G., and Saginur, R.,** Bacitracin in antibiotic-associated colitis and diarrhoea caused by *Clostridium difficile* toxin, *Gastroenterology*, 78, 1584, 1980.
51. **George, W. L., Sutter, V. L., Citron, D., and Finegold, S. M.,** Selective and differential medium for isolation of *Clostridium difficile*, *J. Clin. Microbiol.*, 9, 214, 1979.
52. **Willey, S. and Bartlett, J. G.,** Cultures for *Clostridium difficile* in stools containing a cytotoxin neutralized by *Clostridium sordellii* antitoxin, *J. Clin. Microbiol.*, 10, 880, 1979.
53. **Wilson, K. H., Kennedy, M. J., and Fekety, F. R.,** Use of sodium taurocholate to enhance spore recovery on a medium selective for *Clostridium difficile*, *J. Clin. Microbiol.*, 15, 443, 1982.
54. **Raibaund, P., Ducluzeau, R., Dubos, F., Hudault, S., Bewa, H., and Muller, M. C.,** Implantation of bacteria from the digestive tract of man and various animals into gnotobiotic mice, *Am. J. Clin. Nutr.*, 33, 2440, 1980.
55. **Borriello, S. P. and Honour, P.,** Simplified procedure for the routine isolation of *Clostridium difficile* from faeces, *J. Clin. Pathol.*, 34, 1124, 1981.
56. **Elsden, S. R., Hilton, M. G., and Waller, J. M.,** The end products of the metabolism of aromatic amino acids by Clostridia, *Arch. Microbiol.*, 107, 283, 1976.
57. **Hafiz, S. and Oakley, C. L.,** *Clostridium difficile:* isolation and characteristics, *J. Med. Microbiol.*, 9, 129, 1976.
58. **Borriello, S. P. and Honour, P.,** Concomitance of cytotoxigenic and noncytotoxigenic *Clostridium difficile* in stool specimens, *J. Clin. Microbiol.*, 18, 1006, 1983.
59. **Wilson, K. H., Silva, J., and Fekety, F. R.,** Fluorescent-antibody test for detection of *Clostridium difficile* in stool specimens, *J. Clin. Microbiol.*, 16, 464, 1982.
60. **Hubert, J., Ionesco, H., and Sebald, M.,** Detection de *Clostridium difficile* par isolement sur milieu minimal sélectif et pas immuno fluorescence, *Ann. Microbiol. (Institute Pasteur)*, 132A, 149, 1981.
61. **Borriello, S. P. and Honour, P.,** Detection, isolation and identification of *Clostridium difficile*, in *Antibiotic Associated Diarrhoea and Colitis: Clostridium difficile in Gastrointestinal Disorders*, Borriello, S. P., Ed., Martinus Nijhoff, The Hague, 1984, 37.
62. **Borriello, S. P. and Welch, A. T.,** Detection of *Clostridium difficile* toxins, in *Antibiotic Associated Diarrhoea and Colitis: Clostridium difficile in Gastrointestinal Disorders*, Borriello, S. P., Ed., Martinus Nijhoff, The Hague, 1984, 49.
63. **Borriello, S. P.,** An evaluated micromethod for the detection of *Clostridium difficile* enterotoxin, *Microbios. Lett.*, 7, 25, 1978.
64. **Ryan, R. W., Kwasnik, I., and Tilton, R. C.,** Rapid detection of *Clostridium difficile* toxin in human feces, *J. Clin. Microbiol.*, 12, 776, 1980.
65. **Levine, H. G., Kennedy, M., and LaMont, J. T.,** Counterimmuno electrophoresis vs. cytotoxicity assay for the detection of *Clostridium difficile* toxin, *J. Infect. Dis.*, 145, 398, 1982.
66. **Wu, T. C. and Fung, J. C.,** Evaluation of the usefulness of counterimmunoelectrophoresis for diagnosis of *Clostridium difficile*- associated colitis in clinical specimens, *J. Clin. Microbiol.*, 17, 610, 1983.
67. **Jarvis, W., Nunez-Monteil, O., Thompson, F., Dowell, V., Towns, M., Morris, G., and Hill, E.,** Comparison of bacterial isolation, cytotoxicity assay, and counterimmunoelectrophoresis for the detection of *Clostridium difficile* and its toxin, *J. Infect. Dis.*, 147, 778, 1983.
68. **Kurzynski, T. A., Cembrowski, G. S., and Kimball, J. L.,** The use of CIE for the detection of *Clostridium difficile* toxin in stool filtrates: laboratory and clinical correlation, *Am. J. Clin. Pathol.*, 79, 370, 1983.
69. **Sands, M., Yungbluth, M., and Sommers, H. M.,** The non-value of counterimmunoelectrophoresis for the direct rapid detection of *Clostridium difficile* toxin in stool filtrates, *Am. J. Clin. Pathol.*, 79, 375, 1983.
70. **Welch D. F., Menge, S. K., and Matsen, J. M.,** Identification of toxigenic *Clostridium difficile* by counterimmunoelectrophoresis, *J. Clin. Microbiol.*, 11, 470, 1980.
71. **Yolken, R. H., Whitcomb, L. S., Marien, G., Bartlett, J. G., Libby, J., Ehrich, M., and Wilkins, T.,** Enzyme immunoassay for the detection of *Clostridium difficile* antigen, *J. Infect. Dis.*, 144, 378, 1981.
72. **Giugliano, L. G., Stringer, M. F., and Drasar, B. S.,** Detection of *Clostridium perfringens* enterotoxin by tissue culture and double-gel diffusion methods, *J. Med. Microbiol.*, 16, 233, 1983.
73. **Small, J. D.,** Fatal enterocolitis in hamsters given lincomycin hydrochloride *Lab. Anim. Care*, 18, 411, 1968.
74. **Larson, H. E., Parry, J. V., Price, A. B., Davies, P. R., Dolby, J., and Tyrrell, D. A. J.,** Undescribed toxin in pseudomembranous colitis, *Br. Med. J.*, 1, 1246, 1977.
75. **Bartlett, J. G., Chang, T-W., Moon, N., and Onderdonk, A. B.,** Antibiotic-induced lethal enterocolitis in hamsters: studies with eleven agents and evidence to support the pathogenic role of toxin-producing clostridia, *Am. J. Vet. Res.*, 39, 1525, 1978.
76. **Fekety, R., Silva, J., Toshniwal, R., Allo, M., Armstrong, J., Browne, R., Ebright, J., and Rifkin, G.,** Antibiotic-associated colitis: effects of antibiotics on *Clostridium difficile* and the disease in hamsters, *Rev. Infect. Dis.*, 1, 386, 1979.

77. **Ebright, J. R., Fekety, R., Silva, J., and Wilson, K. W.,** Evaluation of eight cephalosporins in hamster colitis model, *Antimicrob. Ag. Chemother.*, 19, 980, 1981.
78. **Borriello, S. P. and Barclay, F. E.,** *Clostridium difficile* infection — protection by prior colonization with non-pathogenic strains, in *Models of Anaerobic Infection*, Hill, M. J., Ed., Martinus Nijhoff, The Hague, 1984, 239.
79. **Wilson, K. H. and Sheagren, J. N.,** Antagonism of toxigenic *Clostridium difficile* by nontoxigenic *C. difficile*, *J. Infect. Dis.*, 147, 733, 1983.
80. **Browne, R. A., Fekety, R., Silva, J., Boyd, D. I., Work, C. O., and Abrams, G. D.,** The protective effect of vancomycin on clindamycin-induced colitis in hamsters, *Johns Hopkins Med. J.*, 141, 183, 1977.
81. **Bartlett, J. G., Onderdonk, A., and Cisneros, R. L.,** Clindamycin-associated colitis in hamsters: protection with vancomycin, *Gastroenterology*, 73, 772, 1977.
82. **Allo, M., Silva, J., Fekety, R., Rifkin, G. D., and Waskin, H.,** Prevention of clindamycin-induced colitis in hamsters by *Clostridium sordelli* antitoxin, *Gastroenterology*, 76, 351, 1979.
83. **Wilson, K. H., Silva, J., and Fekety, F. R.,** Suppression of *Clostridium difficile* by normal hamster cecal flora and prevention of antibiotic-associated cecitis, *Infect. Immun.*, 34, 626, 1981.
84. **Winans, L., Jr., Thornton, G. B., and Carski, T. R.,** The effect of Lactinex granules on *Clostridium difficile* induced pseudomembranous colitis, *Abst. Ann. Meet. Am. Soc. Microbiol. Abstract A4*, 1980.
85. **Price, A. B., Larson, H. E., and Crow, J.,** Morphology of experimental antibiotic-associated enterocolitis in the hamster: a model for human pseudomembranous colitis and antibiotic-associated diarrhoea, *Gut*, 20, 467, 1979.
86. **Humphrey, C. D., Condon, C. W., Cantey, J. R., and Pittman, F. E.,** Partial purification of a toxin found in hamsters with antibiotic-associated colitis, *Gastroenterology*, 76, 468, 1979.
87. **Price, A. B.,** The pathology of antibiotic-associated colitis, in *Antibiotic Associated Diarrhoea and Colitis: Clostridium difficile in Gastrointestinal Disorders*, Borriello, S. P., Ed., Martinus Nijhoff, The Hague, 1984, 133.
88. **Price, A. B. and Day, D. W.,** Pseudomembranous and infective colitis, in *Recent Advances in Histopathology*, Anthony P. P. and MacSween, R. N. M., Eds., Churchill Livingstone, Edinburgh, 1981, 99.
89. **Price, A. B., Borriello, S. P., and Barclay, F. E.,** unpublished data, 1982.
90. **Bartlett, J. G., Onderdonk, A. B., Asneros, A. B., and Kasper, D. L.,** Clindamycin associated colitis due to toxin producing species of *Clostridium* in hamsters, *J. Infect. Dis.*, 136, 701, 1977.
91. **Chang, T.-W., Bartlett, J. G., Gorbach, S. L., and Onderdonk, A. B.,** Clindamycin-induced enterocolitis in hamsters as a model of pseudomembranous colitis in patients, *Infect. Immun.*, 20, 526, 1978.
92. **Wood, J. S., Bennet, I. L., and Yardley, J. H.,** Staphylococcal enterocolitis in chinchillas, *Bull. Johns Hopkins Hosp.*, 98, 454, 1956.
93. **Rehg, J. E.,** Cecal toxin(s) from guinea pigs with clindamycin-associated colitis, neutralized by *Clostridium sordelli* antitoxin, *Infect. Immun.*, 27, 387, 1980.
94. **Rehg, J. E., Yarbrough, B. A., and Pakes, S. P.,** Toxicity of cecal filtrates from guinea pigs with penicillin-associated colitis, *Lab. Anim. Sci.*, 30, 524, 1980.
95. **Rothman, S. W.,** Presence of *Clostridium difficile* toxin in guinea pigs with penicillin-associated colitis, *Med. Microbiol. Immunol.*, 169, 187, 1981.
96. **Lee, S. P. and Thomsen, L. L.,** Toxin-induced cell membrane injury in guinea pigs given lincomycin, *Pathology*, 14, 317, 1982.
97. **Lowe, B. R., Fox, J. G., and Bartlett, J. G.,** *Clostridium difficile*-associated cecitis in guinea pigs exposed to penicillin, *Am. J. Vet. Res.*, 41, 1277, 1980.
98. **Dabard, J., Dubos, F., Martinet, L., and Ducluzeau, R.,** Experimental reproduction of neonatal diarrhoea in young gnotobiotic hares simultaneously associated with *Clostridium difficile* and other *Clostridium* strains, *Infect. Immun.*, 24, 7, 1979.
99. **Onderdonk, A. B., Cisneros, R. L., and Bartlett, J. G.,** *Clostridium difficile* in gnotobiotic mice, *Infect. Immun.*, 28, 277, 1980.
100. **Czuprynski, C. J., Johnson, W. J., Balish, E., and Wilkins, T.,** Pseudomembranous colitis in *Clostridium difficile*-monoassociated rats, *Infect. Immun.*, 39, 1368, 1983.
101. **Gustafsson, B. E., Midtvedt, T., and Strandberg, K.,** Effects of microbial contamination on the cecum enlargement of germfree rats, *Scand. J. Gastroenterol.*, 5, 309, 1970.
102. **Rehg, J. E. and Lu, Y-S.,** *Clostridium difficile* colitis in a rabbit following antibiotic therapy for Pasteurellosis, *J. Am. Vet. Med. Assoc.*, 179, 1296, 1981.
103. **Katz, L., La Mont, J. T., Trier, J. S., Sonnenblick, E. B., Rothman, S. W., Broitman, S. A., and Rieth, S.,** Experimental clindamycin-association colitis in rabbits. Evidence for toxin-mediated mucosal damage, *Gastroenterology*, 74, 246, 1978.
104. **La Mont, J. T., Sonnenblick, E. B., and Rothman, S.,** Role of clostridial toxin in the pathogenesis of clindamycin colitis in rabbits, *Gastroenterology*, 76, 356, 1979.
105. **Borriello, S. P. and Carman, R. J.,** Association of iota-like toxin and *Clostridium spiroforme* with both spontaneous and antibiotic-associated diarrhoea and colitis in rabbits, *J. Clin. Microbiol.*, 17, 414, 1983.

106. **Borriello, S. P., Carman, R. J., and Price, A. B.,** *Clostridium spiroforme:* its toxin and disease, in *Proc. FEMS Workshop Symp. Bacterial Protein Toxins*, Jeljaszewicz, J. and Alouff, J., Eds., Academic Press, London, 1984, 381.
107. **Hall, J. and O'Toole, E.,** Intestinal flora in newborn infants with description of a new pathogenic organism, *Bacillus difficilis*, *Am. J. Dis. Child.*, 49, 390, 1935.
108. **Snyder, M. L.,** Further studies of *Bacillus difficilis* (Hall and O'Toole), *J. Infect. Dis.*, 60, 223, 1937.
109. **Taylor, N. S., Thorne, G. M., and Bartlett, J. G.,** Separation of an enterotoxin from the cytotoxin of *Clostridium difficile*, *Clin. Res.*, 28, 285, 198.
110. **Banno, Y., Kobayashi, T., Watanabe, K., Ueno, K., and Nozawa, Y.,** Two toxins (D-1 and D-2) of *Clostridium difficile* causing antibiotic-associated colitis: purification and some characterization, *Biochem. Int.*, 2, 629, 1981.
111. **Lyerly, D. M. and Wilkins, T. D.,** Characteristics of the toxins of *Clostridium difficile*, in *Antibiotic Associated Diarrhoea and Colitis: Clostridium difficile in Gastrointestinal Disorders*, Borriello, S. P., Ed., Martinus Nijhoff, The Hague, 1984, 89.
112. **Taylor, N. S. and Bartlett, J. G.,** Partial purification and characterization of a cytotoxin from *C. difficile*, *Rev. Infect. Dis.*, 1, 379, 1979.
113. **Rolfe, R. D. and Finegold, S. M.,** Purification and characterization of *Clostridium difficile* toxin, *Infect. Immun.*, 25, 191, 1979.
114. **Sullivan, N. M., Pellett, S., and Wilkins, T. D.,** Purification and characterization of toxins A and B of *Clostridium difficile*, *Infect. Immun.*, 35, 1032, 1982.
115. **Taylor, N. S., Thorne, G. M., and Bartlett, J. G.,** Comparison of two toxins produced by *Clostridium difficile*, *Infect. Immun.*, 34, 1036, 1981.
116. **Higaki, M., Honda, T., Takeda, Y., and Miwatani, T.,** Detection of *Clostridium difficile* toxin in suckling mice, *FEMS Microbiol. Lett.*, 12, 351, 1981.
117. **Lyerly, D. M., Lockwood, D. E., Richardson, S. H., and Wilkins, T. D.,** Biological activities of toxins A and B of *Clostridium difficile*, *Infect. Immun.*, 35, 1147, 1982.
118. **Libby, J. M., Jortner, B. S., and Wilkins, T. D.,** Effect of the two toxins of *C. difficile* in the antibiotic-associated cecitis in hamsters, *Infect. Immun.*, 36, 822, 1982.
119. **Thelestam, M. and Bronnegard, M.,** Partial purification and cytotoxic effects of *Clostridium difficile* toxin, *Toxicon.*, 17, 192, 1979.
120. **Justus, P. G., Martin, J. L., Goldberg, D. A., Taylor, N. S., Bartlett, J. G., Alexander, R. W., and Mathias, J. R.,** Myoelectric effects of *Clostridium difficile*: motility-altering factors distinct from its cytotoxin and enterotoxin in rabbits, *Gastroenterology*, 83, 836, 1982.
121. **Vesely, D. L., Straub, K. D., Nolan, C. M., Rolfe, R. D., Finegold, S. M., and Monson, T. P.,** Purified *Clostridium difficile* cytotoxin stimulates guanylate cyclase activity and inhibits adenylate cyclase activity, *Infect. Immun.*, 33, 285, 1981.
122. **Gurwith, M. J., Langston, C., and Dunsmore, B.,** Morphological and functional effects of *Clostridium difficile* enterotoxin in tissue culture, *Can. J. Microbiol.*, 28, 100, 1982.
123. **Thelestam, M. and Florin, I.,** Interaction of *Clostridium difficile* toxin B (cytotoxin) with cultured cells, in *Antibiotic Associated Diarrhoea and Colitis: Clostridium difficile in Gastrointestinal Disorders*, Borriello, S. P., Ed., Martinus Nijhoff, The Hague, 1984, 103.
124. **Thelestam, M. and Bronnegard, M.,** Interaction of cytopathogenic toxin from *Clostridium difficile* with cells in tissue culture, *Scand. J. Infect. Dis.*, Suppl. 22, 16, 1980.
125. **Wedel, N., Toselli, P., Pothoulakis, C., Faris, B., Oliver, P., Franzblau, C., and LaMont, T.,** Ultrastructural effects of *Clostridium difficile* toxin B on smooth muscle cells and fibroblasts, *Exp. Cell Res.*, in press.
126. **Borriello, S. P.,** Clostridial Flora of the Gastrointestinal Tract in Health and Disease, *Ph.D. thesis*, University of London, 1981.
127. **Sherertz, R. J. and Sarubbi, F. A.,** The prevalence of *Clostridium difficile* and toxin in a nursery population: a comparison between patients with necrotizing enterocolitis and an asymptomatic group, *J. Pediatr.*, 100, 435, 1982.
128. **Thomas, D. E. M., Fernie, D. S., Malone, M., Bayston, R., and Spitz, L.,** Association between *Clostridium difficile* and enterocolitis in Hirschsprung's disease, *Lancet*, 1, 78, 1982.
129. **Brook, I.,** Isolation of toxin producing *Clostridium difficile* from two children with oxacillin- and dicloxacillin-associated diarrhoea, *Pediatrics*, 65, 1154, 1980.
130. **Donta, S. T., Sullivan, N., and Wilkins, T. D.,** Differential effects of *Clostridium difficile* toxins on tissue-cultured cells, *J. Clin. Microbiol.*, 15, 1157, 1982.
131. **Phillips, K. D. and Rogers, P. A.,** Rapid detection and presumptive identification of *Clostridium difficile* by *p*-cresol production and selective medium, *J. Clin. Pathol.*, 34, 642, 1981.
132. **Marrie, T. J., Furlong, M., Faulkner, R. S., Sidorov, J., Haldane, E. V., and Kerr, E. A.,** *Clostridium difficile:* epidemiology and clinical features, *Can. J. Surg.*, 25, 438, 1982.
133. **Mathan, V. I. and Borriello, S. P.,** unpublished data, 1981.

134. **Nakamura, S., Mikawa, M., Nakshio, S., Takabatake, M., Okado, I., Yamakawa, K., Serikawa, T., Okumura, S., and Nishida, S.,** Isolation of *Clostridium difficile* from the feces and the antibody in sera of young and elderly adults, *Microbiol. Immunol.*, 25, 345, 1981.
135. **Borriello, S. P., Barclay, F. E., and Williams, R.,** unpublished data, 1982.
136. **Miller, R. R. and Jick, H.,** Antibiotic associated colitis, *Clin. Pharm. Ther.*, 22, 1, 1977.
137. **Rogers, T. R., Petrou, M., Lucas, C., Chung, J. T. N., Barrett, A. J., Borriello, S. P., and Honour, P.,** Spread of *Clostridium difficile* among patients receiving non-absorbable antibiotics for gut decontamination, *Br. Med. J.*, 283, 408, 1981.
138. **Rogers, T. R., Borriello, S. P., Honour, P., Petrou, M., and White, S.,** A study of *Clostridium difficile* colonization of immunodeficient patients, in *2nd Int. Symp. Infections Immunocompromised Host*, Easmon, C. S. F. and Gaya, H., Eds., Academic Press, London, 1983, 194.
139. **Malamou-Ladas, H., O'Farrell, S., Nash, J. Q., and Tabaqchali, S.,** Isolation of *Clostridium difficile* from patients and the environment of hospital wards, *J. Clin. Pathol.*, 36, 88, 1983.
140. **Kim, K., DuPont, H. L., and Pickering, L. K.,** Outbreaks of diarrhoea associated with *Clostridium difficile* and its toxin in day-care centres: evidence of person-to-person spread, *J. Pediatr.*, 102, 376, 1983.
141. **Greenfield, C., Szawathowski, M., Noone, P., Burroughs, A., Bass, N., and Pounder, R.,** Is pseudomembranous colitis infectious?, *Lancet*, 1, 371, 1981.
142. **Mills, P. R., Main, A. N. H., Gemmell, C. G., Wright, P. A., Lee, F. D., and Russell, R. I.,** Pseudomembranous colitis as infectious disease, *Lancet*, 1, 552, 1981.
143. **Haji, T. C., Lewis, P. S., and Sanderson, P. J.,** Pseudomembranous colitis: an infectious disease?, *Lancet*, 2, 208, 1981.
144. **Pierce, P. F., Wilson, R., Silva, J., Jr., Garagusi, V. F., Rifkin, G. D., Fekety, R., Nunez-montiel, O., Dowell, V. R., Jr., and Hughes, J. M.,** Antibiotic associated pseudomembranous colitis: an epidemiological investigation of a cluster of cases, *J. Infect. Dis.*, 145, 269, 1982.
145. **Ritchie, D. B. C., Jennings, L. C., Lynn, K. R., Bailey, R. R., and Cook, H. B.,** *Clostridium difficile* associated colitis: cross infection in predisposed patients with renal failure, *N.Z. Med. J.*, 95, 265, 1982.
146. **Mogg, G. A. G., Arabi, Y., Youngs, D., Johnson, M., Bentley, S., Burdon, D. W., and Keighley, M. R. B.,** Therapeutic trials of antibotic-associated colitis, *Scand. J. Infect. Dis.*, Suppl. 22, 41, 1980.
147. **Savage, A. M. and Alford, R. H.,** Nosocomial spread of *Clostridium difficile*, *Infect. Contr.*, 4, 31, 1983.
148. **Wust, J., Sullivan, N. M., Hardegger, U., and Wilkins, T. D.,** Investigation of an outbreak of antibiotic-associated colitis by various typing methods, *J. Clin. Microbiol.*, 16, 1096, 1982.
149. **Borriello, S. P.,** Typing *Clostridium difficile*, in *Antibiotic Associated Diarrhoea and Colitis: Clostridium difficile in Gastrointestinal Disorders*, Borriello, S. P., Ed., Martinus Nijhoff, The Hague, 1984, 71.
150. **Sell, T. L., Schaberg, D. R., and Fekety, R.,** Bacteriophage and bacteriocin typing scheme for *Clostridium difficile*, *J. Clin. Microbiol.*, 17, 1148, 1983.
151. **Mulligan, M. E., George, W. L., Rolfe, R. D., and Finegold, S. M.,** Epidemiological aspects of *Clostridium difficile*-induced diarrhea and colitis, *Am. J. Clin. Nutr.*, 33, 2533, 1980.
152. **Kim, K-H., Fekety, R., Batts, D. H., Brown, D., Silva, J., Jr., and Waters, D.,** Isolation of *Clostridium difficile* from the environment and contacts of patients with antibiotic-associated colitis, *J. Infect. Dis.*, 143, 42, 1981.
153. **Borriello, S. P., Honour, P., Turner, T., and Barclay, F.,** Household pets as a potential reservoir for *Clostridium difficile* infection, *J. Clin. Pathol.*, 36, 84, 1983.
154. **George, W. L., Rolfe, R. D., and Finegold, S. M.,** *Clostridium difficile* and its cytotoxin in feces of patients with antimicrobial agent-associated diarrhea and miscellaneous conditions, *J. Clin. Microbiol.*, 15, 1049, 1982.
155. **Bartlett, J. G., Chang, T-W., Taylor, N. S., and Onderdonk, A. B.,** Colitis induced by *Clostridium difficile*, *Rev. Infect. Dis.*, 1, 370, 1979.
156. **Calver, G.,** An outbreak of *Clostridium difficile* infection, paper presented at Anaerobe Discussion Group Meeting, London, March 1, 1983.
157. **Borriello, S. P., Honour, P., Barclay, F. E., Larson, H. E., and Seal, D. V.,** unpublished data, 1980.
158. **Hafiz, S. L.,** *Clostridium difficile* and Its Toxins, Ph.D. thesis, University of Leads, 1974.
159. **Princewell, T. J. T. and Agba, M. I.,** Examination of bovine faeces for the isolation and identification of Clostridium species, *J. Appl. Bacteriol.*, 52, 97, 1982.
160. **Jones, M. A. and Hunter, D.,** Isolation of *Clostridium difficile* from pigs, *Vet. Rec.*, 112, 253, 1983.
161. **Toshniwal, R., Silva, J., Fekety, R., and Kim, K-H.,** Studies on the epidemiology of colitis due to *Clostridium difficile* in hamsters, *J. Infect. Dis.*, 143, 51, 1981.
162. **McBee, R. H.,** Intestinal flora of some antarctic birds and mammals, *J. Bacteriol.*, 79, 311, 1960.
163. **Borriello, S. P. and Honour, P.,** unpublished data, 1981.

Chapter 8

# CLOSTRIDIA AND HUMAN COLORECTAL CANCER

**M. J. Hill**

## TABLE OF CONTENTS

## I. INTRODUCTION

Large bowel cancer is one of the most common cancers in North America, northwestern Europe, Australia, and the river plate area of South America, but is rare in Africa, Asia, and the Andean countries of South and Central America. In Europe the incidence of the disease is higher in the north than the south, and is also higher in the west than in the east; the same is true of the British Isles.[1] Studies of migrants from southern and eastern Europe or from Japan (all of which have low incidences of colorectal cancer) to the U.S. or to Australia (where the disease is common) show that the risk of the disease is associated with the current area of residence rather than with the area of birth; this indicates that the major factors determining the incidence of the disease in a population are environmental and that genetic factors are not important in this respect. Genetics factors may, however, determine which persons in a uniformly exposed population actually develop the disease.

A mechanism for the development of colorectal cancer has been proposed[2] which includes genetic and environmental factors. On this hypothesis, there is a genetic factor which makes persons "adenoma-prone", an environmental factor (E1) which causes the development of adenomas in adenoma-prone persons but not in others, a further environmental factor (E2) which causes small adenomas (which are unlikely to become malignant) to grow to a large size (and with a high risk of malignancy), and a factor C which causes the development of malignancy.

Environmental factors can be divided into physical and cultural factors. The physical factors include climate, geography, air pollution, etc. all of which are shared by all members of a community; in contrast, the cultural factors include diet, smoking, and the use of other drugs, personal hygiene, etc. which may be unique to an individual and not necessarily shared by a person's neighbors. Studies of, e.g., religious groups in Bombay or California or racial groups in Johannesburg indicate strongly that physical factors are not important and that it is cultural factors which determine the risk of the disease. Most epidemiologists now agree that diet is a major factor in colorectal carcinogenesis. The epidemiology of colorectal cancer has been discussed in depth elsewhere.[3-6]

Searches for carcinogens in the diet have proven to be extremely productive with *N*-nitroso compounds, polycyclic aromatic hydrocarbons (PAHs), aflatoxins, and other mycotoxins, pyrolysis products formed during cooking, alkylhydrazines, etc. having been detected in the diet. However, none of these carcinogens have been found in amounts which might explain the relatively high incidence of colorectal cancer in Western populations; in fact, these carcinogens are usually found in highest amounts in the diet of populations with low incidences of large bowel cancer. In 1969, Aries et al.[7] postulated that the carcinogen or co-carcinogen was produced *in situ* in the large bowel by bacterial action on some benign substrate. In order to test this hypothesis it is necessary to hypothesize the nature of both the substrate and the microbial product.

There is a large body of evidence implicating the bile acids in colorectal carcinogenesis, recently summarized elsewhere;[5,6] this includes studies of populations with a wide range of incidences of colorectal cancer, case-control studies, studies of patients with diseases which confer a high risk of colorectal cancer, and studies of animal models. The possibility that the product of microbial metabolism is an unsaturated bile acid is attractive by analogy with many other families of carcinogenic compounds.

## II. MECHANISMS OF BILE ACID NUCLEAR DEHYDROGENATION

Unsaturated bile acids might be produced in the human colon by bacterial action following four major pathways (Figure 1):

1. Dehydrogenation to introduce a double bond in conjugation with an oxo group.

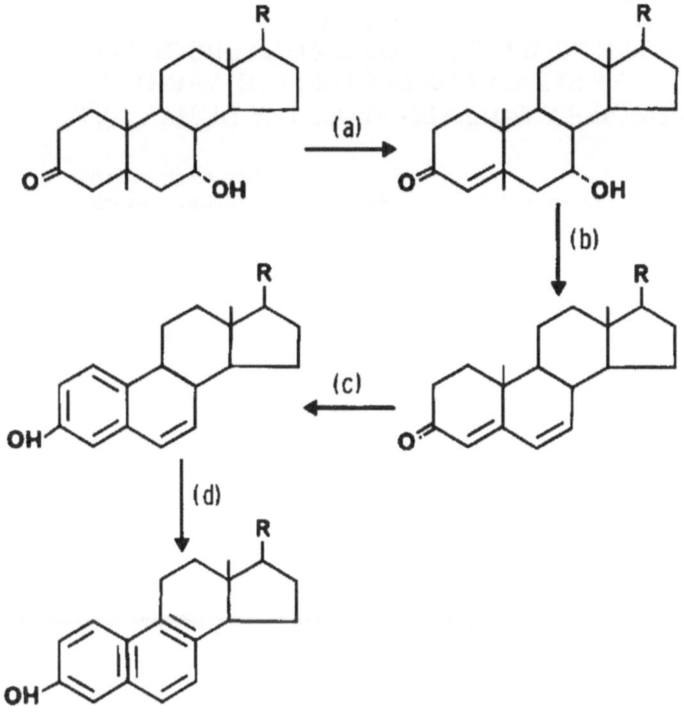


FIGURE 1. The four types of reaction which introduce double bonds into the steroid nucleus, arranged in sequence to give a polycyclic aromatic hydrocarbon.

2. Dehydroxylation, which was shown by Samuelson[8] to be a dehydration reaction involving the transelimination of the elements of water.
3. Removal of the angular methyl groups at C-10 or at C-13 from molecules where ring A contains a 1,4-dien-3-one configuration (or, in the case of the C-13 methyl group, where ring C already contains two double bonds).
4. Dehydrogenation to introduce a double bond in conjugation with other C-C double bonds.

The introduction of a double bond into the steroid nucleus in conjugation with an oxo group (Figure 1a) is of major importance in the pharmaceutical industry in the production of pharmacologically active steroids from more readily available substrates; however, the organisms used are all strictly aerobic and use atmospheric oxygen as the hydrogen acceptor. Since the colon is an anoxic environment and the gut organisms are almost all strictly anaerobic, the reaction is much more difficult; when Goddard et al.[9] screened 1095 strains for their ability to carry out the reaction, only clostridia were found to possess the enzyme (Table 1). Products of the action of the $\Delta^4$-dehydrogenase are present in feces of the rat and rabbit and in the urine of humans. Kallner[10] showed that the 5α bile acids present in rodent feces were produced by gut bacterial action via a 3-oxo-4 cholenoic acid intermediate while Alme et al.[11] showed that a number of urinary bile acids were metabolites produced by gut bacterial action and involving $\Delta^4$-dehydrogenation. The $\Delta^4$-dehydrogenase produced by *Clostridium paraputrificum* was investigated by Aries et al.;[12] the biochemical characteristics of the reaction are summarized in Table 2. The reaction is the first step in the postulated sequential desaturation of the steroid nucleus — the second step being the dehydration of the C6-7 bond (Figure 1c), which is the first step in the 7α-dehydroxylation (Figure 2).

The 7α-dehydroxylation of bile acids is carried out by a wide range of gut bacterial species and genera and has been studied by Midtvedt[13] and by Aries and Hill.[14] The liver synthesizes

**Table 1**
**THE PRODUCTION OF $\Delta^4$-DEHYDROGENASE BY STRAINS OF BACTERIA OF VARIOUS GENERA ISOLATED FROM THE HUMAN GUT**

| Test organisms | No. tested | Percentage producing $\Delta^4$-dehydrogenase |
|---|---|---|
| Nonsporing anaerobic rods | | |
| *Bacteroides fragilis* | 100 | 0 |
| *Bifidobacterium* spp. | 75 | 0 |
| *Eubacterium* spp. | 25 | 0 |
| Facultative anaerobes | | |
| Escherichia coli | 100 | 0 |
| *Streptococcus faecalis* | 100 | 0 |
| Microaerophilic organisms | | |
| *Lactobacillus* spp. | 100 | 0 |
| Spore-forming anaerobes | | |
| Lecithinase-positive | 200 | 1 |
| Lecithinase-negative | 495 | 33 |

Data adapted from Goddard, P., Fernandez, F., West, B., Hill, M. J., and Barnes, P., *J. Med. Microbiol.*, 8, 429, 1975.

**Table 2**
**FACTORS AFFECTING THE ACTIVITY OF $\Delta^4$-DEHYDROGENASE AND $\Delta^1$-DEHYDROGENASE**

| | $\Delta^4$-Dehydrogenase | $\Delta^1$-Dehydrogenase |
|---|---|---|
| pH-optimum | 7—8 | 8—9 |
| Substrate specificity | | |
| Configuration at C-17 | No effect | No effect |
| Substituents at C-12 | No inhibition | No inhibition |
| 7α-Hydroxyl group | No inhibition | No inhibition |
| 7-Oxo-group | Inhibition | Inhibition |
| 5α-Configuration | Inhibition | Inhibition |
| Hydrogen acceptors | Phenazine methosulfate<br>Menadinone | Phanazine methosulfate<br>Menadinone |
| Constitutive/inducible | Inducible | Inducible |
| Cation inhibitors | $Cu^{++}$, $Ca^{++}$ | |
| Other inhibitors | Formaldehyde, methiolate, iodoacetate | |

the primary bile acids (cholic and chenodeoxycholic acid) both of which are 7α-hydroxylated; these undergo enterohepatic circulation involving recovery by active transport from the terminal ileum which is, on average, 98% efficient. The small amount of bile acid lost during each cycle accumulates to a total of 300 to 500 mg/day and during transit through the large intestine they undergo extensive metabolism and almost total dehydroxylation, with less than 10% of the fecal bile acids retaining their 7α-hydroxyl group under normal conditions. The enzyme is widely distributed among the various genera of strictly anaerobic bacteria,[15] is irreversible, and is only produced under extremely anaerobic conditions (certainly more anaerobic than those needed merely to allow growth of the organisms). Although a high proportion of clostridia produce the enzyme, the numerical dominance of the nonsporing anaerobic rods (*Bacteroides* spp., *Bifidobacterium* spp., *Eubacterium* spp.) renders the clostridial enzyme quantitatively insignificant.

The other enzymes capable of introducing double bonds into the steroid nucleus have

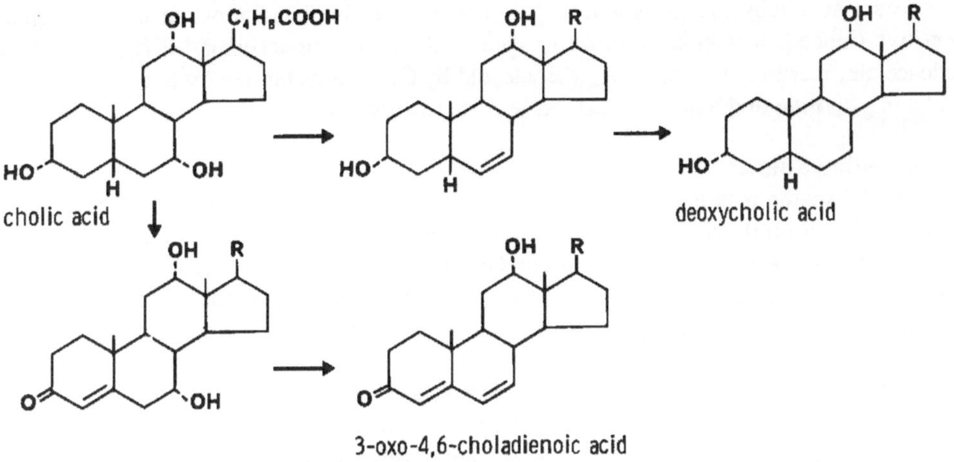


FIGURE 2. The mechanism postulated by Samuelson for the dehydroxylation of cholic acid at C-7. (Adapted from Samuelson, B., *J. Biol. Chem.*, 235, 361, 1960.) The unstable intermediate can be stabilized in conjugation with a 3-ox-4-en configuration. (Adapted from Goddard, P. and Hill, M. J., *Trans. Biochem. Soc.*, 1, 1113, 1973.)

been investigated by Goddard and Hill.[16-18] Because of the slowness of these reactions they are unlikely to be of significance in large bowel carcinogenesis; conversely, because it is so widely distributed, the 7α-dehydroxylase is unlikely to be rate-limiting in the production of unsaturated bile acids and the key reaction is the $\Delta^4$-dehydrogenase. This will now be discussed in much more detail.

## III. THE DEHYDROGENATION OF THE C4-5 BOND IN THE BILE ACID NUCLEUS

### A. Assay Methods

The original assay method described by Aries et al.[12] utilized oxygen as the hydrogen acceptor. Organisms were grown under strictly anaerobic conditions for 3 to 5 days in the presence of an inducer (usually 5β-androstan-3,17-dione). The substrate was then added and the mixture shaken aerobically for 7 days at 37°C. Steroids were then extracted with chloroform and separated by thin-layer chromatography (TLC), after which the unsaturated product could be visualized under UV light. The first development was to use phyloquinone (Vitamin K1) as the hydrogen acceptor,[9] which made the whole assay anaerobic (and so more relevant to studies of the gut); the second was to convert the assay to a microtest[19] which enabled it to be semiautomated and so suitable for use in epidemiological and survey studies.

### B. Biochemistry of the Reaction

The biochemistry of the $\Delta^4$-dehydrogenation reaction is summarized in Table 2. The enzyme has a pH optimum between 7 and 8 and is specific for 5β steroids with an oxo group at C-3. It is inhibited by the presence of an oxo group at C-7 but not by 7α- or 7β-hydroxy groups, by 12α- or 12β-hydroxyl groups, by a 12-oxo group, or by any side chain conformation at C-17. Phenazine methosulfate, a range of quinones, and the nitrate reductase system are all able to act as the hydrogen acceptor but NAD, NADP, FMN, cytochrome C, fumarate, etc. were unable to do so. Fernandez and Hill[20] studied a range of quinones as hydrogen acceptors; the best were the menaquinones (bacterial vitamin K) and phylloquinone (plant vitamin K), followed by the ubiquinones; thus, both the nucleus and the side chain are necessary for optimum activity. The enzyme is inducible by any compound able also to act as a substrate, is produced only under anaerobic conditions but is not oxygen-sensitive (indeed, it is able to utilize oxygen as the hydrogen acceptor). It is inhibited by iodoacetate, merthiolate, and formaldehyde and by $Cu^{++}$ ions but not by a range of anions including fluoride and azide, by EDTA, and by 15 m*M* $Mg^{++}$.

### C. Screening Studies

Table 1 lists the results of screening large numbers of pure strains of gut bacteria grown in Todd Hewitt broth for 5 days for $\Delta^4$-dehydrogenase activity. Only clostridia were found to carry out the reaction. Further studies by Owen[21] using bile acids as the sole carbon source and using very prolonged incubation times showed that under these special conditions some strains of *Bacteroides fragilis* and of *Bifidobacterium* spp. were also able to carry out the reaction, but the significance of this in the context of the human large bowel has still to be established. Within the genus *Clostridium* there was considerable variation between species (Table 3). A high proportion of strains of *C. paraputrificum, C. butyricum,* and *C. tertium,* a smaller proportion of strains of *C. indolis,* and a few strains of *C. rectum C. felsineum, C. cadaveris, C. chauvoei, C. sticklandii, C. sartagoformum,* and *C. inoculum* were able to carry out the reaction but, in general, strains producing lecithinase were inactive in this respect.

Clearly, the production of the enzyme is not restricted to any single clostrial species and so, in assessing the relevance of the reaction to human colorectal carcinogenesis, it was not sensible to attempt to identify the species of clostridia present and then to deduce the $\Delta^4$-dehydrogenase activity. It was more sensible (and much more simple) to screen for production of the enzyme using the microscreening test of Borriello.[19] In this, a total spore count was calculated using the standard serial dilution techniques on egg yolk medium. From the plate containing 100 to 300 colonies, 10 colonies of lecithinase-negative clostridia were selected at random and assayed for NDH activity. This then gives the proportion of persons in which NDH clostridia represent more than 10% of the lecithinase-negative clostridia. In initial studies where very many more colonies were chosen it was noted that the NDH strains were present as a large proportion of the total or were not present at all, and so the carriage rate of the organisms in a population was deemed to be a more appropriate measure than was the mean number of NDH clostridia per gram of stool.

## IV. THE ROLE OF NDH CLOSTRIDIA IN HUMAN BOWEL CANCER

To assess the relevance of NDH clostridia in human colorectal carcinogenesis, the carriage rate of the clostridia was assessed in populations around the world, in case-control studies, and in patients with diseases that put them at high risk of developing large bowel cancer.

### A. Screening of Populations

The recognition of NDH clostridia as a possible risk factor in large bowel carcinogenesis was recognized only after we had already completed our study of the relation between fecal bile acid concentration and colorectal cancer incidence in populations. Consequently, not all of the population studies could be screened for the clostridia. The total number of clostridia per gram of feces was similar for all of the populations (Table 4), but the carriage rate of NDH clostridia was much higher in the populations where large bowel cancer is common (England, Scotland, and the U.S.) than in the countries where the disease is rare (Uganda, India, and Japan). A study of three populations in Hong Kong with varying risks of colorectal cancer found very low carriage rates of NDH clostridia in all three populations.[22] However, those samples had been studied many times between the date of collection (1971) and the date of clostridial study (1975), and the repeated freezing and thawing may have been responsible for the loss of the organisms.

**Table 3**
**THE PRODUCTION OF $\Delta^4$-DEHYDROGENASE BY STRAINS OF VARIOUS CLOSTRIDIAL SPECIES**

| Species | No. of strains tested | Percentage with $\Delta^4$-dehydrogenase |
|---|---|---|
| *C. paraputrificium* | 112 | 93 |
| *C. tertium* | 20 | 90 |
| *C. indolis* | 49 | 31 |
| *C. rectum* | 19 | 11 |
| *C. felsineum* | 32 | 6 |
| *C. cadaveris* | 23 | 9 |
| *C. chauvoei* | 35 | 6 |
| *C. sticklandii* | 16 | 6 |
| *C. sartagoformum* | 31 | 13 |
| *C. inocuum* | 10 | 10 |
| *C. bifermentans* | 100 | 0 |
| *C. perfringens* | 100 | 3 |

**Table 4**
**THE MEAN NUMBER OF CLOSTRIDIA PER GRAM OF FECES AND THE CARRIAGE RATE OF NDC IN VARIOUS POPULATIONS**

| | England | Scotland | U.S. | Uganda | India | Japan | Nigeria |
|---|---|---|---|---|---|---|---|
| No. of persons | 80 | 23 | 34 | 48 | 51 | 46 | 17 |
| Total clostridia | | | | | | | |
| Lec + | 4.2[a] | 4.2 | 3.4 | 4.0 | 4.2 | 4.7 | 5.9 |
| Lec − | 5.7 | 5.6 | 5.5 | 5.2 | 5.7 | 5.6 | 1.1 |
| Total *C. paraputrificum* | 5.0 | 5.2 | 5.0 | 4.2 | ND | 5.1 | ND |
| % Carrying NDC | 35 | 58 | 50 | 7 | 0 | 0 | 0 |

[a] $Log_{10}$ no./g of feces.
[b] ND, not determined.

Data adapted from Borriello, S. P., Drasar, B. S., Tomkins, A., and Hill, M. J., *J. Clin. Pathol.*, 36, 93, 1983.

Later investigation of populations in Denmark, Finland, and Nigeria has given further support to the postulated relationship between the risk of colorectal carcinogenesis and the carriage of NDH clostridia (Table 5).

## B. Case-Control Studies

There have been two studies of the carriage of NDH clostridia in colorectal cancer cases and in control persons. The first was by Hill et al.[23] in which more than 80% of the colorectal cancer cases carried NDH clostria (Table 6) compared with only 43% of the controls. The controls were subdivided into those with large bowel disease other than colorectal cancer, those with upper GI tract disease (including cancer), and those with non-GI disease (e.g., hernia, orthopedic disease, etc.). When controlled for gut pain, fecal blood, constipation, or diarrhea, the difference in carriage rate between colorectal cancer cases and controls remained. In the second study by Blackwood et al.,[24] a similar difference in carriage rate was observed (Table 6); in this study the controls were normal healthy persons living in Glasgow.

In a subsequent extension of our original study, the number of colorectal cancer cases was increased to 88, permitting a subdivision on the basis of subsite and of Dukes classi-

**Table 5**
**THE RELATION BETWEEN THE INCIDENCE OF COLORECTAL CANCER AND CARRIAGE OF NDH CLOSTRIDIA IN VARIOUS POPULATIONS**

| Population | Incidence of cancer[a] Colon | Rectum | Total | % Of persons carrying NDH clostridia | Ref. |
|---|---|---|---|---|---|
| Uganda | 0 | 3.5 | 3.5 | 7 | 1 |
| Nigeria | 2.8 | 3.1 | 5.9 | 0 | 1, 29 |
| Japan | 5.0 | 8.1 | 13.1 | 0 | 1 |
| Finland | 8.3 | 7.5 | 15.8 | 37 | 1 |
| England | 15.8 | 16.2 | 32.0 | 35 | 1 |
| Denmark | 17.1 | 20.8 | 37.9 | 40 | 1 |
| Scotland | 28.2 | 23.3 | 51.5 | 58 | 1 |

[a] Per 100,000 males per annum, age adjusted.

**Table 6**
**CARRIAGE OF NDH CLOSTRIDIA IN CASES OF LARGE BOWEL CANCER AND IN CONTROLS**

| Population group | % Carrying NDH clostridia | Ref. |
|---|---|---|
| Bowel cancer cases | 86 | 23 |
| Controls with GI symptoms | 43 | |
| Bowel cancer cases | 83 | 24 |
| Healthy controls from Glasgow | 58 | |
| Bowel cancer cases | 70 | 19 |
| Healthy controls (England and Wales) | 32 | |

fication (Table 7). The carriage rate of NDH clostridia was high in cancers limited to the mucosa (Dukes A), in those with only local spread (Dukes B), and in those with local metastases (Dukes C), but not in those with metastatic spread to the liver (Dukes C2). When these latter were excluded there was no variation in the carriage rate of NDH clostridia by subsite. In this case-control study, all patients carried clostridia regardless of their diagnosis, and the number of organisms per gram of feces was independent of the diagnosis. Thus, as with the study of populations, there was no relation between carriage of clostridia per se and carriage of NDH clostridia. In those studies where only total clostridia have been monitored, the results showing no difference between colorectal cancer cases and controls (e.g., Moskwitz et al.[25]) in no way contradict the findings of Hill et al.[23] and of Blackwood et al.,[24] and indeed are irrelevant to the problem.

### C. High-Risk Patient Groups

Three groups of patients at high risk of developing colorectal cancer have been studied; these are patients with colorectal adenomas, patients with ulcerative colitis of more than 10 years duration, and patients who have had a large bowel cancer resected.

#### *1. Adenoma Patients*

It has been suggested that almost all colorectal cancers arise in preexisting adenomas and that the risk of malignancy developing in an adenoma increases with the degree of epithelial dysplasia, with the size of the adenoma, and with the "villousness" of the adenoma. The evidence for this has been summarized by Morson[26] and by Morson and Konishi.[27]

### Table 7
### CARRIAGE OF NDH CLOSTRIDIA IN COLORECTAL CANCER CLASSIFIED BY SUBSITE AND BY HISTOLOGICAL STAGE (DUKES CLASSIFICATION)

| Colorectal cancer subdivision | No. of cases | % Carrying NDH clostridia |
|---|---|---|
| All colorectal cancer cases | 88 | 74 |
| Dukes A | 17 | 88 |
| Dukes B | 36 | 72 |
| Dukes C | 25 | 80 |
| Dukes C2 | 10 | 40 |
| All colon cancer cases[a] | 30 | 80 |
| Cecum + asc + transverse | 10 | 80 |
| Descending + sigmoid | 20 | 80 |
| All rectal cancer cases[a] | 38 | 79 |
| Upper third | 13 | 85 |
| Mid third | 12 | 67 |
| Lower third | 13 | 85 |

[a] The analysis by subsite excludes cases classified as Dukes C2.

### Table 8
### RELATIONSHIP BETWEEN TYPE AND SIZE OF POLYP AND CARRIAGE RATE OF NDH CLOSTRIDIA

| | No. of patients | % Carrying NDH clostridia |
|---|---|---|
| All polyp cases | 138 | 38 |
| Adenomas | 114 | 38 |
| Nonadenomas | 24 | 38 |
| Controls | 100 | 37 |
| Adenomas <5 mm diameter | 26 | 6 |
| 5—10 mm diameter | 29 | 39 |
| 10—20 mm diameter | 35 | 50 |
| More than 20 mm diameter | 24 | 69 |

We have studied the fecal analyses of 134 patients with colorectal polyps, of which 110 were adenomas and 24 were nonadenomatous polyps (hamartomas, fibromas, juvenile polyps, or metaplastic polyps). The adenomas were classified by size, subsite, grade of dysplasia, and histological type of adenoma. The carriage of NDH clostridia in the whole group of polyp patients, the nonadenomas, and the adenomas was very similar to that in the control patients (Table 8). When the adenomas were graded by mean diameter, there was a clear correlation between adenoma size and the carriage rate of NDH clostridia. We have interpreted this as meaning that these clostridia have no role in the initial causation of adenomas but may be important in stimulating the subsequent growth of the small adenomas (which rarely become malignant) to a large size (and with a correspondingly high risk of developing malignancy).

There was also a correlation between the degree of dysplasia and carriage of NDH clostridia, but when this was corrected for adenoma size the correlation disappeared. There was no relation between clostridial carriage rate and either adenoma subsite or the degree of villousness in the adenoma either before or after correcting for adenoma size, indicating that the clostridia are probably not involved in the final stage of carcinogenesis.

*2. Ulcerative Colitis*

Patients who have had ulcerative colitis involving the whole colon for more than 10 years are at a high risk of developing a carcinoma and are advised to have a prophylactic colectomy. Those who keep their colon are monitored regularly for early signs of malignancy and so the carcinomas are usually detected at a stage when they can be treated surgically. We have followed 106 such patients; the carriage rate of NDH clostridia in this group of patients is 29% but is 67% in those who subsequently developed severe dysplasia or carcinoma. This indicates that carriage of NDH clostridia is a risk factor (or is associated with a risk factor) for carcinogenesis in these patients.

*3. Colorectal Cancer Patients*

Patients who have had a colorectal cancer removed successfully are at high risk (relative to the general population) of developing a subsequent primary carcinoma in the remaining colon. Those who had an adenoma as well as a carcinoma in the segment removed are at higher risk than those who had only a carcinoma.[28] We have followed 41 colorectal cancer cases postoperatively (81% of whom carried NDH clostridia preoperatively). Immediately after surgery these patients lost their NDH clostridia and at 1 month postoperation, only 29% carried the clostridia. The carriage rate increased to 45% at 3 months and remained at this level for the next 2 years. When a group of patients who had had a colorectal cancer removed at least 12 months prior to analysis were studied, 61% of those who had an adenomas as well as a carcinoma in the resected specimen carried NDH clostridia compared with 38% of those who had only a carcinoma. Once again, this indicates that carriage of NDH clostridia might be a risk factor in large bowel carcinogenesis; because these patients are followed closely and adenomas are removed when detected, none has yet developed a subsequent carcinoma.

## V. OTHER POSSIBLE ROLES OF CLOSTRIDIA IN COLORECTAL CARCINOGENESIS

The mechanisms by which clostridia might be implicated in colorectal carcinogenesis discussed so far in this chapter have been specific to the genus *Clostridium*. They may also, of course, carry out reactions implicated in the causation of the disease which are not specific to clostridia. Two possible mechanisms are worthy of brief mention.

Lithocholic acid sulfate is metabolized by gut bacteria to a range of metabolites, principally lithocholic acid (produced by hydrolysis of the S-O bond), isolithocholic acid (produced by hydrolysis of the C-O bond), and cholenic acid (produced by transelimination of $H_2SO_4$ across the 2-3 bond or the 3-4 bond). It is this latter unsaturated bile acid which is of interest in view of the arguments already discussed for a role for unsaturated bile acids in colorectal carcinogenesis. The metabolism of lithocholic acid sulfate by the human fecal flora has been described by Borriello and Owen,[30] and interestingly, their work shows that it is the NDH clostridia that are the most active organism tested in metabolizing lithocholic acid sulfate to the unsaturated cholenates.

A wide range of amino acid metabolites have been shown to be cocarcinogenic, carcinogenic, or mutagenic in animal or in vitro systems and could therefore be implicated in colorectal carcinogenesis.[6] Clostridia are included in the group of organisms able, for example, to produce volatile phenols from tyrosine, to reduce nitrate to nitrite, and to catalyze the *N*-nitrosation reaction, etc.

## VI. CONCLUSIONS

The data presented in this chapter indicate a possible role for certain clostridia in the causation of large bowel cancer. The evidence is all epidemiological and no good animal model to study the effect of these organisms on colorectal carcinogenesis has been devised. If the relationship between NDH clostria and colorectal cancer were established, there would be profound difficulties in utilizing the information in the prevention of colorectal cancer since we have no clues to the factors determining the carriage (or otherwise) of these organisms and the means which might be used to eradicate them from the gut.

Nevertheless, in the field of human carcinogenesis there are few leads to causation. Study of the role of clostridia presents an intriguing problem in itself and, coincidentally, might give valuable information on the role of the normal gut flora in human disease in general.

## REFERENCES


1. **Doll, R.,** The geographical distribution of cancer. *Br. J. Cancer,* 23, 1, 1969.
2. **Hill, M. J., Morson, B. C., and Bussey, J. H. R.,** Etiology of adenoma-carcinoma sequence in large bowel cancer, *Lancet,* 1, 245, 1978.
3. **Wynder, E. L. and Shigematsu, T.,** Environmental factors in cancer of the colon and rectum, *Cancer,* 20, 1520, 1967.
4. **Correa, P. and Haenszel, W.,** The epidemiology of large-bowel cancer, *Adv. Cancer Res.,* 26, 1, 1978.
5. **Hill, M. J.,** The aetiology of colorectal cancer, in *Recent Advances in Gastrointestinal Pathology,* Wright, R., Ed., W. B. Saunders, London, 1980, 297.
6. **Hill, M. J.,** Metabolic epidemiology of large bowel cancer, in *Gastrointestinal Cancer,* De Cosse, J. and Sherlock, P., Eds., Martinus Nijhoff, The Hague, 1981, 187.
7. **Aries, V. C., Crowther, J. S., Drasar, B. S., Hill, M. J., and Williams, R. E. O.,** Bacteria and the etiology of cancer of the large bowel, *Gut,* 10, 334, 1969.
8. **Samuelson, B.,** Bile acids and steroids. On the mechanism of the biological formation of deoxycholic acid from cholic acid, *J. Biol. Chem.,* 235, 361, 1960.
9. **Goddard, P., Fernandez, F., West, B., Hill, M. J., and Barnes, P.,** The nuclear dehydrogenation of steroids by intestinal bacteria, *J. Med. Microbiol.,* 8, 429, 1975.
10. **Kallner, A.,** The transformation of deoxycholic acid into allodeoxycholic acid in the rat, *Acta Chem. Scand.,* 21, 87, 1967.
11. **Alme, B., Bremmelgaard, A., Sjovall, J., and Thomassen, P.,** Analysis of metabolic profiles of bile acids in urine using a lipophilic anion exchanger and computerised gas liquid chromatography — mass spectrometry, *J. Lipid Res.,* 18, 339, 1977.
12. **Aries, V. C., Goddard, P., and Hill, M. J.,** Degradation of steroids by intestinal bacteria. III. 3-oxo-5β steroid Δ-dehydrogenase and 3-oxo-5β-steroid $\Delta^4$-dehydrogenase, *Biochim. Biophys. Acta,* 248, 482, 1971.
13. **Midtvedt, T.,** 7α-Dehydroxylation of bile acids, *Acta Pathol. Microbiol. Scand.,* 71, 147, 1967.
14. **Aries, V. C. and Hill, M. J.,** Degradation of steroids by intestinal bacteria. II Enzymes catalysing the oxidoreduction of the 3α-, 7α- and 12α-hydroxyl groups in cholic acid, and the dehydroxylation of the 7α-hydroxyl group, *Biochim. Biophys. Acta,* 202, 535, 1970.
15. **Hill, M. J.,** The role of colon anaerobes in the metabolism of bile acids and steroids, and its relation to colon cancer, *Cancer,* 36, 2387, 1975.
16. **Goddard, P. and Hill, M. J.,** Aromatisation of androstan-4-ene-3,17-dione by human intestinal bacteria, *Biochem. J.,* 124, 73, 1971.
17. **Goddard, P. and Hill, M. J.,** Degradation of steroids by intestinal bacteria. IV. The aromatisation of ring A, *Biochim. Biophys. Acta,* 280, 336, 1972.
18. **Goddard, P. and Hill, M. J.,** The dehydrogenation of the steroid nucleus by human gut bacteria, *Trans. Biochem. Soc.,* 1, 1113, 1973.
19. **Borriello, S. P.,** Clostridial Flora of the Gastrointestinal Tract in Health and Disease, Ph.D. thesis, St. Thomas's Hospital Medical School, London, 1981.
20. **Fernandez, F. and Hill, M. J.,** A faecal hydrogen acceptor for clostridial 3-oxo-steroid $\Delta^4$-dehydrogenase, *Trans. Biochem. Soc.,* 6, 376, 1978.
21. **Owen, R. W.,** Anaerobic Degradation of Steroids, Ph.D. thesis, Council for National Academic Awards, London, 1981.
22. **Crowther, J. S., Drasar, B. S., Hill, M. J., MacLennan, R., Magnin, D., Peach, S., and Teoh-Chun, CH.,** Faecal steroids and bacteria and large bowel cancer in Hong-Kong by socio-economic groups, *Br. J. Cancer,* 34, 191, 1976.

23. **Hill, M. J., Drasar, B. S., Williams, R. E. O., Meade, T. W., Simpson, J. E. P., and Morson, B. C.,** Faecal bile acids and clostridia in patients with cancer of the large bowel, *Lancet*, 1, 535, 1975.
24. **Blackwood, A., Murray, W. R., Mackay, C., and Calman, K.,** Faecal bile acids and clostridia in the etiology of colorectal cancer and breast cancer, *Br. J. Cancer*, 38, 175, 1978.
25. **Moskwitz, M., White, C., and Floch, M.,** Diet, faecal bile acids and neutral steroids in carcinoma of the colon, *Dig. Dis. Sci.*, 24, 746, 1977.
26. **Morson, B. C.,** The polyp-cancer sequence in the large bowel, *Proc. R. Soc. Med.*, 67, 451, 1974.
27. **Morson, B. C. and Konishi, F.,** Dysplasia of the colorectum, in *Recent Advances in Gastrointestinal Pathology*, Wright, R., Ed., W. B. Saunders, London, 1980, 331.
28. **Polk, H. C., Spratt, J. S., and Butcher, H. R.,** Frequence of multiple primary malignant neoplasms associated with colorectal carcinoma, *Am. J. Surg.*, 109, 71, 1965.
29. **Borriello, S. P., Drasar, B. S., Tomkins, A., and Hill, M. J.,** Relative carriage rates of nuclear dehydrogenating clostridia in two populations of different colorectal cancer risk, *J. Clin. Pathol.*, 36, 93, 1983.
30. **Borriello, S. P. and Owen, R. W.,** The metabolism of lithocholic acid and lithocholic acid-3α-sulphate by human faecal bacteria, *Lipids*, 17, 477, 1982.

Chapter 9

# HISTOPATHOLOGY OF CLOSTRIDIAL GUT DISEASES IN MAN

**Ashley B. Price**

## TABLE OF CONTENTS

# I. INTRODUCTION

On the assumption that the evidence incriminating clostridial species presented in the accompanying chapters is valid, then there are only two clear-cut patterns of GI disease that allow the diagnostic histopathologist to feel sure he is dealing with a clostridial disease, namely pseudomembranous colitis (PMC) and pneumatosis cystoides intestinalis (PCI). While the evidence for a clostridial etiology of the other conditions discussed, namely necrotizing enterocolitis (NEC), Pig-Bel, botulism, and food poisoning may be indisputable, the actual disease morphology is not diagnostic and merges into what in this chapter is to be grouped together and best described as a "necrotizing enteropathy". Unfortunately, even seeing Gram-positive rods associated with GI lesions is not that helpful to the pathologist as one cannot easily distinguish primary from secondary involvement.

# II. NECROTIZING ENTEROPATHY

This term encompasses a confusing host of conditions such as necrotizing colitis,[1] enteritis necroticans,[2] necrotizing enterocolitis,[3] hemorrhagic enterocolitis[4] and phlegmonous enterocolitis.[5] The terms mostly convey the dominant pathological process of intestinal necrosis, imply a clinical emergency, but do not reflect the etiology. That is not to say these conditions are all due to clostridial organisms but clostridia have been implicated in a proportion. All these entities fall under one umbrella of pathological change which makes it apt to consider them here.

It is not surprising that the morphological picture in these necrotic bowel diseases is blurred as the intestine has only a limited range of responses to a severe insult. This is usually mucosal necrosis in the first instance, for the mucosa is the most sensitive region of the gut wall and furthermore, it is the one in contact with the bowel flora. Once the mucosal barrier is breached for whatever initial reason, secondary invasion by bacteria can occur. Thus, finding bacteria in a gut lesion does not answer the question of whether they are of primary or secondary importance. Intestinal ischemia also causes mucosal necrosis and it is in relation to the role of ischemia that the biggest source of confusion arises. Uncomplicated ischemia results in a necrotic bowel accompanied by a mucosal membrane. Transient ischemia will produce enough in the way of mucosal damage to allow bacterial invasion and its consequences. Thus, to distinguish the role of ischemia vs. bacteria in these necrotic syndromes is exceptionally difficult and much is to be learned from experimental models using gnotobiotic animals. It is made harder in the clinical situation because the majority of patients affected by this range of diseases are of an age, or have associated disease, in which there is cardiovascular impairment. As a further complication, certain clostridial toxins are in themselves vasoconstrictive[6] so that although the etiology may be bacterial, the pathogenesis is ischemic. As an example of the confusion, most textbooks up until current editions carried PMC in the chapter on ischemic bowel disease.[7] Indeed, in severe PMC when there is diffuse confluent mucosal necrosis, the picture is indistinguishable from varieties of occlusive vascular disease. Marston[6] considers the whole group of "necrotizing enteropathies" as examples of acute intestinal failure, the development of which depends on the bacterial flora, the blood flow, and intramural factors, e.g., immunological changes.

As just mentioned, the clinical situation in these conditions is commonly one where the patient is a good candidate for a bacterial or ischemic insult. They are usually elderly, postoperative, have chronic medical disease, or are on therapy for malignant growths. However, a form of necrotizing enteritis, excluding neonatal disease has been described in the young.[8]

FIGURE 1. The typical gross appearance of small bowel in a "necrotizing enteropathy". It is edematous, dilated, and there are patchy areas of incipient gangrene. No abnormal vascular pathology was demonstrated.

**A. Pathology**

In general, the pathology of these "necrotizing enteropathies"[1-6,9] is that of a variable length of small or large bowel which is deeply discolored, black, or purple. The mucosa varies from being frankly gangrenous and covered by a yellow/green membrane to, in mild cases, being simply congested and edematous (Figure 1).

On microscopy, the mucosa has undergone varying degrees of necrosis and hemorrhage. In severe cases this extends through the full thickness of the bowel which may then be paper thin. With less severe involvement, the submucosa, and this is the most striking feature, is greatly expanded due to edema and frequently it is infiltrated by neutrophil polymorphs (Figure 2). The muscle coat is also edematous and usually shows signs of myocytolysis. Thrombotic vessels can be found but they are not a dominant feature.

The absence of dominating vessel occlusion and the presence of the submucosal inflammatory cell infiltrate has been taken as evidence of primary infection[1] in this group of conditions rather than the primary event being ischemia. In some cases clostridial species have been cultured and Gram-positive rods seen in large numbers on the necrotic mucosal surface or to a lesser degree in the submucosa.[1-10]

Whenever a picture similar to the above is found in surgical specimens, a careful search for clostridial species should be made both in the tissue and in the stools. Reflecting the severity of the pathology is the fact that most patients present as abdominal surgical emergencies though a clinical as well as a pathological spectrum exists. In the acute segmental ischemic enteritis described by Welch and Sumitswan,[11] putatively due to *Clostridium welchii*, early diagnosis and antibiotics avoided surgery (Figure 3). In three cases I have personally examined the changes were limited to short lengths of jejunum with florid submucosal edema. The mucosa was still intact with large numbers of polymorphs present in the submucosa along with scanty Gram-positive rods in two of the cases (Figure 2). Stool cultures failed to yield pathogenic or commensal species of clostridia in abnormal numbers.

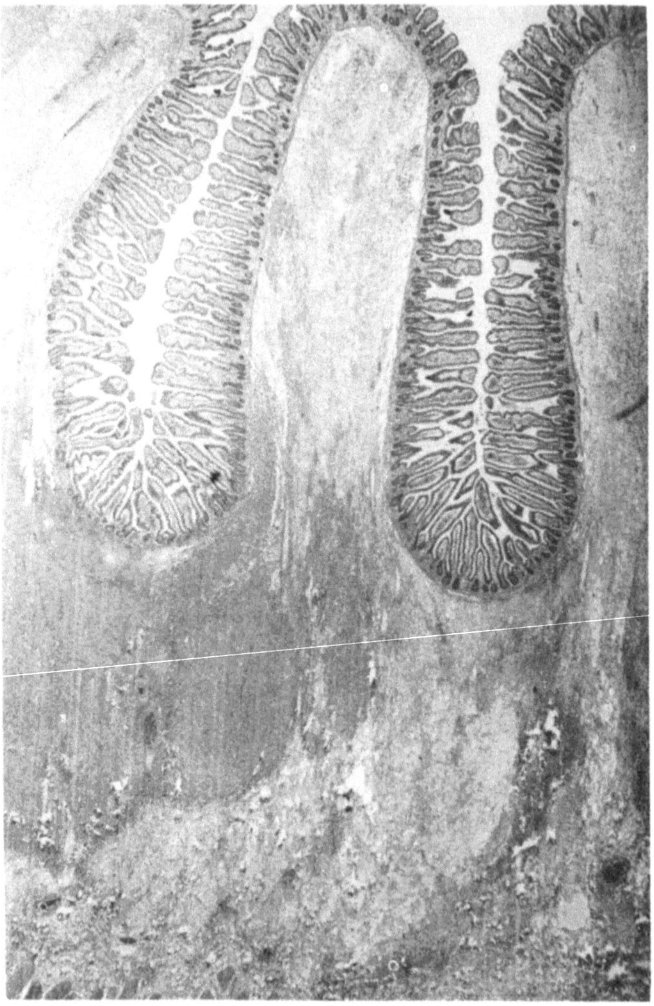

FIGURE 2. Small bowel showing intact mucosa but a hugely expanded submucosa with areas densely infiltrated by acute inflammatory cells. (H.E.; magnification × 10.)

The hallmark of the necrotizing bowel syndromes is hemorrhagic necrosis of the mucosa with intense submucosal edema and inflammation. The bewildering array of names reflects the ignorance surrounding their etiology. It seems likely that ischemia plays some role and, in a proportion, infection by clostridial organisms. Which mechanism has the primary role is not clear. More than likely more than one type of bacteria is involved, the particular invader depending on the bowel flora present in the vicinity of the initiating insult. Clostridia as common intestinal inhabitants which produce such potent toxins are clearly prime suspects and they are already the proven criminal in some cases.

## III. PSEUDOMEMBRANOUS COLITIS (PMC)

*C. difficile* is now thought to be the main cause of morphologically proven PMC.[12,13,14] Its association with antibiotic-associated colitis (defined as the presence of biopsy-proven colitis but in the absence of pseudomembranes) and antibiotic-associated diarrhea (AAD) (defined as diarrhea associated with antibiotic ingestion but normal biopsies) is more tenuous.

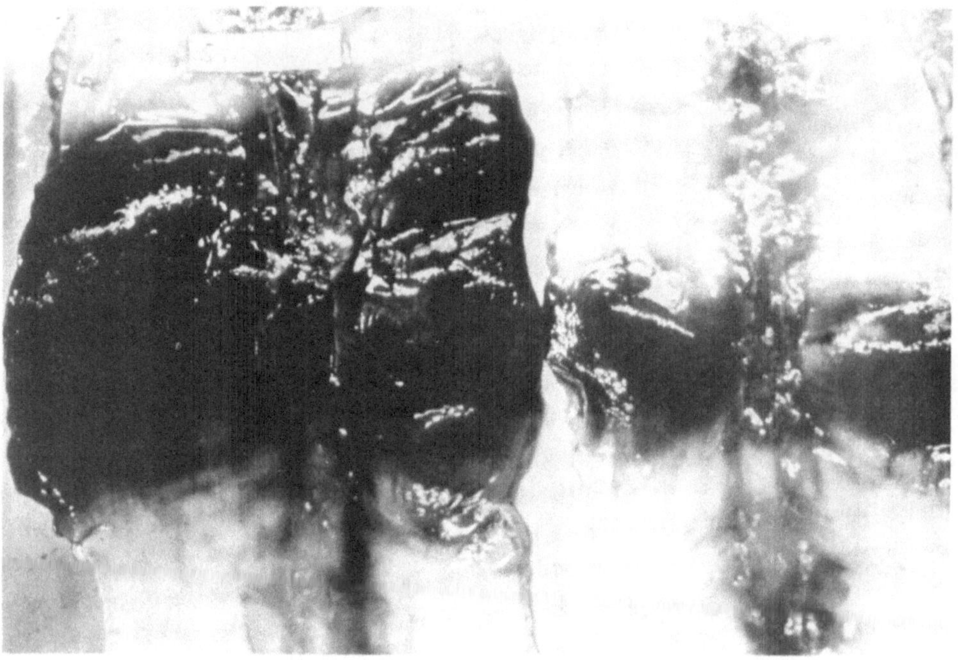

FIGURE 3. Acute segmental ischemic enteritis. (Courtesy Mr. T. Welch.)

In the latter, *C. difficile* was implicated in only 6% of one series.[15] While *C. difficile* is the dominant organism in PMC, in the wider context of AAD there are clearly many other mechanisms to consider.

A pseudomembrane refers to the discrete yellow plaques seen over the colonic mucosa in an established case of PMC[16] (Figure 4). What determines their distribution along the lengths of the colon is not clear but the rectum is involved in some 60 to 70% of cases, making rectal biopsy a diagnostic tool in the majority of instances.[17] Although the disease can be seen in the small intestine, it appears to be limited to the large bowel in association with antibiotics and in the modern era this is the most common presentation. The lesions vary in size from pinpoint up to 1 cm. Much beyond this size they coalesce to produce a confluent necrotic membrane (Figure 4). In the past the condition was seen only at this stage, a stage when, both macro- and microscopically, it is much like any of the other forms of "necrotizing enteropathy" previously described.

Each plaque represents a small group of disrupted crypts capped by fibrin, mucus, and inflammatory debris (Figure 5). Separating the plaques are areas of intact mucosa with only minimal inflammatory changes present. Naturally, these areas diminish as the plaques coalesce. In the earliest recognizable stage of the disease only a tiny intercryptal erosion may be present, "the summit lesion"[16] (Figure 6).

The main changes remain confined to the mucosa and are usually accompanied by submucosal edema. Rarely, toxic megacolon is seen.[18] In confluent disease the necrotic mucosa is replaced by ulcer slough (Figure 7) though the ghost outlines of the crypts may occasionally be recognized, providing a clue to the diagnosis. Microthrombi may be found in the lesions but are not a dominant feature. Gram-positive rods are also only occasionally seen on the surface.

The unpurified toxin of *C. difficile* has been demonstrated to produce epithelial damage in the hamster[19] model, but to date no acceptable pathogenesis has been put forward to explain the unique mucosal eruptions found in PMC. As mentioned, organisms associated

FIGURE 4. This colon is covered by a confluent necrotic membrane, the end result of the stage of individual plaques (inset) coalescing.

with the lesions have not been noted in any significant numbers, indeed, it is not yet known if *C. difficile* is enteroinvasive. Our preliminary observations using a fluorescent antibody suggests it remains on the mucosal surface. The animal model for the disease,[20,21] the hamster in particular, has been very successful from the micribiological point of view. Morphologically however, the disease has been predominantly cecal and while a colitis is easily produced, the discrete plaques are not seen.[20]

In human disease the patterns described form a spectrum of pathology from the small

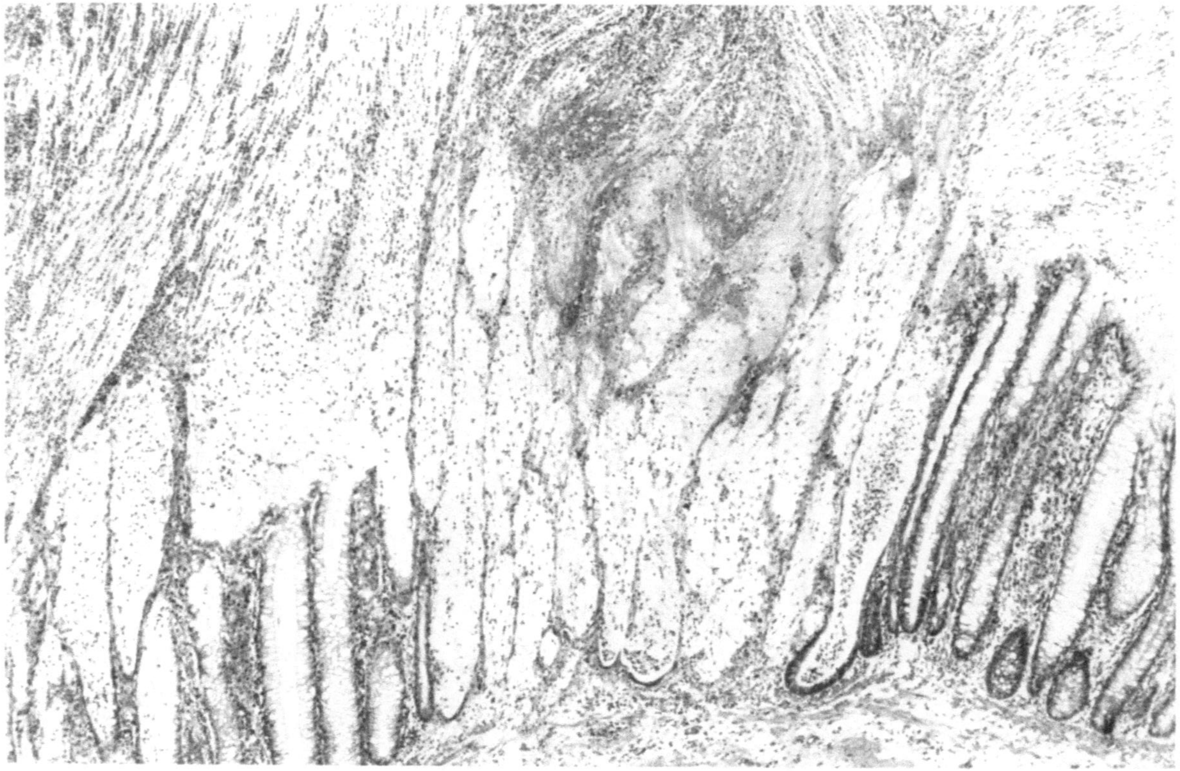

FIGURE 5. The classical group of disrupted crypts with overlying membrane seen in PMC. Normal mucosa is seen to each side. (H.E.; magnification × 112.)

FIGURE 6. The earliest recognizable lesion in PMC is the tiny intercryptal erosion with a spray of inflammatory cells. (H.E.; magnification × 56.)

"summit lesions", through the classical plaques, to the stage of confluent necrosis.[22] What determines the development of the spectrum is not known but there is a significant correlation between high toxin titers and the presence of plaques and between the presence of plaques and the duration of the diarrhea.[23] However, Lishman et al.[24] have reported a group of patients exposed to antibiotics, without diarrhea, yet with toxin titers in the same range as a group with diarrhea. There is clearly much to be discovered in the field interrelating pathology, microbiology, and clinical symptomatology.

## IV. PNEUMATOSIS CYSTOIDES INTESTINALIS (PCI)

The evidence for the role of clostridia in PCI is presented in Chapter 3. However the organisms are implicated,[25] the pathogenesis of pneumatosis in adults needs separate consideration from that in the neonate where in the latter, inflammation and necrosis dominate the morphology.[26]

Adult disease may be divided into primary and secondary types. Secondary disease is usually associated with chronic pulmonary disease or chronic peptic ulceration,[27] and the

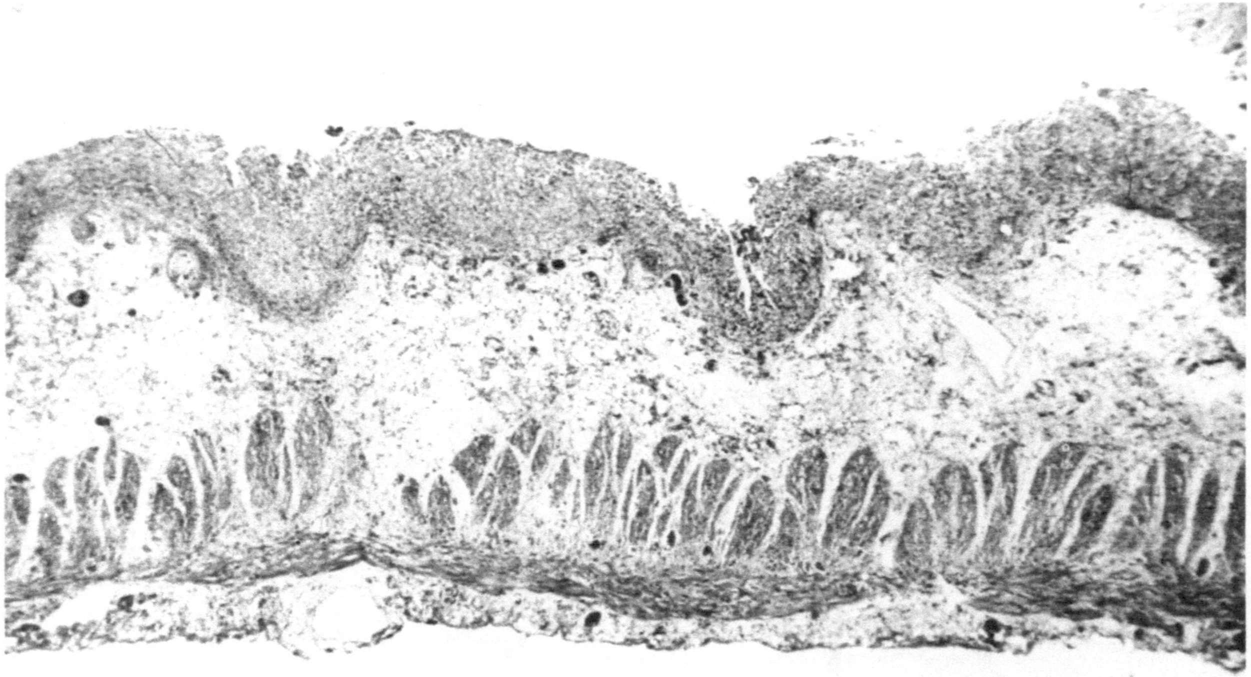

FIGURE 7. Confluent necrosis of the mucosa in PMC. The mucosa is replaced by an ulcer slough at this stage. (H.E.; magnification × 35.)

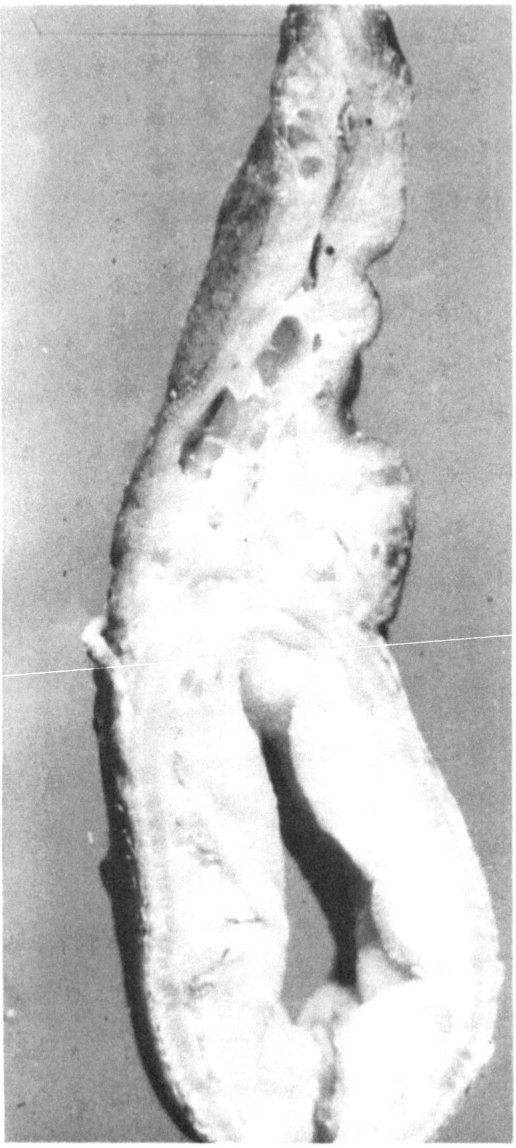

FIGURE 8. The small intestine and mesentery showing the presence of gas-filled cysts (arrows).

lesions predominate on the serosal aspect of the bowel. In primary or idiopathic PCI the submucosa is involved and the bowel has a spongy texture even though nothing may be visible externally (Figure 8). PCI is one cause of colonic polyps,[28] which, when pierced, disappear.

The histology of both primary and secondary PCI is similar, irrespective of whether there is large or small bowel involvement.[29] Characteristic is the gas-filled cyst, lined in part by giant cells, in part by inflammatory cells, and occasionally flat endothelial-like cells (Figure 9). The cysts are similar be they in the serosa or submucosa. Depending on their size, the overlying mucosa may become thin but necrosis and inflammation are absent. If resolution occurs there is gradual fibrosis at that site.

The major complication of PCI relates to pneumoperitoneum if a serosal cyst ruptures,

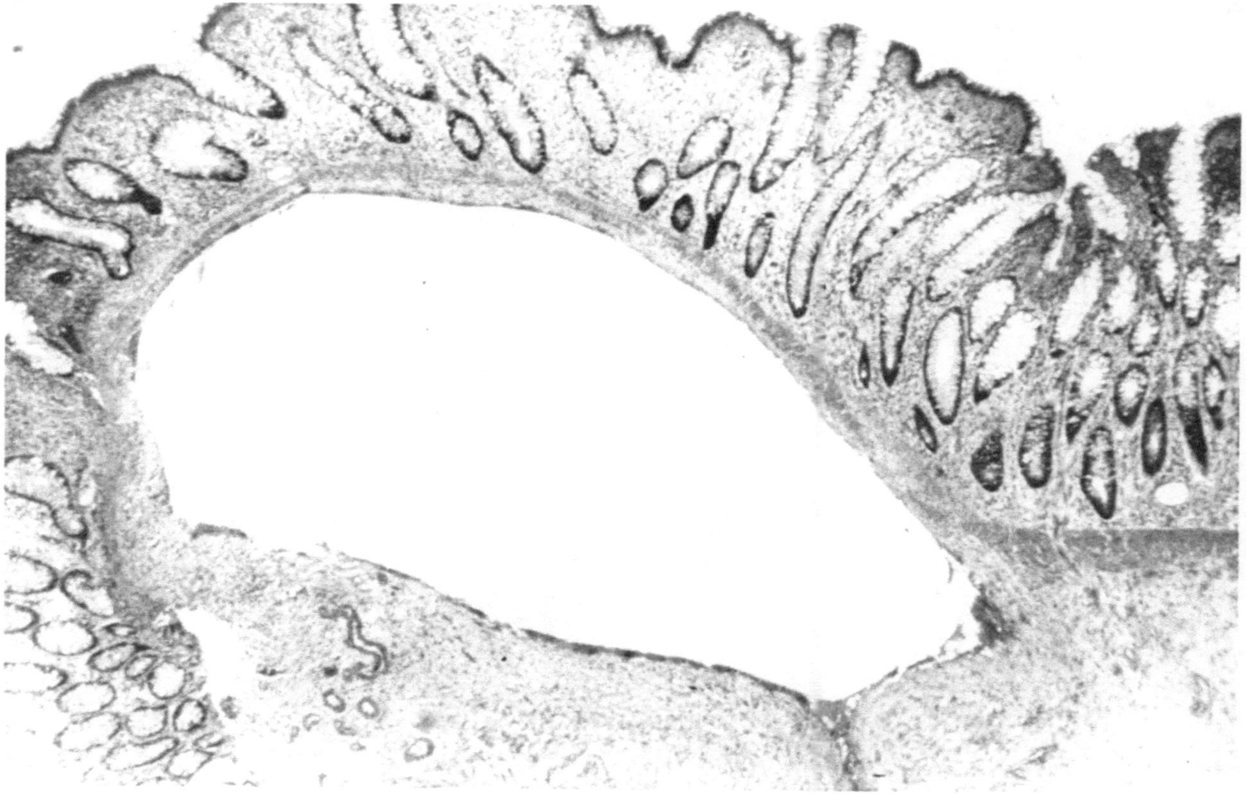

FIGURE 9. Colonic mucosa with a large gas-filled submucosal cyst lined by giant cells. (H.E.; magnification × 112.)

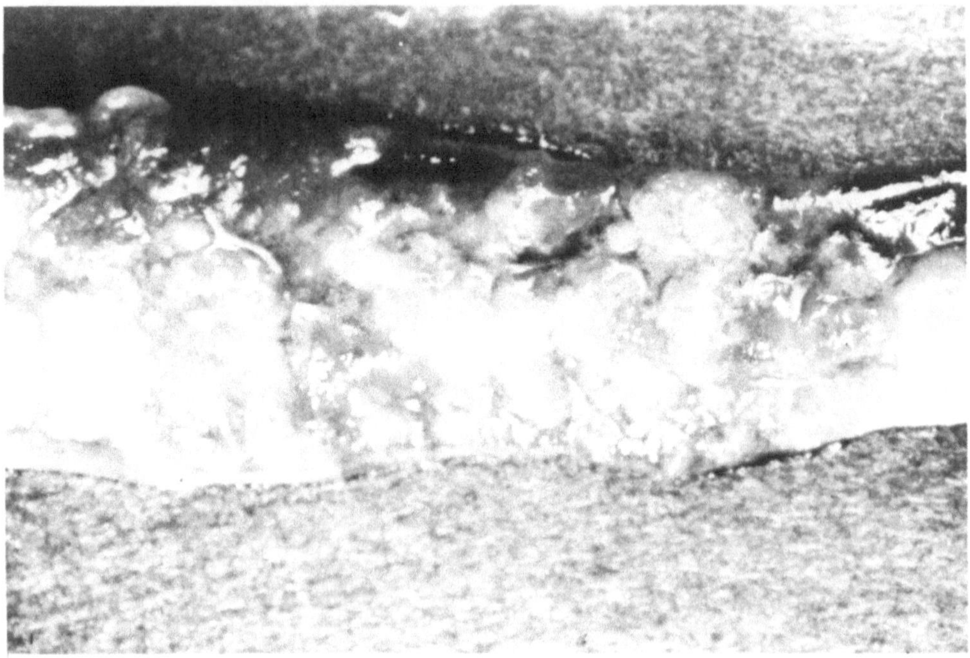

FIGURE 10. The mucosal aspect of a colon from a baby with neonatal NEC. The mucosa has a raised yellow surface due to necrosis and underlying submucosal gas.

and intestinal obstruction if some mucosal cysts grow large enough to occlude the bowel lumen.

The pathogenesis of the cyst clearly depends on the etiology which is considered in Chapter 3. Of some importance, whatever the etiology, is whether the cysts are within lymphatics[30] or are simply in connective tissue. The suggestion that the gas accumulates in distended lymphatics was based on the belief that the cysts have an endothelial lining, but this is not the case as many cysts have no apparent lining apart from the giant cells. Their appearance consequent to trauma and ruptured diverticula[31,32] would also suggest an *in situ* local origin rather than a lymphatic derivation.

## V. NEONATAL NECROTIZING ENTEROCOLITIS

This condition incorporates the basic pathology of PCI accompanied by a severe mucosal inflammatory reaction and necrosis.[26,33] This results in extensive mucosal destruction and perforation is frequent. Any part of the small or large bowel can be involved and the subsequent findings depend on the stage of development of the disease. In established cases the bowel appears congested and thickened, the latter a consequence of edema and the presence of gas-filled cysts. A fibrinous exudate can cover the serosa, especially if a perforation has occurred.

The mucosal aspect of the bowel may be extensively ulcerated or, if intact, have a cobblestone appearance related to the submucosal gas cysts (Figure 10). These cysts in the submucosa vary in size from minute to 0.5 cm to 1 cm in diameter and give the bowel wall an "aerated" appearance.

Histology[26] shows submucosal gas-filled cysts, vascular congestion, edema, and a variable inflammatory infiltrate in the mucosa. In most cases there is extensive mucosal necrosis but this is not invariable (Figure 11). The inflammatory infiltrate is most marked in the submucosa and may be accompanied by small-vessel thrombosis. The gas-filled cysts are usually lined

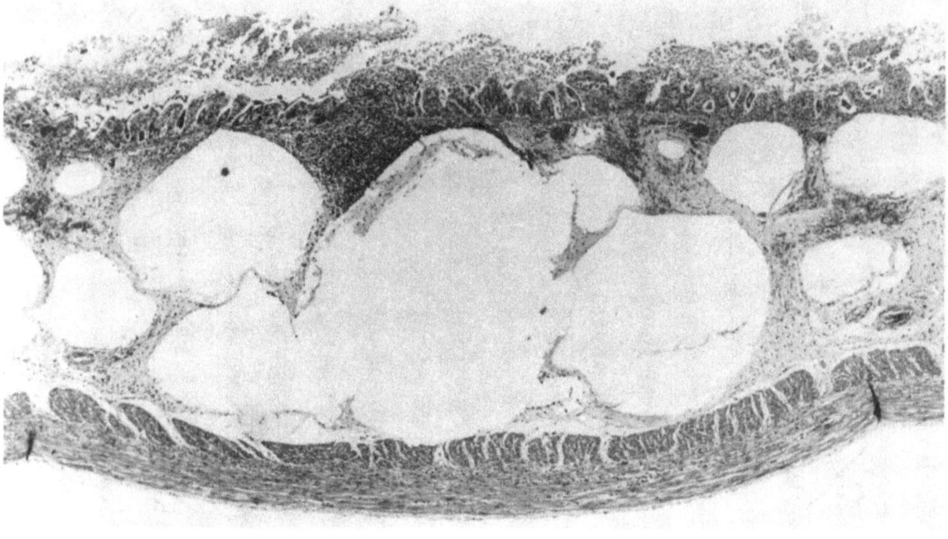

FIGURE 11. Histopathology from the colon seen in Figure 10 showing the mucosal necrosis and huge submucosal gas cysts. (H.E.; magnification × 25.)

by giant cells and occasionally an attenuated endothelium. While Gram-positive rods can frequently be seen in the lumen, they are not a feature in the cysts.

It is now clear that not all cases of NEC show radiologically diagnostic intramural gas and in such cases proctoscopy is useful as it demonstrates a macroscopic colitis.[34] In the cases of Kliegman[35] where no gas was present, the pathology was confirmed on surgical resection or at autopsy as one of widespread focal intestinal necrosis. This picture clearly merges with the other conditions making up the necrotizing enteropathies. Although the term NEC is still applicable morphologically, it may be that the absence of gas has etiological significance. Thus, not all cases of neonatal colitis are of the necrotizing variety and apart from obvious bacterial causes (*Salmonella, Shigella,* and *Campylobactor jejuni*), protein-antigen-induced colitis[36] is being increasingly recognized. In these latter cases where pathology was available, the colitis showed a wide spectrum of inflammatory changes but necrosis was not a feature.

From the observations in the literature it would seem that if clostridia are incriminated in the neonate,[33,37,38] primarily or secondarily, necrotizing colitis (plus or minus pneumatosis) is the likely pathological outcome. However, careful pathological documentation is worthwhile in order to delineate other etiologies.

## VI. TYPHLITIS

Within the category of necrotizing bowel syndromes one can delineate the above entity and its synonyms.[39-41] The etiology of this predominantly right-sided colitis is still uncertain but, as in some instances clostridia have been implicated,[42] it gains entry to this chapter.

The colitis involves the cecum and ascending colon and is invariably associated with a neutropenic state. The latter is usually due to concomitant leukemia/lymphoma, the patients having had a recent regime of chemotherapy. Occasionally it is seen during the course of cyclic neutropenia.[42] This is a rare condition manifest by recurrent episodes of neutropenia.

*C. difficile* and its toxin were associated with a series of 13 cases reported from Cambridge.[43] The association of other clostridia, e.g., *C. septicum* is documented as part of a larger series in which other intestinal pathogens were also implicated.[42]

The pathology is less necrotic than in non-neutropenic necrotizing colitides and yet less

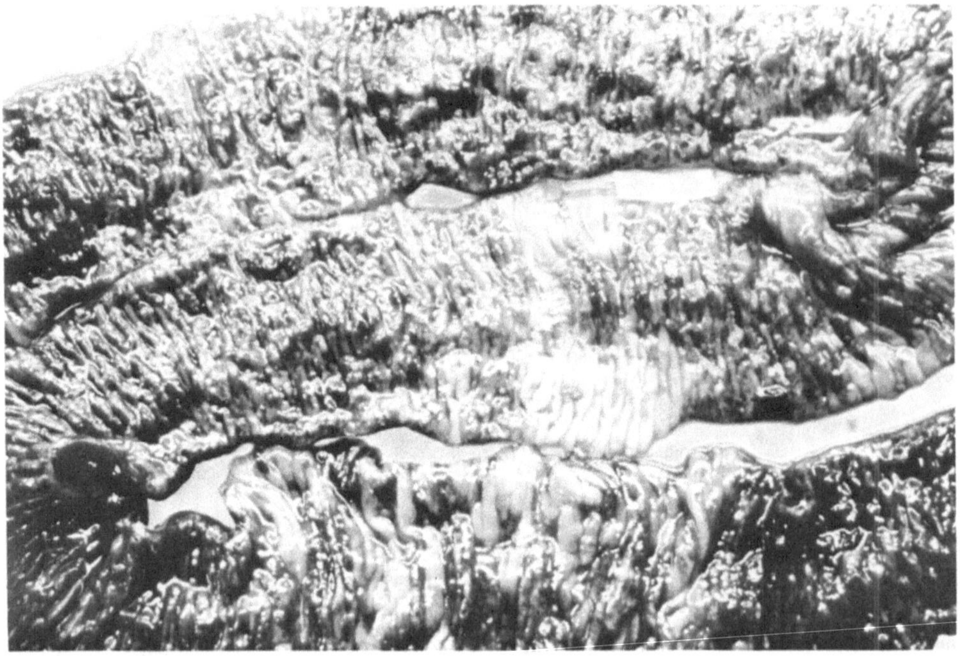

FIGURE 12. The typical appearance of patchy hemorryhage and early gangrene in the small bowel in Pig-Bel.

specific than PMC. Some workers argue that it is the altered immunological state and depressed white cell count that affects the morphology. The cecum is usually dilated, edematous, and variably discolored. The mucosa is congested and shows patchy small foci of ulceration. Submucosal expansion is the predominant feature. It is edematous with increased numbers of inflammatory cells present though with limited numbers of polymorphs. The overlying mucosa shows focal hemorrhage and any ulceration is localized to small areas. The muscle coat also shows some inflammation and can occasionally be necrotic. No membranes are found. In some of the cases implicating clostridial species, clusters of Gram-positive rods have been seen localized to the ulcerated areas of the mucosa.[42,43] Rampling et al.,[43] who isolated *C. difficile* from their series, believe the disease is a form "fruste" of PMC in leukemic patients and in patients with defective white cells. Hopkins and Kushner[42] expressed the attractive idea that this form of NEC is the corollary of the trypsin deficiency in Pig-Bel. In this case, the white cell protease levels are insufficient (due to the neutropenia) to withstand and deal with clostridial exotoxins presented from the lumen.

The potentially fatal condition is important to recognize in patients at risk[44] as the best results are from early surgical intervention. Clostridial species probably account for its etiology in a small percentage; the other causes remain to be delineated.

## VII. NECROTIZING JEJUNITIS (PIG-BEL)

The pathology of Pig-Bel[45] is adequately described in Chapter 5 and is not repeated in detail here. Strictly speaking, it refers to an acute, patchy, necrotizing, and inflammatory disease of the small bowel (Figure 12) affecting natives in the Highlands of New Guinea.[46] This is related to pig feasting and infection by *C. welchii* type C. It is identical to Darmbrand[47] that occurred as an outbreak during the post-war period (1947 to 1948) in northern Germany. Apart from these epidemics, cases with similar pathology (necrotizing jejunitis) have been described in many countries though the microbiological correlation has seldom been complete. It is likely that the acute segmental ischemic enteritis described in Indonesian children[8,11] and the enteritis necroticans of Zeissler and Rassfeld-Sternberg[48] also represent the condition. From the pathology outlined in Chapter 5 it is clear that *C. welchii* type C produces a disease within the spectrum of a "necrotizing enteropathy". The balance between the organism, the intestinal integrity, and other host defenses is interesting.

As well as there being a complete clinical spectrum of severity of disease, there is a similar pathological spectrum.[46] Some patients survive without recourse to surgery and with resolution of the pathology, healing can occur that produces a subsequent intestinal stricture.

*C. welchii* can produce the classical changes of gas gangrene in the bowel when necrosis and gas-filled cysts are apparent.

## VIII. *C. PERFRINGENS* TYPE A FOOD POISONING

Luckily, cases of food poisoning due to *C. welchii* type A seldom require surgical intervention. An enterocolitis has been described and in the few cases that have unfortunately come to autopsy the picture is again one of necrotizing small bowel disease.[49]

## IX. CONCLUSIONS

Apart from PMC and PCI, the pathology of clostridial bowel disease is one dominated by progressive hemorrhagic necrosis and submucosal edema. There is no one distinctive entity except for a large group of acute clinical conditions described in the literature under a variety of guises, all basically variations on the term NEC. It seems likely that this clinical and pathological collection of conditions represent the end result of a disturbed balance between intestinal mucosal integrity, the lumenal flora, and the local blood flow. Variations in each factor account for variations in clinical and morphological presentation.

## REFERENCES


1. **Killingback, M. J. and Lloyd-Williams, K.,** Necrotizing colitis, *Br. J. Surg.*, 49, 175, 1962.
2. **Patterson, M. and Rosenbaum, H. D.,** Enteritis necroticans, *Gastroenterology*, 21, 110, 1952.
3. **Kay, A. W., Richards, R. L., and Watson, A. J.,** Acute necrotizing (pseudomembranous) enterocolitis, *Br. J. Surg.*, 46, 45, 1958.
4. **Williams, L. F., Anastasia, L. F., Hasiokis, C. A., Bosniak, M. A., and Byrne, J. J.,** Non-occlusive mesenteric infarction, *Am. J. Surg.*, 114, 374, 1967.
5. **Rosen, Y. and Hee Won, O. K.,** Phlegmonous enterocolitis, *Dig. Dis.*, 23, 249, 1978.
6. **Marston, A.,** *Intestinal Ischaemia*, Edward Arnold, London, 1977, 81.
7. **Whitehead, R.,** The pathology of intestinal ischaemia, *Clin. Gastroenterol.*, 1(3), 613, 1972.
8. **Headington, J. T., Sathornsumathi, S., Simark, S., and Siyatanond, W.,** Segmental infarcts of the small intestine and mesenteric adenitis in Thai children, *Lancet*, 1, 802, 1967.
9. **Wilson, R. and Qualheim, R. E.,** A form of acute haemorrhagic enterocolitis affecting chronically ill individuals, *Gastroenterology*, 27, 431, 1954.
10. **Blenkinsopp, W. K. and Dupont, P. A.,** Bacteria in necrotizing enterocolitis, *Lancet*, 2, 617, 1977.
11. **Welch, T. P. and Sumitswan, S.,** Acute segmental ischaemic enteritis in Thailand, *Br. J. Surg.*, 62, 716, 1975.
12. **Bartlett, J. G., Chang, T. W., Gurwith, M., Gorbach, S. L., and Onderdonk, A. B.,** Antibiotic-associated pseudomembranous colitis due to toxin-producing clostridia, *N. Engl. J. Med.*, 298, 531, 1978.
13. **George, R. H., Symonds, J. M., Dimock, F., Brown, J. D., Arabi, Y., Shinagawa, N., Keighley, M. R. B., Alexander-Williams, J., and Burdon, D. W.,** Identification of *Clostridium difficile* as a cause of pseudomembranous colitis, *Br. Med. J.*, 1, 65, 1978.
14. **Larson, H. E., Price, A. B., Honour, P., and Borriello, S. P.,** *Clostridium difficile* and the aetiology of pseudomembranous colitis, *Lancet*, 1, 1063, 1978.
15. **Bartlett, J. G., Moon, N., Chang, T. W., Taylor, N., and Onderdonk, A. B.,** Role of *Clostridium difficile* in antibiotic-associated pseudomembranous colitis, *Gastroenterology*, 75, 778, 1978.

16. **Price, A. B. and Davies, D. R.,** Pseudomembranous colitis, *J. Clin. Pathol.*, 30, 1, 1977.
17. **Tedesco, F. J., Corless, J. K., and Brownstein, R. E.,** Rectal sparing in antibiotic-associated pseudomembranous colitis: a prospective study, *Gastroenterology*, 83, 1259, 1982.
18. **Cone, J. B. and Wetzel, W.,** Toxic megacolon secondary to pseudomembranous colitis, *Dis. Colon Rectum*, 25, 478, 1982.
19. **Abrams, G. D., Allo, M., Rifkin, G. D., Fekety, R., and Silva, J.,** Mucosal damage mediated by clostridial toxin in experimental clindamycin-associated colitis, *Gut*, 21, 493, 1980.
20. **Price, A. B., Larson, H. E., and Crow, J.,** Morphology of experimental antibiotic-associated enterocolitis in the hamster: a model for human pseudomembranous colitis and antibiotic-associated diarrhoea, *Gut*, 20, 467, 1979.
21. **Czuprynski, C. J., Johnson, W. J., Balish, E., and Wilkin, T.,** Pseudomembranous colitis in *Clostridium difficile* mono-associated rats, *Infect. Immun.*, 39, 1368, 1983.
22. **Price, A. B.,** Pseudomembranous colitis, in *Recent Advances in Gastrointestinal Pathology*, Wright, R., Ed., W. B. Saunders, London, 1980, chap. 10.
23. **Burdon, D. W., George, F. H., Mogg, G. A. G., Arabi, Y., Thompson, H., Johnson, M., Alexander-Williams, J., and Keighley, M. R. B.,** Faecal toxin and severity of antibiotic-associated pseudomembranous colitis, *J. Clin. Pathol.*, 34, 548, 1981.
24. **Lishman, A. H., Al-Jumaili, I. J., and Record, C. O.,** Spectrum of antibiotic-associated diarrhoea, *Gut*, 22, 34, 1981.
25. **Yale, C. E., Balish, E., and Wu, J. P.,** The bacterial etiology of pneumatosis cystoides intestinalis, *Arch. Surg.*, 109, 89, 1974.
26. **Tait, R. A. and Kealy, W. F.,** Neonatal necrotizing enterocolitis, *J. Clin. Pathol.*, 32, 1090, 1979.
27. **Koss, L. G.,** Abdominal gas cysts (pneumatosis cystoides intestinorum hominis): analysis with report of a case and a critical review of the literature, *Arch. Pathol.*, 53, 523, 1952.
28. **Calne, R. Y.,** Gas cysts of the large bowel simulating multiple polyposis, *Br. J. Surg.*, 47, 212, 1959.
29. **Morson, B. C. and Dawson, I. M. P.,** *Gastrointestinal Pathology*, Blackwell, Oxford, 1979, 433.
30. **Keyting, W. S., McCarver, R. R., Kovarik, J. L., and Daywilt, A. L.,** Pneumatosis intestinalis: a new concept, *Radiology*, 76, 733, 1961.
31. **Werlkin, M. G., Wetchler, B. B., Waye, J. D., and Brown, L. K.,** *Pneumatosis coli* associated with sigmoid volvulus and colonoscopy, *Am. J. Gastroenterol.*, 65, 209, 1976.
32. **Zakhour, H. D. and Clark, R. G.,** Intramural gas cysts in a case of diverticular disease of jejunum, *Histopathology*, 6, 363, 1982.
33. **Kliegman, R. M.,** Neonatal necrotizing enterocolitis: implication for an infectious disease, *Pediatr. Clin. North Am.*, 26, 327, 1979.
34. **Fenton, T. R., Walker-Smith, J. A., and Harvey, D. R.,** Proctoscopy in infancy with reference to its use in necrotizing enterocolitis, *Arch. Dis. Child.*, 56, 121, 1981.
35. **Kliegman, R. M. and Fanaroff, A. A.,** Neonatal necrotizing enterocolitis in the absence of pneumatosis intestinalis, *Am. J. Dis. Child.*, 136, 618, 1982.
36. **Lake, A. M., Whitington, P. F., and Hamilton, S. R.,** Dietary protein-induced colitis in breast fed infants, *J. Pediatr.*, 101, 906, 1982.
37. **Howard, F. M., Flynn, D. M., Bradley, J. M., Noone, P., and Szawatkowski, M.,** Outbreak of necrotizing enterocolitis caused by *Clostridium butyricum*, *Lancet*, 2, 1099, 1977.
38. **Pederson, V. P., Hansen, F. H., Halveg, A. B., Christiansen, E. P., Justesen, T., and Høgh, P.,** Necrotizing enterocolitis of the newborn — is it gas-gangrene of the bowel, *Lancet*, 2, 715, 1976.
39. **Varki, A. P., Armitage, J. O., and Feagler, J. R.,** Typhlitis in acute leukaemia, *Cancer*, 43, 695, 1979.
40. **Freeman, H. J., Rabeneck, L., and Owen, D.,** Survival after necrotizing enterocolitis of leukaemia treated with oral vancomycin, *Gastroenterology*, 81, 791, 1981.
41. **Prolla, J. C. and Kirsner, J. B.,** The gastrointestinal lesions and complications of the leukemias, *Ann. Intern. Med.*, 61, 1084, 1964.
42. **Hopkins, D. G. and Kushner, J. P.,** Clostridial species in the pathogenesis of necrotizing enterocolitis in patients with neutropenia, *Am. J. Hematol.*, 14, 289, 1983.
43. **Rampling, A., Warren, R. E., Berry, P. J., Swirsky, D., Hoggarth, C. E., and Bevan, P. C.,** Atypical *Clostridium difficile* colitis in neutropenic patients, *Lancet*, 2, 162, 1982.
44. **Dosik, G. M., Luna, M., Valdivieso, M., McCredie, K. B., Gehan, E. A., Gil-Extremera, B., Smith, T., and Bodey, G. P.,** Necrotizing colitis in patients with cancer, *Am. J. Med.*, 67, 647, 1979.
45. **Cooke, R.,** The pathology of Pig-bel, *Papua New Guinea Med. J.*, 22, 35, 1979.
46. **Murrell, T. G. C., Roth, L., Egerton, J., Samels, J., and Walker, P. D.,** Pig-bel: enteritis necroticans, *Lancet*, 1, 217, 1966.
47. **Siegmund, H.,** Zur pathologie und pathogenese der als Darmbrand bezeichneten Enteritis necroticans, *Klin. Wochenschr.*, 26, 33, 1948.

48. **Zeissler, J. and Rassfeld-Sternberg, L.,** Enteritis necroticans due to *Clostridium welchii* type F, *Br. Med. J.*, 1, 267, 1949.

49. **Parry, W. M.,** Outbreak of *Clostridium welchii* food-poisoning, *Br. Med. J.*, 2, 1616, 1963.

Chapter 10

# CLOSTRIDIAL DISEASES OF THE GASTROINTESTINAL TRACT IN ANIMALS

**S. P. Borriello and Robert J. Carman**

## TABLE OF CONTENTS

## I. INTRODUCTION

Clostridia can cause a great variety of different diseases both in domestic and wild animals. The diseases caused in livestock are economically very important, as too are those caused in animals held by research or breeding establishments. This chapter is devoted to an overview of the many diseases of the GI tract that can be attributed to clostridia.

With such a variety of different animal species and causative organisms to be described we have adopted the approach of discussing the various diseases under the heading of the known causative organism.

It is hoped that this work will be of interest not only to the veterinarians, but also to those interested in human disease, as frequently information on the etiology and pathogenesis of human disease can be obtained from the study of both natural and induced disease in animals.

## *II. CLOSTRIDIUM PERFRINGENS*

*C. perfringens* (*C. welchii*) is widespread in nature, being found in soil, sewage water, and human and animal intestinal tracts. Excellent reviews of the occurrence and toxigenic nature of the various types of *C. perfringens* already exist.[1-3] The species is subdivided into five major toxigenic types based on the major toxins produced and designated types A to E (Table 1). Due to its ubiquitous nature and toxigenic capability, it is an organism that is ideally suited to take advantage of any compromise in the defense mechanisms of animals and cause disease. It is probably true to say that *C. perfringens* accounts for more GI diseases of animals than many other bacteria. These diseases caused by *C. perfringens* will be dealt with on a toxin type basis.

## III. *CLOSTRIDIUM PERFRINGENS* TYPE A

Of the five toxin types of *C. perfringens* (Table 1), the most frequent in the animal bowel is type A. It is generally held that this is because it is a member of the normal gut flora. It is probable that the majority of published work which does not specify a toxin type does in fact refer to *C. perfringens* type A and we have therefore included such reports where applicable.

**Table 1**
**TOXIN TYPES OF *C. PERFRINGENS***

| Toxin types | Toxins | | | | | | | | | | | |
|---|---|---|---|---|---|---|---|---|---|---|---|---|
| | α | β | γ | δ | ε | η | θ | ι | κ | λ | μ | ν |
| A | ■[a] | –[b] | – | – | – | ±[c] | ± | – | ± | – | ± | ± |
| B | ■ | ■ | ± | ± | ■ | – | +[d] | – | + | + | + | + |
| C | ■ | ■ | ± | + | – | – | + | – | + | – | ± | + |
| D | ■ | – | – | – | ■ | – | + | – | + | + | ± | ± |
| E | ■ | – | – | – | – | – | + | ■ | + | + | ± | ± |

[a] ■ Major toxins used for designation of toxin types.
[b] – Not produced.
[c] ± Produced by some strains.
[d] + Produced by the majority of strains.

In humans, the most common enteric condition caused by type A is food poisoning which is caused by the ingestion of contaminated food and the subsequent production of an enterotoxin in the bowel (see Chapter 6). The enterotoxin differs from the other *C. perfringens* exotoxins by being fabricated intracellularly by sporulating bacteria and by being released when the cells lyse.[4] There are no reports of naturally occurring *C. perfringens*-mediated food poisoning of animals. *C. perfringens* type A enterotoxemia has been reported in animals, although it is by no means certain that the reports are reliable. The isolation of the organism from clinical material, albeit in high numbers, is not a good diagnostic feature as *C. perfringens* type A is a commensal in the bowel. Furthermore, it frequently increases in numbers following minor changes of the diet such as the carbohydrate component.[1] The criteria by which any enterotoxemia should be assessed are the isolation of toxigenic bacteria from gut content, the demonstration of an identical toxin in the gut and, strictly speaking, from the blood as well. Invariably, the test for toxemia is omitted which in most cases, in practical terms, is probably permissible. The difficulty is compounded by the frequent invasion of tissues by *C. perfringens* post-mortem, which, combined with its easy isolation, can create the impression that it was present as a causative organism.

There are numerous reports of spontaneous type A enterotoxemia in animals (Table 2). The major criticism common to many of these is that although *C. perfringens* type A was isolated from animals presenting with the clinical signs of enterotoxemia, toxin was not demonstrated in the gut. Consequently, only the more reliable or more interesting cases will be discussed.

### A. Lamb Enterotoxemia

In 1958, McGowan et al.[5] reported a type A enterotoxemia of young lambs. Known locally in California as yellow lamb disease because of the victims' jaundiced appearance, the clinical symptoms other than icterus were lethargy, occasionally collapse, rapid respiration, and clearly visible hemoglobinuria. The lambs in which both jaundice and hemoglobinuria occurred, with a single exception, all died. Autopsy revealed generalized icterus and pale mucous membranes. The livers were enlarged, yellowish, and friable. Sometimes white foci could be seen on the surface of the liver. Hepatic infarctions were not seen. Renal pathology was extensive. Kidneys were reddish brown, swollen, and soft; red conical zones and infarctions were often encountered. Lesions in other organs did occur but no consistent pattern emerged. In a number of the animals, reddening of the small bowel was noted; alpha toxin was detected in bowel content post-mortem and *C. perfringens* type A was isolated from the anterior ileum. It was possible to reproduce the disease by IV challenge with the cultures of *C. perfringens* isolated.

**Table 2**
***C. PERFRINGENS* TYPE A AND GI DISEASE**

| Condition | Species affected | Isolate shown to be α-toxigenic | α-Toxin in GI tract | Ref. |
|---|---|---|---|---|
| Enterotoxemia | Alpaca | ?[a] | ? | 7 |
| | Cattle | + | + | 111 |
| | Chicken | + | − | 8—10, 112 |
| | Dog | + | + | 6, 113 |
| | Goat | + | − | 114 |
| | Horse | + | − | 11, 12 |
| | Kangaroo | − | − | 115 |
| | Sheep | + | + | 5, 116 |
| | Rabbit | + | − | 117, 118 |
| | Reindeer | + | − | 119 |
| Acute gastric dilatation | Dog | + | − | 15, 120, 121 |
| | Primate | + | − | 16, 17, 122, 123 |
| | Other mammals | − | − | 15 |

[a] ?, Unclear from the literature.

## B. Canine Enterotoxemia

*C. perfringens* type A may cause an enterotoxemia in dogs.[6] A dog suffering from chronic but intermittent diarrhea carried high numbers of the bacterium in its gut and a 1:100 dilution of its loose feces contained a toxin neutralized by anti-alpha toxin. No predisposing factors contributing to either the initial bout of diarrhea or subsequent relapses were established. Full recovery of the dog followed a course of oral metrodiazole and change to a bland diet of chicken and rice.

## C. Alpaca Enterotoxemia

Alpacas are susceptible to type A enterotoxemia[7] at 10 to 14 days old. The susceptibility diminishes as they get older. Death can occur suddenly, with some calves failing to show symptoms. Most diseased animals have diarrhea. Treatment with antibiotics (aureomicin) and a vaccine (analcultivo) have been successful.

## D. Necrotic Enteritis in Fowl

Necrotic enteritis of chickens[8-10] is an enterotoxemia that can be caused by *C. perfringens* type A (see also *C. perfringens* type C). Superficially, it resembles ulcerative enteritis (see *C. colinum*) but in reality the two are quite separate.

Histologically, the condition is characterized by a coagulative necrosis of up to half the thickness of the ileal mucosa. *C. perfringens* type A can be cultured from the ileal contents. To date, toxin has not been demonstrated in the bowel of infected chickens; however, cell-free culture filtrates, containing neutralizable toxins, can reproduce the typical symptoms.[10]

## E. Equine Intestinal Clostridiosis

*C. perfringens* type A in horses can lead to an enterotoxemia known as equine intestinal clostridiosis, which is characterized by acute diarrhea.[11,12] Its diagnosis is based upon the presence of diarrhea and intestinal carriage of *C. perfringens* exceeding $10^4$ colony-forming units (CFU) per gram wet weight of feces. Control healthy animals carry less than 10 CFU per gram.[11,13] Using a clinical isolate of *C. perfringens* type A, Wierup[11] reproduced the disease in horses and detected circulating antibody to alpha toxin more often in the sick horses than in healthy ones. Although no absolute quantitative data on the titers are available, they were higher, though not significantly so, in spontaneously and experimentally infected

**Table 3**
***C. PERFRINGENS* TYPE B AND ENTEROTOXEMIA**

| Species affected | Ref. |
|---|---|
| Lamb | 18, 19, 2 |
| Sheep | 20 |
| Calf | 1 |
| Foal | 1 |
| Goat | 20 |
| Guinea pig | 21 |

horses than in healthy individuals, although the trend throughout was one of low titers. However, there is some indirect evidence from the work of Boyd et al.[14] that low titers of anti-alpha toxin may afford some protection. They studied the active immunization of sheep against gas gangrene and concluded, from the survival of some of their nonvaccinated controls, that titers of less than 0.1 unit/mℓ serum might well provide effective protection. Although it would be unwise to extrapolate from gas gangrene to enterotoxemia, there is at least the inferrence that even low antibody responses to alpha toxin offer some measure of protection.

### F. Acute Gastric Dilatation

Although acute gastric dilatation (AGD) has been reported from man, horses, ruminants, pigs, cats, foxes, mink, rabbits, nutria, guinea pigs, rats, mice, and dogs,[15] it is a disease which has received its most useful attention in nonhuman primates. In AGD the major problem presented by *C. perfringens* is not its toxins, although they have been implicated,[16] but is in fact the excessive gas produced by its metabolism. The disease occurs in four distinct forms; AGD, AGD with volvulus, AGD with rupture, and AGD with both rupture and volvulus.[15] No concurrent predisposing factors have ever been conclusively identified and it now seems likely that AGD is multifactorial in its origins, arising from a combination of gluttony, fermentative bacteria in the stomach, excess carbohydrate, a history of gastric disease, and mismanagement. Van Kruiningen et al.[15] proposed that volvulus occurs as a consequence of reverse peristalsis against the closed cardia of the stomach. Rupture is simply the result of the volume of gas produced by the gastric flora exceeding the capacity of the stomach. Clinical signs of AGD begin with discernible discomfort, inhibited movement, and the loss of appetite. Subsequently, animals adopt a hunched posture, at which stage a firm distended abdomen is apparent by palpation. Bacterial studies of the stomach ingesta in most cases reveal *C. perfringens* in diseased animals but not in healthy individuals.[17] The organism can be isolated from other organs if rupture occurs.

## IV. *CLOSTRIDIUM PERFRINGENS* TYPE B

The role of *C. perfringens* type B in animal enterotoxemia has long been known.[18,19] It has been reported in calves, lambs, adult sheep and goats, and guinea pigs (Table 3). This particular toxin type of *C. perfringens*-associated disease apparently occurs only in the U.K., Europe, South Africa, and the Middle East. It has so far not been found in North America, Australia, or New Zealand.[1] The interested reader is directed to the excellent reviews by Smith[1] and Sterne and Batty.[2]

### A. Lamb Dysentery

Lambs of up to 2 weeks are vulnerable and are infected by ingestion of the bacterium

**Table 4**
***C. PERFRINGENS* TYPE C AND ENTEROTOXEMIA**

| Species affected | Ref. |
|---|---|
| Sheep | 2, 125 |
| Pig | 22, 124 |
| Cow | 125, 126 |
| Horse | 127 |
| Alpaca | 7 |
| Buffalo | 128 |
| Camel | 129 |
| Guinea pig | 21 |
| Mink | 130 |
| Chicken | 24—26, 64 |
| Wildfowl | 131 |

from contaminated teats and pasture. Affected lambs adopt a posture usually described as "tucked up",[2] presumably to provide some relief from the pain. Their feces are semifluid, yellowish, and often flecked with frank blood, at which stage death generally ensues rapidly. At necropsy the major lesion is a severe hemorrhagic inflammation of the ileal submucosa. Gas bubbles are often discernible under the serosal surface of the ileum. Discrete ulcerations, with a tendency to coalesce, forming wide zones of necrosis, are commonly seen. If the ulcerations are severe enough to perforate the ileum, adjacent loops of bowel can become fused together and peritonitis may occur. Although these symptoms are pathognomonic of lamb dysentery, a confirmed diagnosis can only be made by demonstrating beta and epsilon toxins in the bowel content of affected animals.

### B. Sheep and Goat Enterotoxemia

Enterotoxemia of adult goats and sheep which is caused by *C. perfringens* type B is confined to the Middle East. The idiotype of the species involved produces the lethal beta and epsilon toxins typical of type B. However, it fails to produce the proteinase which is typical of lamb dysentery strains.[20]

### C. Guinea Pig Enterotoxemia

Germ-free guinea pigs monoassociated with *C. perfringens* type B(CDC KA41) either via their water supply or already associated neighbors, died rapidly, whereas bacteriologically conventional animals suffered no such fate.[21] Clinical signs were restricted to ruffled fur, lethargy, and flaccid paralysis. Pathological changes in the bowel were only slight. Some animals show signs of hyperemia in areas other than the cecum, which were invariably hemorrhagic and distended with fluid and gas. Mucosal sloughing in the cecum was frequent.

## V. *CLOSTRIDIUM PERFRINGENS* TYPE C

Although *C. perfringens* type C can cause enterotoxemia in a variety of animals species (Table 4), it most commonly affects sheep, lambs, calves, piglets, and fowl. In adult sheep the disease is known as "struck" and although sporadic outbreaks have been reported from a number of countries worldwide, the disease remains of particular economic importance in the Romney Marsh area of Kent, England.

### A. Struck in Sheep

The disease usually occurs mainly during early spring in sheep at pasture, and death of the animal is usually sudden with terminal convulsive episodes. Unlike the situation in lambs, diarrhea is not a feature of the disease in adult sheep.

At autopsy the peritoneal vessels, especially of the ometum, small intestine, and urinary bladder are congested and multiple subperitoneal hemorrhages may also be seen. There may be acute inflammation of the abomasum. There is an acute enteritis with, in most cases, ulceration of the small intestine (which occurs most frequently in the jejunum) and areas of necrosis. The large intestine remains normal. *C. perfringens* type C do not invade the healthy mucosa but are present in already necrotic tissue. The predisposing factors which allow the establishment and multiplication of this *Clostridium* in the GI tract and the uptake, through the gut wall, of the toxins produced are little understood. With respect to toxin uptake it is interesting to note that healthy animals dosed orally with large amounts of toxin remain healthy, and that in many cases of diseased animals a fatal toxemia develops at a time when the intestinal wall appears to be intact.

### B. Enterotoxemia in Lambs and Calves

The symptoms and lesions of the hemorrhagic enterotoxemia seen in lambs and calves are very similar and as such will be discussed together. The disease in lambs is clinically indistinguishable from classical lamb dysentery caused by *C. perfringens* type B (see above). The animals are affected within the first few days of life, frequently within the first 12 hr. Diarrhea, sometimes of sudden onset, is frequently seen. In the most severe forms of the disease there is an acute hemorrhagic enteritis with mucosal necrosis, the most prominent changes occurring in the jejunum and ileum.

### C. Enterotoxemia in Pigs

*C. perfringens* type C has also been incriminated in outbreaks of enterotoxemia in young pigs. The acute form of the disease is seen in pigs within the first 3 days of life and often within the first 24 hr. Although the clinical course of the disease is very short, some animals exhibit signs of dysentery prior to death. At post-mortem, acute inflammation of the small bowel, with the jejunum being particularly affected, and extensive necrosis of the intestinal mucosa can be seen. Many large Gram-positive bacilli are present within the necrotic tissue. It has been shown that *C. perfringens* type C cannot only adhere to the small intestinal villi of pigs, but once attached, can also proliferate, thereby releasing toxins at the site of action.[22] The role of the bacterial toxins in the pathogenesis of this disease of pigs has been confirmed by the reproduction of enterotoxemia by bacteria-free filtrates of *C. perfringens* type C.[23]

### D. Necrotic Enteritis in Fowl

The etiology of this disease was determined by Parish[24-26] and although attributed to *C. perfringens* type F, it is generally accepted that this toxin type should be considered as a subtype of *C. perfringens* type C. The factors that predispose birds to infection are little understood, although it is thought that high-energy diets and gut stasis may be contributory factors, and it is known that in many cases of necrotic enteritis subclinical coccidiosis occurs concurrently. The disease usually starts in flocks of birds that are 2 to 3 weeks old, and has been diagnosed as many as 3 times in 1 grow-out period. Affected birds show loss of condition and appetite, pass blood-stained diarrheic feces, and if made to move either creep or run and fall. Death occurs within 2 to 3 hr of symptoms becoming apparent. The histopathology was described in some detail by Parish.[24]

In general, at post-mortem the birds show signs of dehydration, have water-filled crops, and the liver is dark and congested. A pseudomembrane may be seen covering the mucosal surface of the small intestine or lying free in the lumen; the mucosa at this site will slough

**Table 5**
***C. PERFRINGENS* TYPE D AND ENTERITIS**

| Species affected | Ref. |
|---|---|
| Mouse | 27 |
| Guinea pig | 21 |
| Sheep | 132 |
| Goat | 133 |
| Cow | 134 |
| Buffalo | 128 |
| Camel | 129 |

off and may appear yellowish-white and bile stained. Lesions are particularly noticeable in the distal third of the small intestine. The basic lesion is a fibronecrotic enteritis with cystic dilatation of the deep glands.[24]

A necrotic enteritis of fowl has also been attributed to *C. perfringens* type A (see above).

## VI. *CLOSTRIDIUM PERFRINGENS* TYPE D

This organism can cause disease in a variety of animal species (Table 5). In most cases the disease is associated with concentrated feeding or overeating. The enterotoxemia caused by *C. perfringens* type D is especially important in sheep and is assuming increasing importance in calves. The disease is popularly known as "pulpy kidney" based on the characteristic degeneration of the kidney cortex in lambs. However, these degenerative lesions in the kidney are frequently absent from older sheep and calves. Frequent findings at death are endocardial hemorrhages, areas of congestion in the abomasum and small intestine, congestion and edema of the lungs, and excessive volumes of straw-colored fluid in the pericardium. Although there is no visible GI inflammation at autopsy, a mild enteritis is apparent on microscopic examination. Hyperemia and gas distention are present in short lengths of the small bowel. In calves, superficial desquamation in the intestine is seen. In general, the contents of the bowel are fluid, and diarrhea is a frequent symptom, especially in adult sheep and goats. In the goat the feces may be blood-stained and diarrhea is a symptom of a chronic course of the disease.

*C. perfringens* type D is a normal inhabitant of the soil and may also be isolated from the intestinal tract of healthy sheep in which small amounts of toxin are produced without ensuing disease. The predisposing factors that permit this opportunist pathogen to express its pathogenic potential are little understood, though in sheep at least they are known to be associated with concentrated feeding and overeating. In dry-lot feeding or sudden change to rich pasture, it is possible that substrates such as starch escape metabolism in the rumen, where the ruminal flora is changing in response to the change in diet, and so becomes available for metabolism by *C. perfringens* type D which proliferate rapidly and produce high levels of toxin in the intestine. Additional factors conducive to disease would be oral challenge with the organism and decreased peristalsis which could be induced by substances present in grass. It is highly probable that the normal ruminal flora acts as a barrier to *C. perfringens* type D reaching the intestine but that this barrier is compromised by microbial and physiological changes induced by sudden changes in diet.

In addition to these diseases of normal animals, *C. perfringens* type D has been associated with disease in a cesarean-derived, barrier-sustained mouse colony[27] and has been used to induce a disease in germ-free guinea pigs that resembled that seen to occur in guinea pigs

**Table 6**
***C. PERFRINGENS* TYPE E AND ENTEROTOXEMIA**

| Species affected | Ref. |
|---|---|
| Calf | 28, 29 |
| Sheep | 30 |
| Guinea pig | 31 |
| Rabbit | 34 |

soon after their removal from a germ-free environment.[21] In both of these cases it is probable that *C. perfringens* type D was able to cause disease because the exposed animals had a poorly developed natural GI flora. Exposure of conventional guinea pigs to this organism did not result in disease, whereas exposure of the germ-free animals resulted in disease in every case. Attempts to control the disease in mice included attempts to establish a competing microflora that would exclude *C. perfringens*. Although this approach to disease control failed, it must be noted that the attempts made were limited.

## VII. *CLOSTRIDIUM PERFRINGENS* TYPE E

Not entirely without justification, reviewing authors tend to disregard iota toxigenic *C. perfringens* (known as type E) as a pathogen. There are but four reports (Table 6) of spontaneous infection by the organism and, in the main, they are equivocal. There is to date no record of the isolation of *C. perfringens* type E from humans in either health or disease.

### A. Enterotoxemia of Cattle

Bosworth[28] reported on three cases of scouring in week-old ox-calves from a single herd.[28] All three animals presented with symptoms indicative of enterotoxemia. In the ceca of two, a toxin lethal to mice was demonstrated but not characterized. From one, *C. perfringens* type E was cultured. However, two further strains of iota producing *C. perfringens* were also isolated from the intestines of apparently healthy calves. Gross pathological changes were seen at autopsy in all three animals. The typical features were reddening of the ileum, petechial hemorrhage of the ileal mucosa, and yellowish, watery gut contents. The common histological appearance reported by Bosworth[28] was of central necrosis of some liver lobules with occasionally larger areas of hepatic damage. Lesions in the bowel consisted of intense capillary congestion leaving red cells distributed throughout the mucous membrane. No inflammatory infiltrate was reported. Attempts to reproduce the disease in very young calves failed. He concluded that there was insufficient evidence to absolutely incriminate *C. perfringens* type E as the organism causing enterotoxemia among calves.

Acute necrotic hemorrhagic enteritis was the common feature of iota enterotoxemia in 2 calves under 2 weeks of age[29] in which the pathological changes seen were essentially the same as those reported by Bosworth.[28] The diagnoses were based upon the demonstration of a toxin in the small intestine which was lethal to mice but otherwise uncharacterized, and the isolation of *C. perfringens* type E from the kidneys, though not the bowel, of the calves within 20 min of their deaths. Hemolytic *C. perfringens* isolated from the small intestines proved to be type A, reflecting all too well the problem of selecting appropriate colonies for toxin typing. The outbreak of enterotoxemia reported by Hart and Hooper[29] lasted only 4 weeks. No treatment was attempted although a vaccination program was proposed, as were measures to reduce the level of heat stress suffered by the cattle which were frequently left without shade in temperatures in excess of 35°C.

**B. Lamb Dysentery**

The first description of *C. perfringens* type E by Bosworth[28] made other workers aware of its potential occurrence in the bowel contents and stools of diarrheic animals. Ross et al.[30] examined many such specimens between 1943 and 1947 but found only 4 isolates while investigating an outbreak of lamb dysentery (see *C. perfringens* type B). They identified iota toxin in the mixed cultures from 40 of 100 swabs from the udders of ewes; 7 of the 40 toxin-positive mixed cultures yielded *C. perfringens* type E. No mention was made of the frequency of isolation of type B from the clinical material. However, as none of the ewes were described as being diarrheic and as no etiological role was suggested for type E in the pathogenesis of lamb dysentery, it may be that the isolation of type E was incidental to the disease.

**C. Guinea Pig Enterotoxemia**

It was noted that when adult germ-free guinea pigs were moved to conventional surroundings many of them became ill and subsequently died.[31] Deaths occurred within 2 to 10 days and gross pathology was similar to that described in calves.[28] *C. perfringens* were isolated from diseased animals. An important new observation was that prior to death, the guinea pigs suffered with general flaccid paralysis. Paralysis, particularly of the hind limbs, can be seen in mice following intraperitoneal inoculation with a cell-free filtrate of iota toxin-positive material.[32] Unfortunately, the laboratory diagnosis of iota enterotoxemia did not include the demonstration of toxin in cecal material from the guinea pigs. The toxin assay of isolates of *C. perfringens* was also unsatisfactory. Overnight culture filtrates were lethal to mice but could be neutralized by type A antitoxins. Samples activated by trypsinization were not neutralized with antitoxins to types A, B, and D. Despite not attempting neutralization with types C and E, the authors concluded that the isolates were iota toxigenic type E. Although iota toxin has been reported as being a protoxin requiring trypsin activation,[30] no such treatment was needed for overnight cultures; indeed, it has been shown that after 6 hr of incubation at 37°C all iota protoxin becomes activated despite the absence of trypsin.[30] Consequently, the diagnosis of iota enterotoxemia by Madden et al.[31] may have been hasty. These criticisms notwithstanding, the same workers[21] showed that germ-free guinea pigs monoassociated with *C. perfringens* type E either via the water or by housing the guinea pigs in the same cages as already contaminated animals suffered from a per acute enterotoxemia of a similar nature to the spontaneous disease. Bacteriologically conventional guinea pigs were resistant to the challenge.

**D. Rabbit "Iota"-Enterotoxemia**

The reports of *C. perfringens* type E as the cause of rabbit iota enterotoxemia are reviewed below under the heading, *C. spiroforme* as it is now evident that the disease is due to *C. spiroforme* which produce a toxin that can be neutralized by anti-*C. perfringens* type E iota toxin.[33] However, there is one report[34] of experimentally induced *C. perfringens* type E iota enterotoxemia which is also discussed below.

## VII. *CLOSTRIDIUM SPIROFORME*

Curved and helically coiled Gram-positive organisms have often been seen in the gut of animals. Few are isolated and properly identified. *C. spiroforme* is the only such bacteria connected with a clinical condition (Table 7). It is an organism with a very characteristic helical morphology when grown on agar plates.[35] It was first isolated from the feces of normal adult humans[36] and later correctly identified as *C. spiroforme*.[35] Further isolates were recovered from healthy chickens and humans by Kaneuchi et al.,[35] and Gossling et al.[37] have reported the recovery of similar organisms from the intestines of leopard frogs (*Rana pipens*).

**Table 7**
***C. SPIROFORME* AND ENTEROTOXEMIA**

| Species affected | Ref. |
|---|---|
| Rabbit | 33, 38, 39 |
| Cottontail | 51 |
| Guinea pig | 53 |
| Hamster | 53 |
| Rat | 53 |
| Mouse | 53 |

### A. Rabbit "Iota"-Enterotoxemia

Many of the papers on iota enterotoxemia of rabbits speculate on the role of *C. perfringens* type E in the etiology of the disease. In fact, the disease is now known to be caused by *C. spiroforme*.[33,38,39]

While endeavoring to establish a rabbit model of human antibiotic-associated PMC, Katz et al.[40] chanced upon a previously unreported enterotoxemia following challenge with clindamycin. A rise in total clostridia from $10^5$ to $10^8$ CFU per gram of cecal material was recorded. Coadministration of clindamycin and vancomycin produced no illness. Gross pathology in diarrheic animals was restricted to the ceca which were highly distended with watery contents. Histology showed necrosis of the cecal epithelium and an inflammatory infiltrate in the lamina propria. The same workers reported a toxin from the ceca of scouring animals which was lethal to mice but could be neutralized by *C. perfringens* type E iota antitoxin.[41] However, they were unable to isolate *C. perfringens*, nor was there any evidence of either *C. difficile* or its toxins, the organism and toxin they might have expected to encounter.

Orcutt et al.[34] described the first cases of spontaneous iota enterotoxemia among rabbits of various ages. The common features of the condition were hemorrhagic ceca and the presence of iota toxin. Although they reported the successful cultivation of *C. perfringens*, identified as type E with commercial fluorescent antiserum, they were unable to demonstrate the in vitro production of iota toxin by their isolates.[42] Using a bovine strain of type E *C. perfringens*, iota enterotoxemia could be produced in otherwise healthy rabbits and a formalized toxin was able to protect against the experimental disease.

In 1978, Patton et al.[43] published a report on rabbit enterotoxemia which has become something of a seminal reference. Over a 23-month period, 46 rabbits of all ages died with a presumptive enterotoxemia. Gross pathology and histology was similar to that described by Katz et al.;[40] iota toxin was found in the cecal content of 23 rabbits and *C. perfringens* was isolated from 12. Thereafter the paper is confusing. It fails to state whether or not the isolates were toxin typed, nor does it make clear if any of the 12 animals from which *C. perfringens* was isolated were also toxin-positive. It transpires that the isolates were all type A.[44] Whether or not toxin-positive material was culture-positive remains unclear. Eaton and Fernie[45] described the history of a colony of laboratory rabbits which had suffered a severe epidemic of iota enterotoxemia predominately at weaning. They outlined the differential diagnoses and gave a very full description of the pathognomic symptoms and histology seen in the condition. Studies of the incidence of clostridia and their toxins in the same outbreak were reported by Fernie and Eaton.[46] As in the work reviewed above and subsequent additions to the literature,[47,48] iota toxin could be demonstrated but in no instance was *C. perfringens* cultured. Fernie and Eaton[46] concluded that neutralization of the toxin by iota antitoxin may have resulted from serological cross reactions, an assumption also reached by previous workers.[41]

The possible role of *C. spiroforme* in the etiology of this condition in primarily post-weaning rabbits was first proposed by Carman and Borriello.[38] *C. spiroforme* was isolated from all 36 of both spontaneous and clindamycin-induced iota toxin-positive cecal material and only 2 of 72 (3%) of seemingly healthy, toxin-negative specimens. The clinical material came from several outbreaks, including three reviewed above and from commercial rabbiteries in France. The role of *C. spiroforme* in the disease was given further credance by the demonstration that the organism produced a toxin neutralized by *C. perfringens* type E antitoxin.[33,39] More recently, we have shown that oral administration of *C. spiroforme* to rabbits pretreated with clindamycin induces a fatal enterotoxemia. Adult animals receiving organism or antibiotic alone and kept in isolation survived.[49] A number of workers have shown that bacteria-free filtrates of the cecal contents of diseased rabbits are cytotoxic to cells in tissue culture[41,48] and we have recently shown that cell-free filtrate of *C. spiroforme* are also cytotoxic.[50] It is possible that demonstration of the presence of cytotoxicity and its neutralization by specific antitoxin may be used diagnostically and obviate the need for live animal toxin assays. A diagnosis of iota enterotoxemia was made in two 3-week-old cottontail rabbits presenting with hemorrhagic cecitis.[51] Samples taken within 1 hr of death yielded *C. spiroforme*. The filtered and centrifuged cecal contents injected intracecally into young, healthy cottontails were able to reproduce an identical hemorrhagic cecitis.

The pathogenesis of *C. spiroforme* iota enterotoxemia has yet to be fully elucidated. However, recent evidence would suggest that disease is due to exposure of animals with an incomplete GI flora, either due to age or e.g., antibiotic treatment to toxigenic *C. spiroforme*. It is probably not due to overgrowth of small numbers of the organism normally resident in the gut.[49] These proposals are based on the observations that *C. spiroforme* cannot be isolated from known healthy rabbits,[33] that the organism fails to establish or cause disease when given to normal healthy rabbits, and that antibiotic treatment alone in rabbits housed in isolators fails to induce disease. Both exposure to *C. spiroforme* and some procedure to disrupt the normal gut flora are required to produce disease.[49]

Although no work has been reported on the potential of vaccination against *C. spiroforme* iota enterotoxemia there are ample precedents within the genus to suggest that a formalized toxin vaccine may be effective. More immediately, vancomycin treatment of enterotoxemic animals would seem to be possible. Evidence to this effect is provided by Katz et al.,[40] who were unable to induce clindamycin-associated diarrhea in rabbits given vancomycin at the same time as the latter antibiotic. It is also probable that bacitracin would be useful as it has a similar spectrum of activity to vancomycin, is also poorly absorbed, thereby ensuring high GI levels, and has the advantage of being cheaper than vancomycin.

It is interesting to speculate that *C. spiroforme*-associated enterotoxemia was first seen by Lesbouyries and Berthelon[52] who reported the presence of curved organisms in the ceca of scouring rabbits. Abundant curved bacilli can be seen in the cecal contents of diseased animals and are thought to be the in vivo growth form of *C. spiroforme*.[50]

### B. Guinea Pig Antibiotic-Associated Enterotoxemia

Three guinea pigs given clindamycin orally for 3 days and caged beside rabbits with iota enterotoxemia succumbed to infection by *C. spiroforme*.[53] Previously, culture for the organism had been negative. The course of the disease in the guinea pigs was very similar to that in rabbits. Death followed the first signs of diarrhea within 12 hr; iota toxin was demonstrable in cecal content collected post-mortem. The experimental procedures were refined by isolating guinea pigs to prevent exogenous colonization by *C. spiroforme* following treatment with clindamycin. Guinea pigs held in isolators remained free of disease. After the end of the course of antibiotic (7 days), animals challenged orally with *C. spiroforme*

($10^5$ CFU per animal) became diarrheic within 72 hr; iota toxin and toxigenic *C. spiroforme* were detected in cecal material post-mortem. Controls were negative for both.

### C. Hamster Antibiotic-Associated Enterotoxemia

A fatal enterotoxemia has also been reproduced in the Syrian hamster by a combination of antibiotic pretreatment and challenge with *C. spiroforme*.[53] In all cases a toxin that could be neutralized by *C. perfringens* type E antitoxin could be demonstrated in the cecal contents and *C. difficile* and its associated cytotoxin were shown to be absent.

### D. Red-Leg Partridge "Frothy Gut"

Among the conditions of game birds in the U.K. is an ill-defined syndrome known as "frothy gut". Typically, frothy gut is associated with yellow foamy droppings and high levels of intestinal protozoans.[54] Of interest to the bacteriologist is the observation that, at post-mortem examination of diseased red-leg partridges, the cecal flora is characterized by many helically coiled bacteria. Attempts to show the presence of iota toxin and to culture the organism have been fruitless and their role, if any, in frothy gut pathogenesis is not clear.[55]

### E. Human Disease

As it has been shown that *C. spiroforme* has the potential to cause disease in a variety of animal species, it is of interest to note that eight diarrheic stools from one of the authors (RJC) collected during two different courses of ampicillin and one of oxytetracycline at no time harbored either *C. spiroforme* or iota toxin despite constantly working with the organisms during this time. No incidences of iota enterotoxemia are known in humans. The type strain of *C. spiroforme*, NCTC 11211, (ATCC 29900) which was isolated from humans, is not toxigenic.[33] However, in a study on the effects of clindamycin on the fecal flora of volunteers, *C. ramosum* was noted in the stools of those with diarrhea, as was *C. difficile*.[56] On primary isolation, *C. ramosum* may have many of the morphological features of *C. spiroforme*, which at this time had yet to be described. It is therefore possible that *C. spiroforme* may have unknowlingly been isolated from the stools of subjects with clindamycin-associated diarrhea. The possible role of *C. spiroforme* in diseases of the human GI tract remains to be determined.

## IX. *CLOSTRIDIUM DIFFICILE*

There are only two naturally occurring diseases of the GI tract of animals due to *C. difficile* recorded in the literature. This organism has only recently been shown to be the cause of diarrhea and colitis in man (see Chapter 7) and although the usual course of events is that a *Clostridium* sp. is known to cause disease at this site in a variety of animals long before its involvement in human disease is demonstrated, it may well be that the converse will be shown to be the case with *C. difficile*.

A spontaneous disease due to *C. difficile* has been recently described in the neonatal hare.[57] Neonatal diarrhea in young, conventional hares raised in captivity is common and frequently precedes death of the animal, with a mortality peak at 7 days after birth.[57] A similar disease can be reproduced in young gnotobiotic hares by exposure to *C. difficile*. Diarrhea coincided with the appearance of large numbers of *C. difficile* in fecal material and preceded death by 1 to 3 days. Induction of diarrhea occurs sooner if the animals are diassociated with *C. difficile* and *C. perfringens*, and this may also be the case for *C. difficile* and *C. tertium*. All three organisms are found more frequently in animals with diarrhea than in healthy animals, with all three species being present simultaneously in 80% of diseased animals but in none of the healthy controls. *C. difficile* is found in 100% of diarrheic cases but only 23% of healthy animals.[57]

**Table 8**
**ANIMAL CARRIAGE OF *C. DIFFICILE***

| Animal | Symptoms | Ref. |
|---|---|---|
| Weddel seal | − | 135 |
| Cow | − | 136, 137 |
| Horse | − | 137 |
| Donkey | − | 137 |
| Camel | − | 137 |
| Cat | − | 62 |
| Dog | − | 62 |
| Duck | − | 62 |
| Goose | − | 62 |
| Deer | + | 62 |
| Bear | + | 61 |
| Rat (laboratory) | − | 138 |
| Hare | + | 57 |
| Pig | + | 58, 60 |
| Mouse (laboratory) | − | 139 |

Recently, *C. difficile* was isolated from necrotic lesions in the colon of two pigs affected with a naturally occurring enterocolitis not associated with swine dysentery.[58] In both animals a diphtheritic layer which showed patches of necrosis and ulceration affecting the superficial layers covered the mucosa of the cecum, colon, and rectum. Although *E. coli* serotype E4 and *Salmonella typhimurium* phage type 12 were also isolated, these organisms are infrequently associated with disease in the pig and on these grounds were discounted as playing a major part in the disease process. In addition, *Treponema hyodysenteriae*, the proposed major etiological agent of swine dysentery, could not be shown to be present. These observations and the absence of *C. difficile* in an unspecified number of presumably healthy pigs prompted the authors to suggest that further investigations into the possible pathogenicity of *C. difficile* in pigs should be undertaken, particularly since many growing pigs are given oral antibiotics in the treatment of enteric disease and as growth promoters, a practice which may predispose them to *C. difficile*-mediated enterocolitis. *C. difficile* was also shown to be absent from the feces of healthy pigs[59] and 18 pigs with swine dysentery.[60] Further evidence that *C. difficile* may have played a primary role in this disease is the ability of this organism to cause lesions in the cecum and colon of gnotobiotic pigs after oral challenge.[60]

In the majority of human cases, *C. difficile*-mediated disease is associated with prior antibiotic therapy and there have been a few cases of nonlaboratory animals that have developed this disease after treatment with antibiotics, e.g., an American brown bear which responded to vancomycin therapy.[61]

Asymptomatic carriage of *C. difficile* has been shown in a number of animal species (Table 8) and the potential for disease probably exists in many of them. However, a recent study has shown that even though carriage was higher in animals that had recently received antibiotics than in those which had not, the difference was not statistically significant and carriage of the organism did not appear to be associated with GI disease in either group.[62] So although disease can be readily induced in laboratory animals such as the hamster (see Chapter 7), examples of naturally occurring disease are as yet rare.

## X. *CLOSTRIDIUM SORDELLII*

There have been rare reports of the possible involvement of *C. sordellii* in enterotoxemias of a variety of animals (Table 9), the most recent involving cattle and sheep. In a study of

**Table 9**
***C. SORDELLII*-ASSOCIATED GI DISEASE**

| Species affected | Ref. |
|---|---|
| Cow | 63, 64 |
| Sheep | 65 |
| Horse | 140 |
| Dog | 66 |

bacteria associated with inflammatory lesions of the bovine GI tract, *C. sordellii* was isolated from two animals with GI disease.[63] The one animal was an adult with hemorrhagic diarrhea and the other a calf with necrotic enteritis and soft feces. In both cases congestion and necrosis of the small intestine were present. Pure cultures of an isolate of *C. sordellii* obtained from the adult cow were used to infect calves in a number of experiments designed to confirm the role of this organism in GI disease.[63] Oral challenge of calves with suspensions of *C. sordellii* resulted in two or more episodes of soft feces with mucous and occasional blood over a 10-day period. In all cases, the jejunal and ileal contents were fluid and mucoid and the contents of the large intestine were slightly mucoid. There was inflammation of the jejunum and ileum and inflammation and disruption of the epithelium of the large intestinal mucosa. Calves that were challenged orally with cell-free filtrates of spent *C. sordellii* culture medium passed firm or pasty feces with excess clear mucus containing streaks of fresh blood. Both gross and microscopic changes of the GI tract resembled those induced by challenge with live organism. *C. sordellii* was also isolated from a cow with GI disease by Brooks et al.[64] However, they were unable to produce disease in calves by the oral inoculation of pure cultures of the organism. Hemorrhagic ileitis in sheep has also been attributed to *C. sordellii* infection.[65] In one case, *C. sordellii* was isolated in high numbers from the necrotic hemorrhagic small intestine of an ewe that had suffered from diarrhea for some time prior to death. Filtrates of the intestinal contents were not lethal for mice; however, the toxigenic status of the isolate was not reported.[65]

It is worth mentioning that a toxin neutralized by *C. novyi* and *C. sordellii* antitoxins has been implicated in disease due to experimental strangulation obstruction. It was determined that the lethal toxin of *C. sordellii* was the only active agent in this model and that neutralization by *C. novyi* antitoxin was due to a cross-reaction.[66]

## XI. *CLOSTRIDIUM COLINUM*

This organism has been implicated in an ulcerative enteritis of a variety of avian species (Table 10). As the disease most frequently affects quail it is commonly known as "quail disease". The etiology of this disease remained unknown until the implication of an anaerobic, Gram-positive, sporeforming rod which was given the name *C. colinum*.[67,68] The disease) can affect any part of the GI tract although the ileum and rectum are most commonly affected. As the name suggests, the disease is characterized by the presence of ulcers which originate as small yellow foci with a hemorrhagic border which may coalesce, giving rise to large necrotic areas. The degree and nature of the gross lesions depends on the time between infection and death. In birds dying early, only a hemorrhagic enteritis which is restricted to the proximal intestine is seen. At the other end of the spectrum there are cases of necrosis of the entire intestine resulting in a pseudomembranous-like enteritis with sloughing of mucosa. The sequence of discrete small yellow foci through to a pseudomembranous-type enteritis has some parallels to *C. difficile*-mediated PMC in humans, although ulcers

**Table 10**
***C. COLINUM* AND ULCERATIVE ENTERITIS**

| Species affected | C. colinum isolated | Ref. |
|---|---|---|
| Quail | + | 69, 141—143 |
| Grouse | —[a] | 69, 141, 143 |
| Partridge | —[a] | 69 |
| Turkey | —[a] | 69 |
| Pigeon | —[a] | |
| Pheasant | + | 142, 143 |
| Robin | + | 144 |
| Chicken | + | 142, 145 |

*Note:* [a] = no data.

are not seen in the human situation (Chapter 7). The intestinal lesions in ulcerative enteritis are almost always accompanied by areas of necrosis in the liver. A detailed description of the histopathology of this disease is given by Durant and Doll.[69]

The main evidence implicating *C. colinum* in the etiology of ulcerative enteritis was the isolation of this organism from pathological material from a variety of avian species with natural disease, the induction of disease in healthy animals by oral or intravenous challenge with these strains of *C. colinum*, the reisolation of this organism from the diseased animal, and the ability of these isolates to repeat the above cycle.[70] These organisms have also now been demonstrated in infected tissue by use of a fluorescent antibody test.[71] The pathologic features of ulcerative enteritis strongly suggest the involvement of a toxin; however, to date it has not been possible to show production of a toxin by this organism, or to link a biologically active toxin to the disease.

It is also interesting that the strains of *C. colinum* tested have been resistant to streptomycin in vitro, as this antibiotic has been successfully used to control the naturally occurring disease.[67] The most obvious implication of this finding is that the antibiotic disrupts the microenvironment such that it is unfavorable to the establishment of and expression of pathogenicity by *C. colinum*. There is certainly evidence that in chickens, *C. colinum* alone is not sufficient to induce ulcerative enteritis as it is frequently found associated with intestinal coccidiosis,[72] and attempts to induce the disease in healthy chickens by oral challenge with up to $10^9$ *C. colinum* organisms have consistently failed.[70]

The finding of intestinal ulcers in this disease has attracted the interest of one group of workers currently investigating the etiology of Crohn's disease. However, they failed to demonstrate any involvement of this organism.[73] Although *C. colinum* causes disease in a variety of avian species, it is almost always restricted to gallinaceous birds and has not yet been implicated in disease of any nonavian.

It is important at this point to make a distinction between necrotic enteritis due to *C. perfringens* type C (see above) and ulcerative enteritis in chickens. The distinguishing features of necrotic enteritis are that it involves only the small intestine and skip lesions may occur. At the sites of mucosal necroses, lesions tend to coalesce rather than remain as discrete ulcerations, and save for congestion, liver pathology is rare. It has also been shown that quail inoculated with intestinal scrapings from chickens with necrotic enteritis suffer no ill effects whereas inocula prepared from chickens with ulcerative enteritis caused disease.[74]

## XII. *CLOSTRIDIUM FALLAX*

There has been one report implicating this organism in an hemorrhagic enteritis in young

**Table 11**
*C. CARNIS*

| Host | Disease | Ref. |
|---|---|---|
| Muskrat | Errington's disease | 76, 78—81 |
| Racoon | Errington's disease | 80 |
| Mink | Septicemia | 77, 84, 85 |
| Man | ?[a] | 86 |

[a] ?, Evidence of possible involvement in disease in man derived from one case report only.

chickens.[75] The disease caused high mortality in a flock of 200 9-week-old Black Australorps on a farm in the southern region of Malawi. Death, usually within 3 days of symptoms, followed listlessness, inappetence, and diarrhea. Hemorrhagic enteritis, splenic enlargement, and congestion with petechial hemorrhages and subcutaneous edema were seen at postmortem. The description of this disease shows many similarities to ulcerative enteritis (see above). In particular, GI lesions were restricted to the small bowel, the spleen was enlarged, and hemorrhagic coccidiosis appeared to be a predisposing factor. Death followed within 3 days of initial symptoms and young birds were affected. In view of these similarities it is important to look more closely at the incriminated organism.

*C. fallax* is very similar to *C. colinum*[68] (the etiological agent of ulcerative enteritis), although they are easily differentiated on the basis of the volatile fatty acid end products (VFAs) of carbohydrate metabolism. However, VFAs were not analyzed in this study and *C. colinum* had not yet been described. Unlike the situation with *C. colinum*, oral challenge of healthy birds with *C. fallax* induced disease, although attempts to further passage the disease failed. However, on the basis of the identification manuals and techniques used the possibility that the yet to be described *C. colinum* was the organism isolated and not *C. fallax* must be entertained.

## XIII. *CLOSTRIDIUM CARNIS*

The isolation of *C. carnis* or a very similar organism from muskrats and mink has been infrequently reported.[76,77] Despite the elapse of almost 40 years since the first observations, little progress has been made beyond the attempted speciation of the isolates and some informed speculation upon the association of *C. carnis* with diarrhea. The field is ripe for research as is testified by the small literature on the subject (Table 11).

Errington's disease is an infectious and fatal disease of muskrats. It was first described by Errington in 1946[76] as an epizootic of muskrats in the midwestern U.S. The major features of the disease were hemorrhagic enteritis in one or more areas of the intestines, focal liver necroses, and a high mortality. Clinical symptoms have not been recorded. Only rarely were mink and racoons feeding upon dead muskrats victims to similar pathologies. Lord et al.[78,79] reported very fully on the gross pathological changes and the histological appearance of the viscera; their findings were largely in accord with those of Errington.[76,80] Of interest was the report of a degeneration and necrosis of the submucosal epithelium without a cellular infiltrate. Lord et al.[78] also described a thick necrotic membrane of degenerated epithelial cells, leukocytes, and erythrocytes emeshed in fibrin. A terminally sporing bacteria was seen in the sites of liver necrosis and within the intestinal submucosa. A species of *Clostridium*, subsequently given a tentative identification of *C. carnis*, was seen in Gram-stained smears of affected organs and was cultured from the same specimens. The isolates were not pathogenic for guinea pigs or mice.[79] However, muskrats housed in cages previously inhabited by diseased animals died, even following decontamination with 5% creolin. No deaths resulted from exposure of healthy animals to uncontaminated cages.

In a short but useful review of Errington's disease, Karstad[81] notes criticisms of the conclusions drawn by Lord et al.[78,79] that *C. carnis* was the causal organism of this disease — the major one being that clostridia can be abundant and harmless in the bowel, liver, kidneys, and environment of muskrats. This is clearly true but is perhaps not an altogether valid reason for disregarding *C. carnis* or a similar bacterium as the organism responsible for Errington's disease. Further reservations about the role of *C. carnis* were expressed by Spalatin et al.[82] who attributed lesions similar to those of Errington's disease to a species of *Chlamydia*.

Much work remains to be done on Errington's disease. Notable among the outstanding projects are the investigation of toxins in the cecal content of diseased muskrats; the occurrence of both disease and organism in species such as the racoon[80] and lastly, the studies into the relationship between Tyzzer's Disease and Errington's disease.[83]

Although not strictly within the realms of this work, mention should be made of three reports of two instances of the isolation of *C. carnis* from the livers of ranched mink in Denmark and the U.S.[77,84,85] In none of these reports was there mention of pathology in the gut. Lesions were confined to the liver and kidneys and were consistent with septicemia.

Experimental work has shown that *C. carnis* is pathogenic for many animal species[85] including the cat, dog, rabbit, mouse, sparrow, frog, and axolotl. The organism has also been isolated from an 8-week-old boy who had diarrhea and was admitted to hospital dehydrated and in severe shock.[86] The organism was isolated in pure culture from the blood. The authors state that *C. carnis* was not recognized in feces because anaerobic culture was not performed. However, it should be noted that *C. carnis* is a microerophile capable of growth under aerobic conditions and should have been present on the medium incubated aerobically. Although the infant had diarrhea, barium studies of the GI tract carried out during convalescence revealed no abnormality. The authors infer in conclusion that this case removes any question that *C. carnis* is pathogenic for man.

## XIV. *CLOSTRIDIUM SEPTICUM*

*C. septicum* causes a disease in sheep known as "braxy", and usually affects lambs during the first year of life, especially those on farms in mountainous areas. The disease is rare in the Americas and Australia, occurring most commonly in Scotland, Iceland, and Scandinavia.

The first signs of illness are loss of appetite and signs of abdominal pain, and sheep with braxy are usually found dead or die within a few hours of being noticed ill. The presumed pathogenesis of braxy is that lambs ingest frozen food which abrades and chills the abomasum, thus permitting invasion of *C. septicum* and multiplication of the organism in the mucosa and submucosa, ultimately leading to a fatal bacteremia and toxemia. In some cases infection appears to originate at the bite points of nematodes. Because the predisposing factors are so poorly understood, it has not been possible to reproduce the disease experimentally. However, the facts that *C. septicum* can be isolated in pure culture from lesions in the abomasum and that the disease can be prevented by immunization against *C. septicum* is strongly indicative of a primary etiological role for this organism. In addition, *C. septicum* produce at least four toxins and is pathogenic for a variety of laboratory animals. Typical lesions noted in sheep with braxy are areas of hemorrhagic inflammation in the abomasum wall. The inflammation may be diffuse or present as many sharp foci. There may be some submucosal gas and edema, and usually an excessive quantity of blood-stained fluid is to be found in the peritoneal cavity.

## XV. CLOSTRIDIAL INFECTIONS EMANATING FROM THE GUT

Some species of clostridia which inhabit the bowel of animals, though not causing enterotoxemia per se are toxigenic and do give rise to frequently fatal conditions once they have crossed the gut wall barrier. In these instances the pathological changes within the gut are of less consequence than the ensuring bacteremia and toxemia. *C. novyi, C. haemolyticum*, and *C. botulinum* are such organisms. They will be discussed here very briefly. The interested reader is directed to a fuller review by Buxton and Fraser.[87]

***Clostridium novyi*** — *C. novyi* type B is the cause of necrotic hepatitis (black disease, bradsot) of sheep, of a very similar condition of cattle, and of a septicemic disease of chelonians.[88] Necrotic hepatitis is characterized by a bacteremia, a toxemia, and associated enterotoxic symptoms. Immature forms of parasites, usually the fluke *Fasciola hepatica*, migrate to the liver where they burrow, creating localized damaged tissue in which *C. novyi* type B multiplies and extends the area of necroses by release of toxin. A bacteremia and toxemia ensue. The source of the infection is thought to be ingested spores, some of which cross the mucous membranes, probably in phagocytes, remaining as latent infection in organs such as the liver. Although relatively difficult to grow, *C. novyi* can be isolated. However, direct fluorescent antibody tests on infected tissues have facilitated diagnoses. If these are unavailable it may be best to demonstrate *C. novyi* toxin in pleural and pericardial effusions.

***Clostridium hemolyticum*** — The disease of cattle, bacillary hemoglobinuria (red water disease), is caused by *C. hemolyticum*. At autopsy, one can see an anemic infarct surrounded by an area of congestion. As with necrotic hepatitis, the nidus of the lesion is created by damage following fluke penetration into the liver and germination of latent spores. Instances of red water disease have also followed liver biopsy.[89] Hemolysis occurs following the formation of the infarct and hemolyzed blood enters the bowel, particularly the colon. Hemoglobinuria is clearly apparent, the bladder containing urine often said to be the color of port wine. Frequently the kidneys are covered in petechia and occasionally the ileum can be extremely hemorrhagic. *C. hemolyticum* is even harder to isolate than *C. novyi* type B. The best diagnostic methods are fluorescent antibody tests on the lesions in infected organs. Liver extracts or effusions from the pericardia of diseased cattle will contain neutralizable β-lecithinase. The absence of alpha toxin distinguishes red water disease from necrotic hepatitis.

***Clostridium botulinum*** — It is generally accepted that botulism in animals is due to ingestion of preformed toxin, and it is of interest that some carrion-eating animals appear to be resistant to the toxins of this organism.[90] However, it has recently been shown that botulism can occur in human infants due to ingestion of *C. botulinum* and absorption of toxin formed in the gut, and animal models have been developed to mimic this form of infant botulism (see Chapter 2). In the light of these observations it is probably worth reevaluating the premise that botulism in animals is always due to ingestion of preformed toxins.

## XVI. TYZZER'S DISEASE

In 1917, Tyzzer described an epizootic disease of fatal consequence in laboratory mice. The condition, manifesting as a diarrhea, has since become widely known as Tyzzer's disease and the causative organism, described as a pleomorphic, Gram-negative, subterminally sporulating rod[91] has been called *Bacillus piliformis*. It is in laboratory animals that much of the research into Tyzzer's disease has been done. However, progress has been hampered by the inability of workers to culture the organism in synthetic medium. Tyzzer[92] himself claimed to have briefly grown an organism resembling, in morphology, the bacillus he had seen in clinical material from which he drew his initial inoculum. There have been three

**Table 12**
***BACILLUS PILIFORMIS* AND TYZZER'S DISEASE**

| Species affected | Ref. |
|---|---|
| Mouse | 91—93, 99, 101, 107, 146—150 |
| Guinea pig | 151—153 |
| Hamster | 154—157 |
| Muskrat | 108—110 |
| Rabbit | 158, 159 |
| Cottontail | 160, 161 |
| Gerbil | 162—166 |
| Cyote | 102 |
| Rhesus monkey | 167 |
| Cat | 168—171 |
| Horse | 172—177 |
| Dog | 178, 179 |
| Grey fox | 180 |
| Rat | 181—184 |

more recent reports of successful in vitro growth in accellular media.[93-95] Doubt remains about the relevance of the two earlier reports.[91] The organism remains as a poorly characterized pathogen and is not included in the approved list of bacterial names[96] or in the 8th edition of *Bergey's Manual of Determinative Bacteriology*.[97,98] Confusion exists regarding the classification of Tyzzer's *Bacillus*[91] and there is no firm evidence published to date to place it in the genus *Bacillus*. The gram-negative staining of the organism reported by Fries[91] in his excellent review of Tyzzer's disease is by no means uncommon among the clostridia and it is no more suggestive of *Bacillus* sp. than *Clostridium* sp. The organism is sensitive to penicillin[99,100] which is typical of clostridia, and Tyzzer claimed to have grown the bacterium in an anaerobic atmosphere.[92] In addition, the organism is spore-forming with contaminated bedding remaining infected for over 12 months.[92] Although spontaneous epidemics are frequent, commoner still is the activation of either latent infections or the creation of disease in previously naive hosts following stress, such as challenge with cercaria of *Schistoma mansoni*[101] and immunosuppressives.[102] Stress as an inducer of clostridial disease is a well-documented phenomena. Lastly, the clostridia have a pedigree as enteric pathogens hardly matched by the *Bacillus* genus. For these reasons we have included a description of Tyzzer's disease in this review of clostridial diseases, as it is possible that the organisms belong to this genus.

There are as yet no reports of Tyzzer's disease in humans. However, there are serological data[103] to show that inapparent infection by *B. piliformis* may exist. Furthermore, the possibility that Tyzzer's disease is in some way related to Crohn's disease of humans has been discussed by Mayberry et al.[104] Spontaneous outbreaks have been reported in many animal species (Table 12). A detailed review of Tyzzer's disease host by host is beyond the scope of this chapter; consequently, only mice and muskrats as hosts for *B. piliformis* are discussed. The former is included because it provides the greatest wealth of typical information, the latter because of the controversy surrounding the similarity between Tyzzer's and Errington's disease (see *C. carnis*). For reviews of the clinical presentations of Tyzzer's disease in other species the reader should refer to Fries.[91]

Mice with Tyzzer's disease present with diarrhea lasting from 48 to 72 hr followed by loss of appetite and subsequent weight loss. Death occurs 4 to 6 days after onset. The course of the disease is shorter and mortality higher in younger animals.[91] At post-mortem the perianal region is usually soiled with diarrheic feces. Internal gross pathology can be discerned only in the liver, where there are visible pale targets like focal lesions of usually only 2-mm diameter.

The number of lesions can be highly variable. Larger lesions appear during chronic cases of the disease. Histologically, a liver lesion comprises of focus of necrosis frequently with an infiltrate of polymorphonuclear lymphocytes, and an outer zone of hepatocytes showing degenerative changes. Extreme liver pathology is characterized by cloudy swelling of the parenchyma.[91]

*B. piliformis* is best visualized by staining by the Warthin Starry silver impregnation technique.[105] There are also immunofluorescent methods for the detection of the organism itself and of antibodies raised to it.[106] Livers from infected animals generally contain clumps of *B. piliformis* lying in parallel in otherwise healthy hepatic cells on the edge of the necrotic foci. Cecal and anterior colonic epithelial cells are likewise involved, albeit to a lesser degree.[91] However, in younger infected mice in which liver lesions were unapparent, intestinal lesions are consistently seen.[107]

Experimental infections with Tyzzer's disease are initiated by the intracerebral inoculation of young mice with a liver suspension from a spontaneously infected animal. Thereafter, brain macerates are the inocula of choice. Affected animals show irritability, paralysis, convulsions, and finally, death. The brain at post-mortem is likely to be free of gross pathological changes but histology should reveal areas of liquefaction necrosis, often associated with infiltrates of polymorphonuclear cells. *B. piliformis* is to be found in very large numbers at the outer limit of necroses.[91]

There have been two documented instances of Tyzzer's disease in muskrats seemingly caused by predisposing factors enhancing a latent infection. These initiators were experimental infection with *Schistosomatium douthitii*[108] and sulfaquinoxaline-medicated feed accompanying changes in husbandry.[109] In both outbreaks, onset of disease was rapid and animals were frequently dead within 12 hr of the first symptoms. At post-mortem the findings were very similar to those in mice reported by Fries.[91] Histology confirmed a diagnosis of Tyzzer's disease; organisms resembling *B. piliformis* were seen but could not be cultured.

Wobeser et al.[83] reported the histological diagnosis of Tyzzer's disease among free living muskrats. Following comment upon the similarity between the gross pathology, histological changes, and observed spore-forming bacilli to those seen in Errington's disease,[78,79] they proposed that the two diseases may be a single entity. Retrospective study of material gathered by Errington strengthened these suspicions.[110]

As has already been stated in regard to *C. carnis*, much work remains to be done on the relationship between Tyzzer's and Errington's disease. While with respect to *B. piliformis*, there is a wealth of work outstanding regarding the taxonomy and culture of the organism.

## REFERENCES


1. **Smith, L. D. S.,** *The Pathogenic Anaerobic Bacteria*, 2nd ed., Charles C Thomas, Springfield, 1975, chap. 7.
2. **Sterne, M. and Batty, I.,** *Pathogenic Clostridia*, Butterworths, London, 1975, chap. 2.
3. **Willis, A. T.,** *Anaerobic Bacteriology: Clinical and Laboratory Practice*, 3rd ed., Butterworths, London, 1977, chap. 7.
4. **Niilo, L.,** The effect of enterotoxin of *Clostridium welchii* (*perfringens*) type A on fowls, *Res. Vet. Sci.*, 20, 225, 1976.
5. **McGowan, B., Moulton, J. E., and Rood, S. E.,** Lamb losses associated with *Clostridium perfringens*, *J. Am. Vet. Med. Assoc.*, 133, 219, 1958.
6. **Carman, R. J. and Lewis, J. C. M.,** Recurrent diarrhoea in a dog associated with *Clostridium perfringens* type A, *Vet. Rec.*, 112, 342, 1983.

7. **Moro, M. and Guerrero, C.,** La alpaca enfermedades infecciosas y parasitarias, *Bol. Divulg. I.V.I.T.A.*, 8, 1, 1971.
8. **Al-Sheikhly, F. and Truscott, R. B.,** The pathology of necrotic enteritis of chickens following infusion of broth cultures of *Clostridium perfringens* into the duodenum, *Avian Dis.*, 21, 230, 1977.
9. **Al-Sheikhly, F. and Truscott, R. B.,** The pathology of necrotic enteritis of chickens following infusion of crude toxins of *Clostridium perfringens* into the duodenum, *Avian Dis.*, 21, 241, 1977.
10. **Al-Sheikhly, F. and Truscott, R. B.,** The interaction of *Clostridium perfringens* and its toxins in the production of necrotic enteritis of chickens, *Avian Dis.*, 21, 250, 1977.
11. **Wierup, M.,** Equine intestinal clostridiosis. An acute disease in horses associated with high intestinal counts of *Clostridium perfringens* type A, *Acta Vet. Scand.*, Suppl. 62, 1, 1977.
12. **Wierup, M. and Dilietro, J. A.,** Bacteriologic examination of equine fecal flora as a diagnostic tool for equine intestinal clostridiosis, *Am. J. Vet. Res.*, 42, 2167, 1981.
13. **White, G. and Prior, S. D.,** Comparative effects of oral administration of trimethoprim/sulphadiazine or oxytetracycline on the faecal flora of horses, *Vet. Rec.*, 111, 316, 1982.
14. **Boyd, N. A., Thompson, R. O., and Walker, P. D.,** The prevention of experimental *Clostridium novyi* and *C. perfringens* gas gangrene in high-velocity missile wounds by active immunisation, *J. Med. Microbiol.*, 5, 467, 1972.
15. **Van Kruiningen, H. J., Gregoire, K., and Meuten, D. J.,** Acute gastric dilatation: a review of comparative aspects, by species, and a study in dogs and monkeys, *J. Am. Anim. Hosp. Assoc.*, 10, 294, 1974.
16. **Newton, W. M., Beamer, P. D., and Rhoades, H. E.,** Acute bloat syndrome in stumptailed macaques (*Macaca arctoides*): a report of four cases, *Lab. Anim. Sci.*, 21, 193, 1971.
17. **Bennett, B. T., Cuasay, L., Welsh, T. J., Beluhan, F. Z., and Schofield, L.,** Acute gastric dilatation in monkeys: a microbiologic study of gastric contents, blood and feed, *Lab. Anim. Sci.*, 30, 241, 1980.
18. **Gaiger, S. H. and Dalling, T.,** Bacillary dysentery in lambs. A note on some recent research into the etiology and source of infection, *J. Comp. Pathol. Ther.*, 36, 120, 1923.
19. **Mason, J. H., Ross, H. E., and Dalling, T. B.,** Paludis (McEwen) — a comparison with the lamb dysentery bacillus, *J. Comp. Pathol. Ther.*, 44, 258, 1931.
20. **Brooks, M. E. and Entessar, F.,** Anomalous *Clostridium welchii* Type B strains isolated in Iran, *Br. Vet. J.*, 113, 506, 1957.
21. **Horton, R. E., Madden, D. L., and McCullough, N. B.,** Pathogenicity of *Clostridium perfringens* for germ-free pigs after oral ingestion, *Appl. Microbiol.*, 19, 314, 1970.
22. **Arbuckle, J. B. R.,** The attachment of *Clostridium welchii* (*C. perfringens*) type C to intestinal villi of pigs, *J. Pathol.*, 106, 65, 1972.
23. **Field, H. I. and Goodwin, R. F. W.,** The experimental reproduction of enterotoxaemia in piglets, *J. Hyg.*, 57, 81, 1959.
24. **Parish, W. E.,** Necrotic enteritis in the fowl (Gallus Gallus Domesticus). I. Histopathology of the disease and isolation of a strain of *Clostridium welchii, J. Comp. Pathol.*, 71, 377, 1961.
25. **Parish, W. E.,** Necrotic enteritis in the fowl. II. Examination of the causal *Clostridium welchii, J. Comp. Pathol.*, 71, 394, 1961.
26. **Parish, W. E.,** Necrotic enteritis in the fowl. III. The experimental disease, *J. Comp. Pathol.*, 71, 405, 1961.
27. **Clapp, H. W. and Graham, W. R.,** An experience with *Clostridium perfringens* in cesarean derived barrier sustained mice, *Lab. Anim. Care*, 20, 1081, 1970.
28. **Bosworth, T. J.,** On a new type of toxin produced by *Clostridium welchii, J. Comp. Pathol.*, 53, 245, 1943.
29. **Hart, B. and Hooper, P. T.,** Enterotoxaemia of calves due to *Clostridium welchii* Type E, *Aust. Vet. J.*, 43, 360, 1967.
30. **Ross, H. E., Warren, M. E., and Barnes, J. M.,** *Clostridium welchii* iota toxin: its activation by trypsin. *J. Gen. Microbiol.*, 3, 148, 1949.
31. **Madden, D. L., Horton, R. E., and McCullough, N. B.,** Spontaneous infection in ex-germfree guinea pigs due to *Clostridium perfringens, Lab. Anim. Care*, 20, 454, 1970.
32. **Carman, R. J.,** unpublished data, 1982.
33. **Borriello, S. P. and Carman, R. J.,** Association of toxigenic *Clostridium spiroforme* with iota toxin positive enterotoxaemia in rabbits, *J. Clin. Microbiol.*, 17, 414, 1983.
34. **Orcutt, R. P., Foster, H. L., and Jonas, A. M.,** *Clostridium perfringens* Type E enterotoxemia as the cause of acute diarrhoeal death or "hemorrhagic typhlitis" in rabbits *Proc. Am. Assoc. Lab. Anim. Sci.*, 100(Abstr.), 1978.
35. **Kaneuchi, C., Miyazato, T., Shinjo, T., and Mitsuoka, T.,** Taxonomic study of helically coiled, spore forming anaerobes isolated from the intestines of humans and other animals: *Clostridium cocleatum* sp. nov. and *Clostridium spiroforme* sp. nov., *Int. J. Syst. Bacteriol.*, 29, 1, 1979.
36. **Moore, W. E. C. and Holdeman, L. V.,** Human fecal flora: the normal flora of 20 Japanese-Hawaiians, *Appl. Microbiol.*, 27, 961, 1974.

37. **Gossling, J., Loesche, W. J., and Nace, G. W.,** Large intestine bacterial flora of non-hibernating and hibernating leopard frogs, *(Rana pipiens), Appl. Environ. Microbiol.*, 44, 59, 1982.
38. **Carman, R. J. and Borriello, S. P.,** Observation on an association between *Clostridium spiroforme* and *Clostridium perfringens* Type E iota enterotoxaemia in rabbits, *Eur. J. Chemother. Antibiot.*, 2, 143, 1982.
39. **Carman, R. J. and Borriello, S. P.,** *Clostridium spiroforme* isolated from rabbits with diarrhoea, *Vet. Rec.*, 111, 461, 1982.
40. **Katz, L., LaMont, J. T., Trier, J. S., Sonnenblick, E. B., Rothman, S. W., Broitman, S. A., and Rieth, S.,** Experimental clindamycin-associated colitis in rabbits. Evidence for toxin-mediated mucosal damage, *Gasteroenterology*, 74, 246, 1978.
41. **LaMont, J. T., Sonnenblick, E. B., and Rothman, S.,** Role of clostridial toxin in the pathogenesis of clindamycin colitis in rabbits, *Gastroenterology*, 76, 356, 1979.
42. **Orcutt, R. P.,** Personal communication, 1981.
43. **Patton, N. M., Holmes, H. T., Riggs, R. J., and Cheeke, P. R.,** Enterotoxemia in rabbits, *Lab. Anim. Sci.*, 28, 536, 1978.
44. **Patton, N. M.,** personal communication, 1980.
45. **Eaton, P. and Fernie, D. S.,** Enterotoxaemia involving *Clostridium perfringens* iota toxin in a hysterectomy-derived colony, *Lab. Anim.*, 14, 347, 1980.
46. **Fernie, D. S. and Eaton, P.,** The demonstration of a toxin resembling *Clostridium perfringens* iota toxin in rabbits with enterotoxaemia, *FEMS Microbiol. Lett.*, 8, 33, 1980.
47. **Baskerville, M., Wood, M., and Seamer, J. H.,** *Clostridium perfringens* type E enterotoxaemia in rabbits, *Vet. Rec.*, 107, 18, 1980.
48. **Rehg, J. E. and Pakes, S. P.,** Implication of *Clostridium difficile* and *Clostridium perfringens* iota toxins in experimental lincomycin-associated colitis of rabbits, *Lab. Anim. Sci.*, 32, 253, 1982.
49. **Carman, R. J. and Borriello, S. P.,** Infectious nature of *Clostridium* spiroforme mediated rabbit enterotoxaemia, *Vet. Microbiol.*, 9, 497, 1984.
50. **Carman, R. J. and Borriello, S. P.,** Laboratory diagnosis of *Clostridium spiroforme* mediated diarrhoea (iota enterotoxaemia) of rabbits, *Vet. Rec.*, 113, 184, 1983.
51. **Evans, R. H.,** personal communication, 1982.
52. **Lesbouyries, M. M. and Berthelon, M.,** Entéro-toxémie du lapin, *Bull. Acad.*, 1, 74, 1936.
53. **Carman, R. J. and Borriello, S. P.,** Animal models of *Clostridium spiroforme* mediated diarrhoea (iota enterotoxaemia) of rabbits, in *Models of Anaerobic Infection*, Hill, M. J., Ed., Martinus Nijhoff, The Hague, 1984, 77.
54. **Norton, C. C.,** Coccidia of the pheasant, *Folia Vet. Lat.*, 6, 218, 1976.
55. **Carman, R. J., Beer, J. V., and Wharton, A.,** unpublished data, 1982.
56. **Sutter, V. L. and Finegold, S. M.,** The effect of antimicrobial agents on human faecal flora: studies with cephalexin, cyclacillin and clindamycin, in, *The Normal Microbial Flora of Man*, Skinner, F. A. and Carr, J. G., Eds., Academic Press, London, 1974, 229.
57. **Dabard, J., Dubos, F., Martinet, L., and Ducluzeau, R.,** Experimental reproduction of neonatal diarrhoea in young gnotobiotic hares simultaneously associated with *Clostridium difficile* and other *Clostridium* strains, *Infect. Immun.*, 24, 7, 1979.
58. **Jones, M. A. and Hunter, D.,** Isolation of *Clostridium difficile* from pigs, *Vet. Rec.*, 112, 253, 1983.
59. **Honour, P. and Borriello, S. P.,** unpublished data, 1983.
60. **Lysons, R. J. and Hall, G. A.,** *Clostridium difficile* in the gnotobiotic pig, *Eur. J. Chemother. Antibiot.*, 2, 133, 1982.
61. **Orchard, J. L., Fekety, R., and Smith, J. R.,** Antibiotic-associated colitis due to *C. difficile* in a Kodiak bear, *Am. J. Vet. Res.*, 44, 1547, 1983.
62. **Borriello, S. P., Honour, P., Turner, T., and Barclay, F.,** Household pets as a potential reservoir for *Clostridium difficile* infection, *J. Clin. Pathol.*, 36, 84, 1983.
63. **Al-Mashat, R. R. and Taylor, D. J.,** Production of diarrhoea and enteric lesions in calves by the oral inoculation of pure cultures of *Clostridium sordellii*, *Vet. Rec.*, 112, 141, 1983.
64. **Brooks, M. E., Sterne, M., and Batty, R. W.,** Occurrence of *Clostridium sordellii* in Great Britain, *Vet. Rec.*, 68, 121, 1956.
65. **Al-Mashat, R. R. and Taylor, D. J.,** *Clostridium sordellii* in enteritis in adult sheep, *Vet. Rec.*, 112, 19, 1983.
66. **Bornside, G. H., Floyd, C. E., and Cohn, I.,** *Clostridium sordellii* in strangulation obstruction: further studies of serial changes of intestinal contents, *J. Surg. Res.*, 4, 233, 1964.
67. **Peckham, M.,** Quail disease (ulcerative enteritis), in *Infections and Parasitic Diseases of Wild Birds*, Davis, J. W., Anderson, R. C., Karstad, L., and Trainer, D. O., Eds. Iowa State University Press, Ames, 1971, chap. 11.
68. **Berkhoff, G. A., Campbell, S. G., Naylor, H. B., and Smith, L. D. S.,** Etiology and pathogenesis of ulcerative enteritis ("quail disease"). Characterization of the causative anaerobe, *Avian Dis.*, 18, 195, 1974.

69. **Durant, A. J. and Doll, E. R.,** Ulcerative enteritis in quail, *Mo. Agric. Exp. Stn. Res. Bull.*, 325, 1941.
70. **Berkhoff, G. A. and Campbell, S. G.,** Etiology and pathogenesis of ulcerative enteritis ("quail disease"). The experimental disease, *Avian Dis.*, 18, 205, 1974.
71. **Berkhoff, G. A. and Kanitz, C. L.,** Fluorescent antibody test in a diagnosis of ulcerative enteritis, *Avian Dis.*, 20, 525, 1976.
72. **Davis, R. P.,** Ulcerative enteritis in quail and poultry, in *Proc. 20th West. Poultry Dis. Conf.*, 8, 1972.
73. **Mayberry, J. F.,** personal communication, 1982.
74. **Davis, R. B., Brown, J., and Dawe, D. L.,** Quail-biological indicators in the differentiation of ulcerative and necrotic enteritis of chicken, *Poult. Sci.*, 50, 737, 1971.
75. **Ellwood, D. C. and Halliwell, R. W.,** *Clostridium fallax* infection in poultry, *Trop. Anim. Health Prod.*, 5, 202, 1973.
76. **Errington, P. L.,** Special report on muskrat disease, *Q. Rep. Iowa Coop. Wildl. Res. Unit*, 34, 1946.
77. **Sompolinsky, D.,** En enzooti forarsaget af Clostridium carnis hos mink, *Skand. Vet. Tidskr.*, 38, 506, 1948.
78. **Lord, G. H., Todd, A. C., and Kabat, C.,** Studies on Errington's disease in muskrats. I. Pathological changes, *Am. J. Vet. Res.*, 17, 303, 1956.
79. **Lord, G. H., Todd, A. C., Kabat, C., and Mathiak, H.,** Studies on Errington's disease in muskrats. II. Etiology, *Am. J. Vet. Res.*, 17, 307, 1956.
80. **Errington, P. L.,** *Muskrat Populations*, Iowa State University Press, Ames, 1963, 160.
81. **Karstad, L. H.,** Errington's disease, in *Infectious Diseases of Wild Animals*, Davis, J. W., Karstad, L. H., and Trainer, D. O., Eds., Iowa State University Press, Ames, 1970, chap. 31.
82. **Spalatin, J., Fraser, C. E. O., Connell, R., Hanson, R. P., and Berman, D. T.,** Agents of psittocosis-lymphogranuloma venereum group isolated from muskrats and snowshoe hares in Saskatchewan, *Can. J. Comp. Med. Vet. Sci.*, 30, 260, 1966.
83. **Wobeser, G., Hunter, D. B., and Daoust, P-Y.,** Tyzzer's disease in muskrats: occurrence in free-living animals, *J. Wildl. Dis.*, 14, 325, 1978.
84. **Gianforte, E. M. and Brown, R., Jr.,** Isolation of an organism resembling *Clostridium carnis* from mink, *Am. J. Vet. Res.*, 19, 254, 1958.
85. **Sompolinsky, D.,** Etude d'une souche de *Plectridium carnis* (Klein) Prévot isolée d'une enzootíe danoise du vison, *Ann. Inst. Past.*, 79, 204, 1950.
86. **Wort, A. J. and Ozere, R. L.,** Characteristics of a strain of *Clostridium carnis* causing septicaemia in a young infant, *J. Clin. Pathol.*, 29, 1011, 1976.
87. **Buxton, A. and Frazer, G.,** *Animal Microbiology*, Vol. 1, Blackwell, Oxford, 1977, chap. 20.
88. **Urbain, A., Dechambre, E., and Piette, G.,** Septicémie due a *Clostridium oedematiens*, type A, sur les tortues de la mènagerie des reptiles du Muséum, *Bull. Museum*, 23, 247, 1951.
89. **Olander, J. J., Hughes, J. P., and Biberstein, E. L.,** Bacillary hemoglobinuria: induction by liver biopsy in naturally and experimentally infected animals, *Pathol. Vet.*, 3, 421, 1966.
90. **Kalmbach, E. R.** American vultures and the toxin of *Clostridium botulinum*, *J. Am. Vet. Med. Assoc.*, 94, 187, 1939.
91. **Fries, A. S.,** *Tyzzer's Disease and the Importance of Inapparent Infection in Biomedical Research*, Royal Veterinary and Agricultural University, Copenhagen, 1981, 1.
92. **Tyzzer, E. E.,** A fatal disease of the Japanese waltzing mouse caused by a spore-bearing bacillus (*Bacillus piliformis*, N. sp.), *J. Med. Res.*, 37, 307, 1917.
93. **Kanazawa, K. and Imai, A.,** Pure culture of the pathogenic agent of Tyzzer's disease of mice, *Nature (London)*, 184, 1810, 1959.
94. **Simon, P. C.,** Isolation of *Bacillus piliformis* from rabbits, *Can. Vet. J.*, 18, 46, 1977.
95. **Kawamura, S., Taguchi, F., Ishida, T., Nakayama, M., and Fujwara, K.,** Growth of Tyzzer's organism in primary monolayer culture of mouse hepatocytes, *J. Gen. Microbiol.*, 129, 277, 1983.
96. **Skerman, V. B. D., McGowan, V., and Sneath, P. H. A., Eds.,** Approved list of bacterial names, *Int. J. Syst. Bacteriol*, 30, 225, 1980.
97. **Gibson, T. and Gordon, R. E.,** *Bacillus*, in *Bergey's Manual of Determinative Bacteriology*, 8th ed., Buchanan, R. E. and Gibbons, N. E., Eds., Williams & Wilkins, Baltimore, 1974, 529.
98. **Smith, L. D. S. and Hobbs, G.,** *Clostridium* in *Bergey's Manual of Determinative Bacteriology*, 8th ed., Buchanan, R. E. and Gibbons, N. E., Eds., Williams & Wilkins, Baltimore, 1974, 551.
99. **Craigie, J.,** "*Bacillus piliformis*" (Tyzzer) and Tyzzer's disease of the laboratory mouse. I. Propagation of the organism in embryonated eggs, *Proc. R. Soc. London Biol.*, 165, 61, 1966.
100. **Thunert, A.,** Therapy and prophylaxis of Tyzzers disease (in vivo antibiogram of three strains of *B. piliformis* ), *Z. Versuchstierk*, 24, 206, 1982.
101. **Schneck, G.,** Therapieversuche gegen Tyzzer's disease bei Mäusen, *Wien. Tierätz. Monatschr.*, 56, 243, 1969.
102. **Marler, R. J. and Cooke, J. E.,** Tyzzer's disease in two coyotes, *J. Am. Vet. Med. Assoc.*, 169, 940, 1976.

103. **Fries, A. S.,** Antibodies to *Bacillus piliformis* (Tyzzer's disease) in sera from man and other species, in *Animal Quality and Models in Biomedical Research,* Proc. 7th Symp. Int. Comm. Lab. Anim., Fischer, Stuttgart, 1980.
104. **Mayberry, J. F., Rhodes, J., and Heatley, R. V.,** Infections which cause ileocolic disease in animals: are they relevant to Crohn's disease?, *Gastroenterology,* 78, 1080, 1980.
105. Armed Forces Institute of Pathology, *Manual of Histologic and Special Staining Technics,* McGraw Hill, N.Y., 1960.
106. **Fries, A. S.,** Studies on Tyzzer's disease: application of immunofluorescence for detection of *Bacillus piliformis* and for demonstration and determination of antibodies to it in sera from mice and rabbits, *Lab. Anim.,* 11, 69, 1977.
107. **Mullink, J. W. M. A.,** Tyzzer's disease. Intestinal lesions in both SPF and conventional mice, *Z. Versuchstierkd.,* 10, 271, 1968.
108. **Karstad, L., Lusis, P., and Wright, D.,** Tyzzer's disease in muskrats, *J. Wildl. Dis.,* 7, 96, 1971.
109. **Chalmers, G. A. and MacNeil, A. C.,** Tyzzer's disease in wild-trapped muskrats in British Columbia, *J. Wildl. Dis.,* 13, 114, 1977.
110. **Wobeser, G., Barnes, H. J., and Pierce, K.,** Tyzzer's disease in muskrats: re-examination of specimens of hemorrhagic disease collected by Paul Errington, *J. Wildl. Dis.,* 15, 525, 1979.
111. **Köhler, B. and Ständer, R.,** Untersuchungen zum Vorkommen von *Clostridium perfringens* und seiner Toxine und zur *Clostridium-perfringens* -Enterotoxamie beim Rind, *Arch. Exp. Vet. Med.,* 27, 29, 1973.
112. **Al-Sheikhly, F. and Al-Saieg, A.,** Role of coccidia in the occurrence of necrotic enteritis of chickens, *Avian Dis.,* 24, 324, 1980.
113. **Prescott, J. F., Johnson, J. A., Patterson, J. H., and Bulm, W. S.,** Haemorrhagic gastroenteritis in the dog associated with *Clostridium welchii, Vet. Rec.,* 103, 116, 1978.
114. **Russell, W. C.,** Type A enterotoxemia in captive wild goats, *J. Am. Vet. Med. Assoc.,* 157, 643, 1970.
115. **English, A. W.,** Enterotoxaemia in a juvenile eastern grey kangaroo (*Macropus giganteus*), *Vet. Rec.,* 102, 243, 1978.
116. **Paliwal, O. P., Krishna, L., and Kulshreshta, S. B.,** Enterotoxaemic jaundice in sheep, *Ind. J. Anim. Sci.,* 45, 656, 1977.
117. **Kunstvyr, I., Matthiesen, I., and Matthiesen, T.,** Acute enteritis in rabbits: spontaneous infection by *Clostridium perfringens* type A., *Z. Versuchstierkd.,* 17, 57, 1975.
118. **Tung, Y. F., Chen, H. F., Wong, G. H., and Chen, Y. C.,** A preliminary report on the causal agent of a clostridial dysentery in rabbits and the methods of its control, *J. Appl. Rabbit Res.,* 4, 81, 1981.
119. **Kummeneje, K. and Bakken, G.,** *Clostridium perfringens* enterotoxaemia in reindeer, *Nord. Vet. Med.,* 25, 196, 1973.
120. **Caywood, D., Teague, H. D., Jackson, D. A., Levitt, M. D., and Bond, J. H., Jr.,** Gastric gas analysis in the canine gastric dilatation — volvulus syndrome, *J. Am. Anim. Hosp. Assoc.,* 13, 459, 1977.
121. **Warner, N. S. and Van Kruiningen, H. J.,** The incidence of clostridia in the canine stomach and their relationship to acute gastric dilatation, *J. Am. Anim. Hosp. Assoc.,* 14, 618, 1978.
122. **Christie, R. J. and King, R. E.,** Acute gastric dilatation and rupture in *Macaca arctoides* associated with *Clostridium perfringens, J. Med. Primatol.,* 10, 263, 1981.
123. **Soave, O. A.,** Observations on acute gastric dilatation in non-human primates, *Lab. Anim. Sci.,* 28, 331, 1978.
124. **Kohler, B., Zabke, J., Reiher, K., and Rummler, H. J.,** Necrotic enteritis (Clostridium perfringens type C enterotoxaemia) among unweaned piglets in large pig-breeding units. V. Control, *Arch. Exp. Vet.,* 33, 621, 1979.
125. **Lauerman, L. H., Jensen, R., and Pierson, R. E.,** *Clostridium perfringens* type C enterotoxaemia in feedlot cattle and sheep, *Proc. Am. Assoc. Vet. Lab. Diagnost.,* 20, 363, 1977.
126. **Szemeredi, G., Sola, A., and Lon-So, E.,** Acute gastroenteritis caused by *Clostridium perfringens* in calves, *Mag. Allatorv. Lapja,* 31, 713, 1976.
127. **Dickie, C. W., Klinkerman, D. L., and Petrie, R. J.,** Enterotoxemia in two foals, *J. Am. Vet. Assoc.,* 173, 306, 1978.
128. **Awad, F. I., Farrag, I., Shawkat, M. E., and Abeid, M. H.,** Studies on enterotoxaemia in young buffalo-calves, *Egypt. J. Vet. Sci.,* 14, 25, 1977.
129. **Ipatenko, N. G.,** Infections enterotoxaemia of camels, in *Little-Known Contagious Diseases of Animals,* 2nd ed., Orlov, F. M., Izdatel'stuo Kolos, U.S.S.R., 1973, 248.
130. **Hagen, K., Rienke, S., and Gorham, J.,** Acute enteritis of unknown origin in mink, *Natl. Fur. News.,* 41, 17, 1969.
131. **Jasmin, A. N., Cooperrider, D. E., and Powell, C. P.,** Enterotoxaemia of wild-fowl due to *Clostridium perfringens* Type C, *J. Wildl. Dis.,* 8, 79, 1972.
132. **Pyakural, S. and Singh, N. B.,** Initial studies on "six months disease" in sheep, *Vet. Rec.,* 98, 49, 1976.
133. **Chakrabarty, A. K., Dutta, B. M., and Sharma, A. K.,** Enterotoxaemia in goats due to *Clostridium perfringens* Type D (feed contamination), *Ind. Vet. J.,* 57, 195, 1980.

134. **Kohler, B. and Stander, R.,** Untersuchungen zum Vorkommen von *Clostridium perfringens* und seiner Toxine undzur *Clostridium-perfringens* Enterotoxamie beim rind, *Arch. Exp. Vet.*, 27, 29, 1973.
135. **McBee, R. H.,** Intestinal flora of some antarctic birds and mammals, *J. Bacteriol.*, 79, 311, 1960.
136. **Princewell, T. J. T. and Agba, M. I.,** Examination of bovine faeces for the isolation and identification of *Clostridium* species, *J. Appl. Bacteriol.*, 52, 97, 1982.
137. **Hafiz, S. L.,** *Clostridium difficile* and its Toxins, University of Leeds, Ph.D. thesis, 1974.
138. **Honour, P. and Borriello, S. P.,** unpublished observations, 1982.
139. **Carman, R. J. and Borriello, S. P.,** unpublished observations, 1983.
140. **Hibbs, C. N., Johnson, D. R., Reynolds, K., and Harrington, R.,** *Clostridium sordellii* isolated from foals, *Vet. Med. Small Anim. Clin.*, 72, 256, 1977.
141. **Harris, A. H.,** An outbreak of ulcerative enteritis amongst Bobwhite quail (*Colinus virginianus*), *Vet. Rec.*, 73, 11, 1961.
142. **Berkhoff, G. A., Campbell, S. G., and Naylor, H. B.,** Etiology and pathogenesis of ulcerative enteritis ("Quail Disease"). Isolation of the causative anaerobe, *Avian Dis.*, 18, 186, 1974.
143. **Buss, I. O., Conrad, R. D., and Reilly, J. R.,** Ulcerative enteritis in the pheasant, blue grouse and californian quail, *J. Wildl. Manag.*, 22, 446, 1958.
144. **Winterfield, R. W. and Berkhoff, G. A.,** Ulcerative enteritis in robins, *Avian Dis.*, 21, 328, 1977.
145. **Ononiwu, J. C., Prescott, J. F., Carlson, H. C., and Julian, R. J.,** Ulcerative enteritis caused by *Clostridium Colinum* in chickens, *Can. Vet. J.*, 19, 226, 1978.
146. **Gard, S.,** *Bacillus piliformis* infection in mice and its prevention, *Acta Pathol. Microbiol. Scand.*, 54, 129, 1944.
147. **Gowen, W. J. and Schott, R. G.,** Genetic predisposition to *Bacillus piliformis* infection among mice, *J. Hyg.*, 33, 370, 1933.
148. **Kaneko, J., Fujita, H., Matsuyama, S., Kojima, H., Asakura, H., Nakamura, Y., and Kodama, T., cited in Fujiwara, K., Takagaki, Y., Naiki, M., Maejima, K., and Tajima, Y.,** Tyzzer's disease in mice, *Jpn. J. Exp. Med.*, 34, 59, 1964.
149. **Meshorer, A.,** Tyzzer's disease in laboratory mice, *Refu. Vet.*, 31, 41, 1974.
150. **Tuffery, A. A.,** The laboratory mouse in Great Britain. IV. Intercurrent infection (Tyzzer's disease), *Vet. Rec.*, 68, 511, 1956.
151. **McLeod, C. G., Stookey, J., Harrington, D. G., and White, J. D.,** Intestinal Tyzzer's disease and spirochetosis in a guinea pig, *Vet. Pathol.*, 14, 229, 1977.
152. **Sparrow, S. and Naylor, P.,** Naturally occurring Tyzzer's disease in guinea pigs, *Vet. Rec.*, 102, 288, 1978.
153. **Zwicker, G. M., Dagle, G. E., and Adee, R. R.,** Naturally occurring Tyzzer's disease and intestinal spirochetosis in guinea pigs, *Lab. Anim. Sci.*, 28, 193, 1978.
154. **Nakayama, M., Saegusa, J., Itoh, K., Kiuchi, Y., Tamura, T., Ueda, K., and Fujiwara, K.,** Transmissible enterocolitis in hamsters caused by Tyzzer's organism, *Jpn. J. Exp. Med.*, 45, 33, 1975.
155. **Takasaki, Y., Oghiso, Y., Sato, K., and Fujiwara, K.,** Tyzzer's disease in hamsters, *Jpn. J. Exp. Med.*, 44, 267, 1974.
156. **Zook, B. C., Albert, E. N., and Rhorer, R. G.,** Tyzzer's disease in the Chinese hamster (*Cricetulus griseus*), *Lab. Anim. Sci.*, 27, 1033, 1977.
157. **Zook, B. C., Huang, K., and Thorer, R. G.,** Tyzzer's disease in Syrian hamster, *J. Am. Vet. Med. Assoc.*, 171, 833, 1977.
158. **Allen, A. M., Ganaway, J. R., Moore, T. D., and Kinard, R. F.,** Tyzzer's disease syndrome in laboratory rabbits, *Am. J. Pathol.*, 46, 859, 1965.
159. **Prescott, J. F.,** Tyzzer's disease in rabbits in Britain, *Vet. Rec.*, 100, 285, 1977.
160. **Evans, R. H.,** personal communication, 1983.
161. **Ganaway, J. R., McReynolds, R. S., and Allen, A. M.,** Tyzzer's disease in free-living cottontail rabbits ( *Sylvilagus floridanus* ) in Maryland, *J. Wildl. Dis.*, 12, 545, 1976.
162. **Carter, G. R., Whitenack, D. L., and Julius, L. A.,** Natural Tyzzer's disease in Mongolian gerbils ( *Meriones unguiculatus* ), *Lab. Anim. Care*, 19, 648, 1969.
163. **Port, C. D., Richter, W. R., and Moise, S. M.,** Tyzzler's disease in the gerbil (*Meriones unguiculatus*), *Lab. Anim. Care*, 20, 109, 1970.
164. **Port, C. D., Richter, W. R., and Moise, S. M.,** An ultrastructural study of Tyzzer's disease in the Mongolian gerbil (*Meriones* unguiculatus), *Lab. Invest.*, 25, 81, 1971.
165. **Strittmatter, J.,** Eliminierung der Tyzzer's Krankheit bei der Mongolischen Rennimaus (*Meriones unguiculatus*) durch Ammenaufzuch mit Mäusen, *Z. Versuchstierkd.*, 14, 209, 1972.
166. **White, D. J. and Waldron, M. M.,** Naturally-occurring Tyzzer's disease in the gerbil, *Vet. Rec.*, 85, 111, 1969.
167. **Niven, J. S. F.,** Tyzzer's disease in laboratory animals, *Z. Versuchstierkd.*, 10, 168, 1968.
168. **Bennett, A. M., Huxtable, C. R., and Love, D. N.,** Tyzzer's disease in cats experimentally infected with feline leukaemia virus, *Vet. Microbiol.*, 2, 49, 1977.

169. **Goto, N., Oghiso, Y., Lee, Y.-S., Takahashi, R., and Fujwara, K.,** Fine structure of the Tyzzer's organism in the feline liver, *Jpn. J. Exp. Med.*, 44, 373, 1974.
170. **Kubokawa, K., Kubo, M., Takasaki, Y., Oghiso, Y., Sato, K., Lee, Y.-S., Goto, N., Takahashi, R., and Fujiwara, K.,** Two cases of feline Tyzzer's disease, *Jpn. J. Exp. Med.*, 43, 413, 1973.
171. **Schneck, G.,** Tyzzer's disease in an adult cat, *Vet. Med. Small Anim. Clin.*, 70, 155, 1975.
172. **Hall, W. C. and Van Kruiningen, H. J.,** Tyzzer's disease in a horse, *J. Am. Vet. Med. Assoc.*, 164, 1187, 1974.
173. **Harrington, D. D.,** Naturally-occurring Tyzzer's disease (*Bacillus piliformis* infection) in horse foals, *Vet. Rec.*, 96, 59, 1975.
174. **Harrington, D. D.,** *Bacillus piliformis* infection (Tyzzer's disease) in two foals, *J. Am. Vet. Med. Assoc.*, 168, 58, 1976.
175. **Pulley, L. T. and Shively, J. N.,** Tyzzer's disease in a foal, *Vet. Pathol.*, 11, 203, 1974.
176. **Swerczek, T. W., Crowe, M. W., Prickett, M. E., and Bryans, J. T.,** Focal bacterial hepatitis in foals: preliminary report, *Mod. Vet. Pract.*, 54, 66, 1973.
177. **Thompson, G. W., Wilson, R. W., Hall, E. A., and Physick, S. P.,** Tyzzer's disease in the foal: case reports and review, *Can. Vet. J.*, 18, 41, 1977.
178. **Poonacha, K. B. and Smith, H. L.,** Naturally occurring Tyzzer's disease as a complication of distemper and mycotic pneumonia in a dog, *J. Am. Vet. Med. Assoc.*, 169, 419, 1976.
179. **Qureshi, S. R., Carlton, W. W., and Orlander, H. J.,** Tyzzer's disease in a dog, *J. Am. Vet. Med. Assoc.*, 168, 602, 1976.
180. **Stanley, S. M., Flatt, R. E., and Daniels, G. N.,** Naturally occurring Tyzzer's disease in the gray fox, *J. Am. Vet. Med. Assoc.*, 173, 1173, 1978.
181. **Stedham, M. A. and Bucci, T. J.,** Tyzzer's disease in the rat, *Lab. Invest.*, 20, 604, 1969.
182. **Stedham, M. A. and Bucci, T. J.,** Spontaneous Tyzzer's disease in a rat, *Lab. Anim. Care*, 20, 743, 1970.
183. **Yamada, A., Osada, Y., Takayama, S., Akimoto, T., Ogawa, H., Oshima, Y., and Fujiwara, K.,** Tyzzer's disease syndrome in laboratory rats treated with adrenocorticotropic hormone, *Jpn. J. Exp. Med.*, 39, 505, 1969.
184. **Fries, A. S. and Svendsen, O.,** Studies on Tyzzer's disease in rats, *Lab. Anim.*, 12, 1, 1978.

Chapter 11

# NEWLY DESCRIBED CLOSTRIDIAL DISEASES OF THE GASTROINTESTINAL TRACT: *CLOSTRIDIUM PERFRINGENS* ENTEROTOXIN-ASSOCIATED DIARRHEA AND NEUTROPENIC ENTEROCOLITIS DUE TO *CLOSTRIDIUM SEPTICUM*

**S. P. Borriello**

## TABLE OF CONTENTS

# I. INTRODUCTION

There have been two major developments in the field of clostridia and gastrointestinal disease during the last year.[1,2] Both are sufficiently recent as to have not been included in the preceding chapters, and in a way highlight the degree of current activity in this aspect of microbiology. An up-to-date account of these two diseases, which are *Clostridium perfringens* enterotoxin-associated diarrhea and neutropenic enterocolitis due to *Clostridium septicum*, will be presented.

# II. *CLOSTRIDIUM PERFRINGENS* ENTEROTOXIN-ASSOCIATED DIARRHEA

## A. Introduction

Although *C. difficile* is now a well established cause of antibiotic-associated colitis and diarrhea (see Chapter 7), there are still many cases of this type of diarrhea for which an etiological agent cannot be found. There are, of course, a number of antimicrobial mediated changes in the gut that would account for this problem, in some cases, which would not involve a specific pathogen.[3] However, pathogenic microorganisms must account for some of the cases, especially those where there is histological evidence of infection or blood in the stool. It is now known that some of these cases are due to enterotoxigenic strains of *C. perfringens*.[1]

## B. Pathogenesis

During a recent investigation of fecal samples from patients with antibiotic-associated diarrhea it was shown that a cytotoxin, other than that of *C. difficile*, was present. The toxin was identified as *C. perfringens* enterotoxin by neutralization tests in tissue culture using specific antitoxin and by an enzyme-linked immunoassay.[1] In all of these cases very high numbers of an enterotoxigenic strain of *C. perfringens* were present in the stool.[2] Since this description of 10 patients a further 16 have been detected.[4] An investigation of all 26 patients has shown that in the majority of cases the *C. perfringens* enterotoxin diarrhea was associated with recent administration of antibiotics (Table 1), with only four of the patients definitely not having received antibiotics in the recent past. There were five patients on whom an antibiotic history could not be obtained. This association with antibiotics confirms the observations made on a smaller group of patients.[1] The apparent association with the penicillins noted in the earlier study[1] is also still apparent in the extended series of patients (Table 1). It is of interest to note that blood and mucous was present in the feces of 7 (27%) of these 26 patients. Histological analysis of rectal biopsies obtained at sigmoidoscopy have either been indicative of nonspecific colitis (two cases) or entirely normal (three cases).[1,5] Two of these cases had blood in the stool, one of which had normal rectal mucosa on histological examination. However, it is important to note that rectal sparing has been described in pseudomembranous colitis due to *C. difficile*;[6] so by analogy the excretion of blood and mucous in this patient with normal rectal mucosa may be indicative of more proximal colonic involvement.

There is some preliminary evidence that disease may be due to small bowel colonization by *C. perfringens*. This is based on the observations that *C. perfringens* enterotoxin will not induce net secretion in the ligated rabbit cecum[7] but will cause secretion in the small bowel, and that metronidazole treatment of a patient resulted in resolution of diarrhea and disappearance of toxin in the feces but had little effect on the concentration of the enterotoxigenic serotype of *C. perfringens* in the stool.[8] It is possible that this well-absorbed antibiotic had an effect on colonization of the small bowel and that it was removal of the *C. perfringens* from this site that effected the cure.

Although it is still uncertain whether the disease is due to overgrowth of small numbers

**Table 1**
**ANTIBIOTICS ASSOCIATED WITH 21 CASES OF *C. PERFRINGENS* ENTEROTOXIN DIARRHEA**

| Antibiotic | Alone | In combination | Total |
|---|---|---|---|
| Amoxacillin | 2 | 2 | 4 |
| Co-trimoxazole | 2 | 1 | 3 |
| Flucloxacillin | 2 | 0 | 2 |
| Trimethoprim | 2 | 0 | 2 |
| Penicillin | 1 | 1 | 2 |
| Cefuroxime | 1 | 1 | 2 |
| Ampicillin | 1 | 0 | 1 |
| Nystatin | 1 | 0 | 1 |
| Cephalexin | 0 | 1 | 1 |
| Tetracycline | 0 | 1 | 1 |
| Gentamycin | 0 | 1 | 1 |
| Yes, but unknown | — | — | 2 |
| None | — | — | 4 |

of enterotoxigenic *C. perfringens* that may be normally resident in the gut or is a true infection, recent epidemiological investigations imply that *C. perfringens* may behave as a true infectious agent in much the same way as *C. difficile*.

It is pertinent to comment on the possibility that these patients suffered from food poisoning. Although food as a fomite has not been excluded, the findings are not typical for food poisoning in so far as that most of the cases were sporadic and involved different serotypes, even when cases came from the same hospital, and that the diarrhea was prolonged and there was occasional visible blood. Additional arguments against these being cases of food poisoning, such as recurrence after specific chemotherapy, have been proposed elsewhere.[1]

## C. Treatment

There is very little information available on treatment. Most of the patients described by Borriello et al.[1] had self-limiting diarrhea and even though some of them experienced protracted diarrhea, none of the patients died from this illness. One of the patients was treated with vancomycin but relapsed after treatment.[1] The same serotype of *C. perfringens* was the cause of the relapse. One patient improved without further complication after treatment with codeine.[1] Since then it has proven necessary to treat three patients due to the severity of their diarrhea. All three responded within 36 hrs to treatment with metronidazole. In the one patient studied in detail it was shown that there was a very good relationship between starting metronidazole therapy and the concomitant resolution of symptoms and decrease in the numbers of *C. perfringens* and titers of enterotoxin in the stool.[8]

## D. Epidemiology

During routine screening of fecal specimens from patients with diarrhea two clusters of diarrhea due to enterotoxigenic *C. perfringens* were detected.[9] During a 12-month period, from October 1982 to August 1983, two of 270 stool samples analyzed had *C. perfringens* enterotoxin and large numbers of enterotoxigenic *C. perfringens* present. Both cases were sporadic and involved different serotypes (Table 2). However, during a 15-week period, from September 1983 to January 1984, there were cases of this disease which occurred as two clusters of five patients (Table 2). Apart from one case the disease was restricted to two geriatric wards. In the first instance there were five cases of diarrhea due to *C. perfringens* serotype 41 over a 9-week period, with four cases occurring on the same ward. The other five cases clustered over a 6-week period and involved three cases of diarrhea due to *C.*

**Table 2**
**INCIDENCE OF DIARRHEA DUE TO ENTEROTOXIGENIC STRAINS OF *C. PERFRINGENS* OVER A 20-MONTH-STUDY PERIOD**

| Time period | No. of cases | Wards | Serotypes involved (no. of patients) |
|---|---|---|---|
| October 1982 — August 1983 | 2 | A, B | 36, TW/51 |
| September 1983 — November 1983 | 5 | C, D | 41 |
| December 1983—January 1984 | 5 | C, E | 24(2), 27(3) |
| February 1984—May 1984 | 3 | F.G.H | 53, nontypable (2) |

**Table 3**
**SURFACES FOUND TO BE POSITIVE FOR AN OUTBREAK STRAIN OF *C. PERFRINGENS* DURING A CLUSTER OF CASES**

| | |
|---|---|
| Toilet seat | Radiator |
| Hand rails | Radio |
| Under bed | Baths |
| Bed sheets | Shelves |
| Bed frame | Sink |
| Bedside table | Chair arm rest |

*perfringens* serotype 24 and two due to serotype 27 (Table 2). A full epidemiological investigation conducted during the serotype 41 outbreak resulted in the following findings:

1. The causative serotype of *C. perfringens* could be readily isolated from inanimate surfaces in the immediate environment of a patient with the disease.
2. *C. perfringens* could rarely be isolated from similar areas in the immediate environment of a patient with diarrhea due to *C. difficile*.
3. The outbreak serotype of *C. perfringens* could be isolated from the hands of patients with the disease and occasionally from the staff dealing with these patients.
4. The outbreak serotype of *C. perfringens* was rarely present in the feces of asymptomatic patients on the same ward as the patients with this disease when sampled during the outbreak period.
5. The outbreak serotype of *C. perfringens* was isolated from 81% of surfaces on the outbreak ward but only 24% of similar surfaces on control wards. The surface investigated included those on common user objects (Table 3).
6. The isolation rate from the environment of the outbreak serotypes decreased as the diarrhea resolved.
7. The outbreak serotype was recovered from surfaces in the sluice room associated with the outbreak ward but not from the sluice rooms on control wards.

### E. Conclusions

It is evident from the findings that the serotype of *C. perfringens* causing the disease in the first cluster of cases could be readily isolated from inanimate surfaces associated with patients with the disease, and that the potential for cross-infection exists. It is likely that enterotoxigenic strains of *C. perfringens* can cause cross-infection in much the same way as that seen with *C. difficile*.

It is too early to speculate on the spectrum of disease that may be caused by this organism, but it is already evident that some people suffer from protracted severe diarrhea necessitating the administration of intravenous fluids, whereas others have a milder and shorter period of diarrhea. In addition, the presence of blood and mucous in the stools of some patients implies a more severe form of the disease is present.

We have, also, recently seen a patient with an exacerbation of ulcerative colitis which was associated with infection by enterotoxigenic *C. perfringens* and the presence of *C. perfringens* entertoxin in the stool.

The information available to date would implicate enterotoxigenic *C. perfringens* as an additional cause of antibiotic-associated diarrhea, as well as some cases of diarrhea not linked to recent antibiotic usage. It is hoped that a fuller picture of the disease will emerge in the near future.

## III. NEUTROPENIC ENTEROCOLITIS AND *CLOSTRIDIUM SEPTICUM*

### A. Introduction

One of the types of gastrointestinal lesion that has been described in patients with leukemia is necrotizing enterocolitis.[10] This probably describes a spectrum of disease, and it is thought that pseudomembranous colitis, ischemic colitis, and neutropenic enterocolitis are three separate distinct entities[11] that have all at some time been included under this term. Neutropenic enterocolitis is a distinct clinical and pathological syndrome in which patients often present with symptoms resembling appendicitis. Until the recent work implicating *C. septicum*,[2,12] the etiology of this disease was unknown.

### B. Pathogenesis

The lesion in neutropenic enterocolitis is usually located in the cecum, which shows pronounced edema and wall thickening, with necrosis and hemorrhage. Gas formation is not a feature of the condition. It has frequently been noted that septicemia due to *C. septicum* may occur in patients with neutropenia, leukemia, and large bowel malignancy.[2,12-17] Four of these reports are recent descriptions of seven cases of neutropenic colitis in which there was an association between *C. septicum* septicemia and characteristic ileocecal disease.[2,12,16,17] It was this association that led Rifkin to suggest that *C. septicum* was responsible for the formation of the enterocolitic lesion.[12] This attractive hypothesis was supported by the work of King and colleagues[2] who were able to examine bowel resection material taken at surgery performed early in the course of the illness in three patients. In all three cases *C. septicum* alone were shown to be present in the bowel wall by specific immunofluorescence staining techniques.

Although there is now an increasing amount of evidence to implicate *C. septicum* in the etiology of neutropenic colitis the evidence is still far from conclusive. It is known that *C. septicum* seems to be exceptionally capable of taking advantage of compromise in host defense mechanisms and translocating across the gut wall. So, on this basis it could be argued that the findings are a consequence of the disease as opposed to the cause. An argument against this is that *C. septicum* is not a common component of the fecal flora of healthy Western subjects (see Chapter 1). However, there is evidence that it may be present in recoverable numbers more frequently in patients with colon cancer (see Chapter 1) and that it is frequently present in the normal appendix.[18] Although it is possible to argue that the association between *C. septicum* and neutropenic enterocolitis is a consequence of the disease, it is difficult to believe that this toxigenic microorganism does not play a role in the etiology.

Evidence of the possible pathogenic nature of *C. septicum* is to be found in a disease of sheep called 'braxy' (see Chapter 10), in which *C. septicum* can be isolated in pure culture from regions of hemorrhagic inflammation in the abomasum. This disease can be prevented by immunization against *C. septicum*. Although there is no animal model of neutropenic enterocolitis, it is interesting to note that a toxin that could be neutralized by antitoxin to *Clostridium histolyticum* was found in the fecal supernatants of guinea pigs with clindamycin-associated enterocolitis.[19] The interest here is that no results on the isolation of the incriminated microorganism were mentioned and that the major toxin produced by *C. histolyticum* can be neutralized by antisera to *C. septicum* as the two organisms synthesize toxins with components possessing common antigenic properties.[20] It is possible, therefore, that the disease noted could have been due to *C. septicum*.

## C. Treatment

In the main treatment has been restricted to surgery,[21] though bowel rest and fluid therapy coupled to massive administration of broad spectrum antibiotics has also been successfully employed.[22] The implication of *C. septicum* in the disease may serve to stimulate a more rational nonsurgical approach to treatment. The use of oral vancomycin and parenteral antibiotics active against anaerobes would seem appropriate. By analogy to the *C. septicum* mediated disease of 'braxy' in sheep it may be possible to actively immunize at risk patients.

## D. Conclusions

Although not conclusive, the evidence to date strongly implicates *C. septicum* in the etiology of neutropenic enterocolitis. The development of an animal model would greatly facilitate research in confirming the role of *C. septicum* in this disease. It should be possible to isolate the organism from the feces of suspect patients, and may be possible to use tissue culture techniques to detect the cytotoxic toxins of *C. septicum* in fecal material as an aid in diagnosis.

# REFERENCES


1. **Borriello, S. P., Larson, H. E., Welch, A. R., Barclay, F., Stringer, M. F., and Bartholomew, B. A.**, Enterotoxigenic *Clostridium perfringrens*: a possible cause of antibiotic-associated diarrhoea, *Lancet*, 1, 305, 1984.
2. **King, A., Rampling, A., Wight, D. G., D., and Warren, R. E.**, Neutropenic enterocolitis due to *Clostridium septicum* infection, *J. Clin. Pathol.*, 37, 335, 1984.
3. **Borriello, S. P.**, Bacteria and gastrointestinal secretion and motility, *Scand. J. Gastroenterol.*, 19(Suppl. 93), 115, 1984.
4. **Borriello, S. P., Barclay, F. E., and Welch, A. R.**, unpublished data, 1984.
5. **Borriello, S. P., Williams, R. K. T., and Price, A. P.**, unpublished data, 1984.
6. **Tedesco, F. J., Corless, J. K., and Brownstein, R. E.**, Rectal sparing in antibiotic-associated pseudomembranous colitis: a prospective study, *Gastroenterology*, 83, 1259, 1982.
7. **McDonel, J. L. and Demers. G. W.**, *In vivo* effects of enterotoxin from *Clostridum perfringens* Type A in the rabbit colon: binding vs. biological activity, *J. Infect. Dis.*, 145, 490, 1982.
8. **Borriello, S. P., and Williams, R. K. T.**, Treatment of *C. perfringens* exterotoxin associated diarrhea with metronidazole, *J. Infect.*, in press.
9. **Borriello, S. P., Barclay, F. E., Welch, A. R., Stringer, M. F., and Watson, G. N.**, unpublished data, 1983.
10. **Steinberg, D., Gold, J., and Brodin, A.**, Necrotizing enterocolitis in leukaemia, *Arch. Intern. Med.*, 131, 538, 131.
11. **Dosik, G. M., Luna, M., Valdivieso, M., McCredie, K. B., Gehan, E. A., Gil-Extremera, B., Smith, T. L., and Bodey, G. P.**, Necrotizing colitis in patients with cancer, *Am. J. Med.*, 67, 646, 1979.
12. **Rifkin, G. D.**, Neutropenic enterocolitis and *Clostridium septicum* infection in patients with agranulocytosis, *Arch. Intern. Med.*, 140, 834, 1980.
13. **Alpern, R. J. and Dowell, V. R.**, *Clostridium septicum* infections and malignancy, *JAMA*, 209, 385, 1969.

14. **Koransky, J. R., Stargel, M. D., and Dowell, V. R.,** *Clostridium septicum* bacteraemia. Its clinical significance, *Am. J. Med.*, 66, 63, 1979.
15. **Katlic, M. R., Derkac, W. M., and Coleman, W. S.,** *Clostridium septicum* infections and malignancy, *Ann. Surg.*, 193, 361, 1981.
16. **Bignold, L. P. and Harvey, H. P. B.,** Necrotizing enterocolitis associated with invasion by *Clostridium septicum* complicating cyclic neutropenia, *Aust. N.Z. J. Med.*, 9, 426, 1979.
17. **Hopkins, D. G. and Kushner, J. P.,** Clostridial species in the pathogenesis of necrotizing enterocolitis in patients with neutropenia, *Am. J. Hematol.*, 14, 289, 1983.
18. **Lanz, O. and Tavel, E.,** Bacteriologie de l'appendicite, *Rev. Chir.*, 30, 43, 1904.
19. **Knoop, F. C.,** Clindamycin-associated enterocolitis in guinea pigs: evidence for a bacterial toxin, *Infect. Immun.*, 23, 31, 1979.
20. **Zemlyanitskaya, E. P., Yermakova, M. P., and Matveyev, K. I.,** Investigation of common and specific components of *Cl. septicum* and *Cl. histolyticum* toxins, *J. Hyg. Epidemiol. Microbiol. Immunol.*, 22, 108, 1978.
21. **Ikard, R. W.,** Neutropenic typhlitis in adults, *Arch. Surg.*, 116, 943, 1981.
22. **Shaked, A., Shinar, E., and Freund, H.,** Neutropenic typhlitis: a plea for conservation, *Dis. Colon Rectum*, 26, 351, 1983.

# INDEX

## A



## B



## C

## D



## E

## H



## I



## J



## K



## L



## M

## N



## O



## P

## Q



## R



## S



## T

U



V



Y